AF293736

James E. Humphreys

Linear Algebraic Groups

Springer-Verlag   New York Heidelberg Berlin

James E. Humphreys

Associate Professor of Mathematics and Statistics
University of Massachusetts
Amherst, Massachusetts 01002

Managing Editors

P. R. Halmos

Indiana University
Department of Mathematics
Swain Hall East
Bloomington, Indiana 47401

C. C. Moore

University of California
at Berkeley
Department of Mathematics
Berkeley, California 94720

AMS Subject Classification
20G15

Library of Congress Cataloging in Publication Data

Humphreys, James E
    Linear algebraic groups.
    (Graduate texts in mathematics; v. 21)
    Bibliography: p. 233
    1. Linear algebraic groups. I. Title. II. Series.

QA171.H83   512'.2   74–22237

Softcover reprint of the hardcover 1st edition 1975

ISBN 978-1-4684-9445-7   ISBN 978-1-4684-9443-3 (eBook)
DOI 10.1007/978-1-4684-9443-3

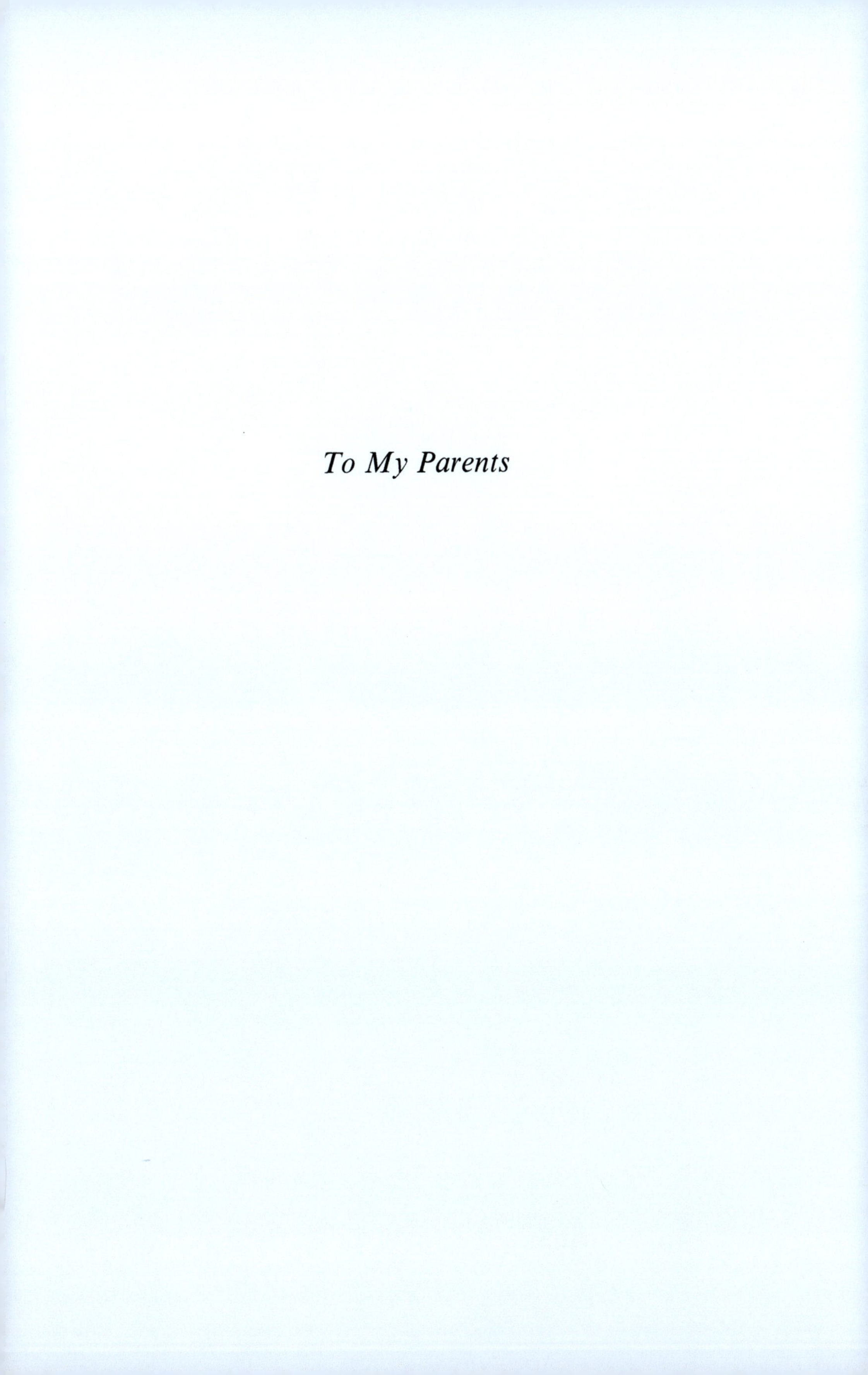

To My Parents

Preface

Over the last two decades the Borel–Chevalley theory of linear algebraic groups (as further developed by Borel, Steinberg, Tits, and others) has made possible significant progress in a number of areas: semisimple Lie groups and arithmetic subgroups, p-adic groups, classical linear groups, finite simple groups, invariant theory, etc. Unfortunately, the subject has not been as accessible as it ought to be, in part due to the fairly substantial background in algebraic geometry assumed by Chevalley [8], Borel [4], Borel, Tits [1]. The difficulty of the theory also stems in part from the fact that the main results culminate a long series of arguments which are hard to "see through" from beginning to end. In writing this introductory text, aimed at the second year graduate level, I have tried to take these factors into account.

First, the requisite algebraic geometry has been treated in full in Chapter I, modulo some more-or-less standard results from commutative algebra (quoted in §0), e.g., the theorem that a regular local ring is an integrally closed domain. The treatment is intentionally somewhat crude and is not at all scheme-oriented. In fact, everything is done over an algebraically closed field K (of arbitrary characteristic), even though most of the eventual applications involve a field of definition k. I believe this can be justified as follows. In order to work over k from the outset, it would be necessary to spend a good deal of time perfecting the foundations, and then the only rationality statements proved along the way would be of a minor sort (cf. (34.2)). The deeper rationality properties can only be appreciated after the reader has reached Chapter X. (A survey of such results, without proofs, is given in Chapter XII.)

Second, a special effort has been made to render the exposition transparent. Except for a digression into characteristic 0 in Chapter V, the development from Chapter II to Chapter XI is fairly "linear", covering the foundations, the structure of connected solvable groups, and then the structure, representations and classification of reductive groups. The lecture notes of Borel [4], which constitute an improvement of the methods in Chevalley [8], are the basic source for Chapters II–IV, VI–X, while Chapter XI is a hybrid of Chevalley [8] and SGAD. From §27 on the basic facts about root systems are used constantly; these are listed (with suitable references) in the Appendix. Apart from §0, the Appendix, and a reference to a theorem of Burnside in (17.5), the text is self-contained. But the reader is asked to verify some minor points as exercises.

While the proofs of theorems mostly follow Borel [4], a number of improvements have been made, among them Borel's new proof of the normalizer theorem (23.1), which he kindly communicated to me.

I had an opportunity to lecture on some of this material at Queen Mary College in 1969, and at New York University in 1971–72. Several colleagues have made valuable suggestions after looking at a preliminary version of the manuscript; I especially want to thank Gerhard Hochschild, George Seligman, and Ferdinand Veldkamp. I also want to thank Michael J. DeRise for his help. Finally, I want to acknowledge the support of the National Science Foundation and the excellent typing of Helen Samoraj and her staff.

James E. Humphreys

Conventions

$K^* =$ multiplicative group of the field K
char $K =$ characteristic of K
char exp $K =$ characteristic exponent of K, i.e., max $\{1, \text{char } K\}$
det $=$ determinant
Tr $=$ trace
Card $=$ cardinality
$\amalg$ $=$ direct sum

Table of Contents

V.   Characteristic 0 Theory   87

VI.   Semisimple and Unipotent Elements   95

VII.   Solvable Groups   109

VIII.   Borel Subgroups   133

IX.   Centralizers of Tori   147

X. Structure of Reductive Groups   163

XI. Representations and Classification of Semisimple Groups   188

XII.   Survey of Rationality Properties   217

Linear Algebraic Groups

Chapter I

Algebraic Geometry

0. Some Commutative Algebra

Algebraic geometry is heavily dependent on commutative algebra, the study of commutative rings and fields (notably those arising from polynomial rings in many variables); indeed, it is impossible to draw a sharp line between the geometry and the algebra. For reference, we assemble in this section some basic concepts and results (without proof) of an algebraic nature. The theorems stated are in most cases "standard" and readily accessible in the literature, though not always encountered in a graduate algebra course.

We shall give explicit references, usually by chapter and section, to the following books:

L = S. Lang, Algebra, Reading, Mass.: Addison-Wesley 1965.

ZS = O. Zariski, P. Samuel, Commutative Algebra, 2 vol., Princeton: Van Nostrand 1958, 1960.

AM = M. F. Atiyah, I. G. Macdonald, Introduction to Commutative Algebra, Reading, Mass.: Addison-Wesley 1969.

J = N. Jacobson, Lectures in Abstract Algebra, vol. III, Princeton: Van Nostrand 1964.

There are of course other good sources for this material, e.g., Bourbaki or van der Waerden. We remark that [AM] is an especially suitable reference for our purposes, even though some theorems there are set up as exercises.

All rings are assumed to be commutative (with 1).

0.1 *A ring R is **noetherian** $\Leftrightarrow$ each ideal of R is finitely generated $\Leftrightarrow$ R has ACC (ascending chain condition) on ideals $\Leftrightarrow$ each nonempty collection of ideals has a maximal element, relative to inclusion. Any homomorphic image of a noetherian ring is noetherian.* [L, VI §1] [AM, Ch. 6, 7]. **Hilbert Basis Theorem:** *If R is noetherian, so is $R[\mathsf{T}]$ (polynomial ring in one indeterminate). In particular, for a field K, $\mathsf{K}[\mathsf{T}_1, \mathsf{T}_2, \ldots, \mathsf{T}_n]$ is noetherian.* [L, VI §2] [ZS, IV §1] [AM, 7.5].

0.2 *If K is a field, $\mathsf{K}[\mathsf{T}_1, \ldots, \mathsf{T}_n]$ is a UFD (unique factorization domain).* [L, V §6].

0.3 **Weak Nullstellensatz:** *Let K be a field, $\mathsf{L} = \mathsf{K}[x_1, \ldots, x_n]$ a finitely generated extension ring of K. If L is a field, then all x_i are algebraic over K.* [L, X §2] [ZS, VII §3] [AM, 5.24; Ch. 5, ex. 18; 7.9].

0.4   *Let* L/K *be a field extension. Elements* $x_1, \ldots, x_d \in$ L *are* **algebraically independent** *over* K *if no nonzero polynomial* $f(T_1, \ldots, T_d)$ *over* K *satisfies* $f(x_1, \ldots, x_d) = 0$. *A maximal subset of* L *algebraically independent over* K *is called a* **transcendence basis** *of* L/K. *Its cardinality is a uniquely defined number, the* **transcendence degree** tr. $\deg._K$ L. *If* L $= K(x_1, \ldots, x_n)$, *a transcendence basis can be chosen from among the* x_i, *say* $x_1, \ldots, x_d$. *Then* $K(x_1, \ldots, x_d)$ *is* **purely transcendental** *over* K *and* L/$K(x_1, \ldots, x_d)$ *is (finite) algebraic.* [L, *X* §1] [ZS, *II* §12] [J, *IV* §3].

Lüroth Theorem: *Let* L $=$ K(T) *be a simple, purely transcendental extension of* K. *Then any subfield of* L *properly including* K *is also a simple, purely transcendental extension.* [J, *IV* §4]. *(Remark: The proof in* J *is not quite complete, so reference may also be made to B. L. van der Waerden, Modern Algebra, vol. I, New York: F. Ungar 1953, p. 198.)*

0.5   *Let* E/F *be a finite field extension. There is a map* $N_{E/F}$: E $\rightarrow$ F, *called the* **norm**, *which induces a homomorphism of multiplicative groups* E* $\rightarrow$ F*, *such that* $N_{E/F}(a)$ *is a power of the constant term of the minimal polynomial of* a *over* F, *and in particular,* $N_{E/F}(a) = a^{[E:F]}$ *whenever* $a \in$ F. *To define the norm, view* E *as a vector space over* F. *For each* $a \in$ E, $x \mapsto ax$ *defines a linear transformation* E $\rightarrow$ E; *let* $N_{E/F}(a)$ *be its determinant.* [L, *VIII* §5] [ZS, *II* §10].

0.6   *Let* R $\supset$ S *be an extension of rings. An element* $x \in R$ *is* **integral** *over* S $\Leftrightarrow x$ *is a root of a monic polynomial over* S $\Leftrightarrow$ *the subring* $S[x]$ *of* R *is a finitely generated S-module* $\Leftrightarrow$ *the ring* $S[x]$ *acts on some finitely generated S-module* V *faithfully (i.e.,* $y.V = 0$ *implies* $y = 0$). R *is* **integral** *over* S *if each element of* R *is integral over* S. *The* **integral closure** *of* S *in* R *is the set (a subring) of* R *consisting of all elements of* R *integral over* S. *If* R *is an integral domain, with field of fractions* F, R *is said to be* **integrally closed** *if* R *equals its integral closure in* F. [L, *IX* §1] [ZS, *V* §1] [AM, *Ch. 5*].

0.7   **Noether Normalization Lemma:**   *Let* K *be an infinite field,* R $=$ $K[x_1, \ldots, x_n]$ *a finitely generated integral domain over* K *with field of fractions* F, $d =$ tr. $\deg._K$ F. *Then there exist elements* $y_1, \ldots, y_d \in R$ *such that* R *is integral over* $K[y_1, \ldots, y_d]$ *(and the* y_i *are algebraically independent over* K). [L, *X* §4] [ZS, *V* §4] [AM, *Ch. 5, ex. 16*].

0.8   **Going Up Theorem:**   *Let* R/S *be a ring extension, with* R *integral over* S. *If* P *is a prime (resp. maximal) ideal of* S, *there exists a prime (resp. maximal) ideal* Q *of* R *for which* $Q \cap S = P$. [L, *IX* §1] [ZS, *V* §2] [AM, *5.10, 5.11*].

Extension Theorem: *Let* R/S *be an integral extension,* K *an algebraically closed field. Then any homomorphism* $\varphi: S \rightarrow$ K *extends to a homomorphism* $\varphi': R \rightarrow$ K. *If* $x \in R, a \in$ K, φ *can first be extended to a homomorphism* $R[x] \rightarrow$ K *sending x to a (then be further extended to R, R being integral over $R[x]$), provided $f(x) = 0$ implies $f_\varphi(a) = 0$ for $f(\mathsf{T}) \in \mathsf{S}[\mathsf{T}]$ ($f_\varphi(\mathsf{T})$ the polynomial over K gotten by applying φ to each coefficient of $f(\mathsf{T})$). [L, IX §3] [AM, Ch. 5] [J, Intro., IV].*

0.9 *Let $P_1, \ldots, P_n$ be prime ideals in a ring R. If an ideal lies in the union of the P_i, it must already lie in one of them. [ZS, IV §6, Remark p. 215].*

0.10 *Let S be a* **multiplicative set** *in a ring R ($0 \notin S, 1 \in S, a, b \in S \Rightarrow ab \in S$). The* **generalized ring of quotients** *$S^{-1}R$ is constructed using equivalence classes of pairs $(r, s) \in R \times S$, where $(r, s) \sim (r', s')$ means that for some $s'' \in S$, $s''(rs' - r's) = 0$. The (prime) ideals of $S^{-1}R$ correspond bijectively to the (prime) ideals of R not meeting S. In case R is an integral domain, with field of fractions F, $S^{-1}R$ may be identified with the set of fractions r/s in F. In general, the canonical map $R \to S^{-1}R$ (sending r to the class of $(r, 1)$) is injective only when S contains no zero divisors. For example, take $S = \{x^n | n \in \mathbb{Z}^+\}$ for x not nilpotent, to obtain $S^{-1}R$, denoted R_x; R is a subring of R_x provided x is not a zero divisor. Or take $S = R - P$, P a prime ideal. Then $S^{-1}R$ is denoted R_P and is a* **local ring** *(i.e., has a unique maximal ideal PR_P, consisting of the nonunits of R_P). The prime ideals of R_P correspond naturally to the prime ideals of R contained in P. If R is an integrally closed domain, then so is R_P. If R is noetherian, so is R_P. If M is a maximal ideal, the fields R/M and R_M/MR_M are naturally isomorphic, and the inclusion $R \to R_M$ induces a vector space isomorphism of M/M^2 onto $MR_M/(MR_M)^2$. [L, II §3] [AM, Ch. 3].*

0.11 **Nakayama Lemma:** *Let R be a ring, M a maximal ideal, V a finitely generated R-module for which $V = MV$. Then there exists $x \notin M$ such that $xV = 0$. In particular, if R is local (with unique maximal ideal M), x must be a unit and therefore $V = 0$. [AM, 2.5, 2.6] [L, IX §1].*

0.12 *The* **Krull dimension** *of a (noetherian) local ring R is the maximum length k of a chain of prime ideals $0 \subsetneq P_1 \subsetneq P_2 \subsetneq \cdots \subsetneq P_k \subsetneq R$. If this equals the minimum number of generators of the maximal ideal M of R, R is called* **regular.** **Theorem:** *A regular local ring is an integral domain, integrally closed (in its field of fractions). [AM, Ch. 11] [ZS, VIII §11; cf. Appendix 7].*

0.13 *Let I be an ideal in a noetherian ring R, and let $P_1, \ldots, P_t$ be the minimal prime ideals containing I. The image of $P_1 \cap \cdots \cap P_t$ in R/I is the nilradical of R/I, a nilpotent ideal. In particular, for large enough n, $P_1^n P_2^n \cdots P_t^n \subset (P_1 \cap \cdots \cap P_t)^n \subset I$. [AM, 7.15] [L, VI §4].*

0.14 *A field extension E/F is* **separable** *if either char $\mathsf{F} = 0$, or else char $\mathsf{F} = p > 0$ and the pth powers of elements $x_1, \ldots, x_n \in \mathsf{E}$ linearly independent over F are again so. This generalizes the usual notion when E/F is finite. $\mathsf{E} = \mathsf{F}(x_1, \ldots, x_n)$ is* **separably generated** *over F if E is a finite separable* *extension of a purely transcendental extension of* F. *For finitely generated extensions* E/F, *"separably generated" is equivalent to "separable", and* E/F *is automatically separable when* F *is perfect. If* $F \subset L \subset E$, E/F *separable, then* L/F *is separable. If* $F \subset L \subset E$, E/L *and* L/F *separable, then* E/F *is separable* [ZS, *II* §13] [L, *X* §6] [J, *IV* §5].

0.15 *A* **derivation** $\delta : E \to L$ (E *a field,* L *an extension field of* E), *is a map which satisfies* $\delta(x + y) = \delta(x) + \delta(y)$ *and* $\delta(xy) = x\,\delta(y) + \delta(x)\,y$. *If* F *is a subfield of* E, δ *is called an* **F-derivation** *if in addition* $\delta(x) = 0$ *for all* $x \in F$ (*so* δ *is F-linear). The space* $\mathrm{Der}_F(E, L)$ *of all F-derivations* $E \to L$ *is a vector space over* L, *whose dimension is* tr. deg.$_F$ E *if* E/F *is separably generated.* E/F *is separable if and only if all derivations* $F \to L$ *extend to derivations* $E \to L$ (L *an extension field of* E). *If* char $E = p > 0$, *all derivations of* E *vanish on the subfield* E^p *of* p^{th} *powers.* [ZS, *II* §17] [J, *IV* §7] [L, *X* §7].

1. Affine and Projective Varieties

In this section we consider subsets of affine or projective space defined by polynomial equations, with special attention being paid to the way in which geometric properties of these sets translate into algebraic properties of polynomial rings. K always denotes an algebraically closed field, of arbitrary characteristic.

1.1. Ideals and Affine Varieties

The set $K^n = K \times \cdots \times K$ will be called **affine n-space** and denoted $\mathbf{A}^n$. By **affine variety** will be meant (provisionally) the set of common zeros in $\mathbf{A}^n$ of a finite collection of polynomials. Evidently we have in mind curves, surfaces, and the like. But the collection of polynomials defining a geometric configuration can vary quite a bit without affecting the geometry, so we aim for a tighter correspondence between geometry and algebra. As a first step, notice that the ideal in $K[T] = K[T_1, \ldots, T_n]$ generated by a set of polynomials $\{f_\alpha(T)\}$ has precisely the same common zeros as $\{f_\alpha(T)\}$. Moreover, the Hilbert Basis Theorem (0.1) asserts that each ideal in $K[T]$ has a finite set of generators, so every ideal corresponds to an affine variety. Unfortunately, this correspondence is not $1-1$: e.g., the ideals generated by T and by T^2 are distinct, but have the same zero set $\{0\}$ in $\mathbf{A}^1$. We shall see shortly how to deal with this phenomenon.

Formally, we can assign to each ideal I in $K[T]$ the set $\mathscr{V}(I)$ of its common zeros in $\mathbf{A}^n$, and to each subset $X \subset \mathbf{A}^n$ the collection $\mathscr{I}(X)$ of all polynomials vanishing on X. It is clear that $\mathscr{I}(X)$ is an ideal, and that we have inclusions:

$$X \subset \mathscr{V}(\mathscr{I}(X)),$$
$$I \subset \mathscr{I}(\mathscr{V}(I)).$$

Of course, neither of these need be an equality (examples?). Let us examine more closely the second inclusion. By definition, the **radical** $\sqrt{I}$ of an ideal I is $\{f(\mathsf{T}) \in \mathsf{K}[\mathsf{T}] \,|\, f(\mathsf{T})^r \in I \text{ for some } r \geqslant 0\}$. This is easily seen to be an ideal, including I. If $f(\mathsf{T})$ fails to vanish at $x = (x_1, \ldots, x_n)$, then $f(\mathsf{T})^r$ also fails to vanish at x for each $r \geqslant 0$. From this it follows that $\sqrt{I} \subset \mathscr{I}(\mathscr{V}(I))$, which refines the above inclusion. Indeed, we now get equality—a fact which is crucial but not at all intuitively obvious.

Theorem (Hilbert's Nullstellensatz). *If I is any ideal in* $\mathsf{K}[\mathsf{T}_1, \ldots, \mathsf{T}_n]$, *then* $\sqrt{I} = \mathscr{I}(\mathscr{V}(I))$.

Proof. In view of the finite generation of I, the theorem is equivalent to the statement: "Given $f(\mathsf{T}), f_1(\mathsf{T}), \ldots, f_s(\mathsf{T})$ in $\mathsf{K}[\mathsf{T}]$, such that $f(\mathsf{T})$ vanishes at every common zero of the $f_i(\mathsf{T})$ in $\mathbf{A}^n$, there exist $r \geqslant 0$ and $g_1(T), \ldots,$

$$g_s(T) \in \mathsf{K}[\mathsf{T}] \text{ for which } f(\mathsf{T})^r = \sum_{i=1}^{s} g_i(\mathsf{T}) f_i(\mathsf{T})."$$

We show first that this statement follows from the assertion:

$$(*) \qquad\qquad \text{If } \mathscr{V}(I) = \emptyset \quad \text{then} \quad I = \mathsf{K}[\mathsf{T}].$$

(Notice that this is just a special case of the theorem, since only the ideal $\mathsf{K}[\mathsf{T}]$ can have $\mathsf{K}[\mathsf{T}]$ as radical!) Indeed, given $f(\mathsf{T}), f_1(\mathsf{T}), \ldots, f_s(\mathsf{T})$ as indicated, we can introduce a new indeterminate T_0 and consider the collection of polynomials in $n + 1$ indeterminates, $f_1(\mathsf{T}), \ldots, f_s(\mathsf{T}), 1 - \mathsf{T}_0 f(\mathsf{T})$. These have no common zero in $\mathbf{A}^{n+1}$, thanks to the original condition imposed on $f(\mathsf{T})$, so (*) implies that they generate the unit ideal. Find polynomials $h_i(\mathsf{T}_0, \ldots, \mathsf{T}_n)$ and $h(\mathsf{T}_0, \ldots, \mathsf{T}_n)$ for which $1 = h_1(\mathsf{T}_0, \mathsf{T}) f_1(\mathsf{T}) + \cdots + h_s(\mathsf{T}_0, \mathsf{T}) f_s(\mathsf{T}) + h(\mathsf{T}_0, \mathsf{T})(1 - \mathsf{T}_0 f(\mathsf{T}))$. Then substitute $1/f(\mathsf{T})$ for T_0 throughout, and multiply both sides by a sufficiently high power $f(\mathsf{T})^r$ to clear denominators. This yields a relation of the desired sort.

It remains to prove (*), or equivalently, to show that a proper ideal in $\mathsf{K}[\mathsf{T}]$ has at least one common zero in $\mathbf{A}^n$. (In the special case $n = 1$, this would follow directly from the fact that K is algebraically closed.) Let us attempt naively to construct a common zero. By Zorn's Lemma, I lies in some maximal ideal of $\mathsf{K}[\mathsf{T}]$, and common zeros of the latter will serve for I as well; so we might as well assume that I is maximal. Then the residue class ring $\mathsf{L} = \mathsf{K}[\mathsf{T}]/I$ is a *field*; K may be identified with the residue classes of scalar polynomials. If we write t_i for the residue class of T_i, it is clear that $\mathsf{L} = \mathsf{K}[t_1, \ldots, t_n]$ (the smallest subring of L containing K and the t_i). Moreover, the n-tuple $(t_1, \ldots, t_n)$ is by construction a common zero of the polynomials in I. If we could identify L with K, the t_i could already be found inside K. But K is algebraically closed, so for this it would be enough to show that the t_i are *algebraic* over K, which is precisely the content of (0.3). $\square$

The Nullstellensatz ("zeros theorem") implies that the operators $\mathscr{V}, \mathscr{I}$ set up a $1-1$ correspondence between the collection of all **radical ideals** in $\mathsf{K}[\mathsf{T}]$ (ideals equal to their radical) and the collection of all affine varieties in $\mathbf{A}^n$.

Indeed, if $X = \mathscr{V}(I)$, then $\mathscr{I}(X) = \mathscr{I}(\mathscr{V}(I)) = \sqrt{I}$, so that X may be recovered as $\mathscr{V}(\mathscr{I}(X))$ (I and $\sqrt{I}$ having the same set of common zeros). On the other hand, if $I = \sqrt{I}$, then I may be recovered as $\mathscr{I}(\mathscr{V}(I))$. Notice that the correspondences $X \mapsto \mathscr{I}(X)$ and $I \mapsto \mathscr{V}(I)$ are *inclusion-reversing*. So the noetherian property of $\mathsf{K}[\mathsf{T}]$ implies DCC (descending chain condition) on the collection of affine varieties in $\mathbf{A}^n$.

Examples of radical ideals are *prime* (in particular, *maximal*) ideals. We shall examine in (1.3) the varieties corresponding to prime ideals. For the moment, just consider the case $X = \mathscr{V}(I)$, I maximal. The Nullstellensatz guarantees that X is nonempty, so let $x \in X$. Clearly $I \subset \mathscr{I}(\{x\}) \subsetneq \mathsf{K}[\mathsf{T}]$, so $I = \mathscr{I}(\{x\})$ by maximality, and $X = \mathscr{V}(I) = \mathscr{V}(\mathscr{I}(\{x\})) = \{x\}$. On the other hand, if $x \in \mathbf{A}^n$, then $f(\mathsf{T}) \mapsto f(x)$ defines a homomorphism of $\mathsf{K}[\mathsf{T}]$ onto K, whose kernel $\mathscr{I}(\{x\})$ is maximal because K is a field. Thus the points of $\mathbf{A}^n$ correspond $1\text{--}1$ to the maximal ideals of $\mathsf{K}[\mathsf{T}]$.

A **linear variety** through $x \in \mathbf{A}^n$ is the zero set of linear polynomials of the form $\sum a_i(\mathsf{T}_i - x_i)$. This is just a vector subspace of $\mathbf{A}^n$ if the latter is viewed as a vector space with origin x. From the Nullstellensatz (or linear algebra!) we deduce that any linear polynomial vanishing on such a variety is a K-linear combination of the given ones.

1.2. Zariski Topology on Affine Space

If K were the field of complex numbers, $\mathbf{A}^n$ could be given the usual topology of complex n-space. Then the zero set of a polynomial $f(\mathsf{T})$ would be closed, being the inverse image of the closed set $\{0\}$ in $\mathbb{C}$ under the continuous mapping $x \mapsto f(x)$. The set of common zeros of a collection of polynomials would equally well be closed, being the intersection of closed sets. Of course, complex n-space has plenty of other closed sets which are unobtainable in this way, as is clear already in case $n = 1$.

The idea of topologizing affine n-space by decreeing that the **closed** sets are to be precisely the affine varieties turns out to be very fruitful. This is called the **Zariski topology**. Naturally, it has to be checked that the axioms for a topology are satisfied: (1) $\mathbf{A}^n$ and $\emptyset$ are certainly closed, as the respective zero sets of the ideals (0) and $\mathsf{K}[\mathsf{T}]$. (2) If I, J are two ideals, then clearly $\mathscr{V}(I) \cup \mathscr{V}(J) \subset \mathscr{V}(I \cap J)$. To establish the reverse inclusion, suppose x is a zero of $I \cap J$, but not of I or J. Say $f(\mathsf{T}) \in I, g(\mathsf{T}) \in J$, with $f(x) \neq 0, g(x) \neq 0$. Since $f(\mathsf{T})g(\mathsf{T}) \in I \cap J$, we must have $f(x)g(x) = 0$, which is absurd. This argument implies that finite unions of closed sets are closed. (3) Let I_α be an arbitrary collection of ideals, so $\sum_\alpha I_\alpha$ is the ideal generated by this collection. Then it is clear that $\bigcap_\alpha \mathscr{V}(I_\alpha) = \mathscr{V}(\sum_\alpha I_\alpha)$, i.e., arbitrary intersections of closed sets are closed.

What sort of topology is this? Points are closed, since $x = (x_1, \ldots, x_n)$ is the only common zero of the polynomials $\mathsf{T}_1 - x_1, \ldots, \mathsf{T}_n - x_n$. But the Hausdorff separation axiom fails. This is evident already in the case of $\mathbf{A}^1$, where the proper closed sets are precisely the finite sets (so no two nonempty open sets can be disjoint). The reader who is accustomed to spaces with good separation properties must therefore exercise some care in reasoning about the Zariski topology. For example, the DCC on closed sets (resulting from Hilbert's Basis Theorem) implies the ACC on open sets, or equivalently, the maximal condition. This shows that $\mathbf{A}^n$ is a *compact* space. But in the absence of the Hausdorff property, one cannot use sequential convergence arguments or the like; for this reason, one sometimes uses the term **quasicompact** in this situation, reserving the term "compact" for compact Hausdorff spaces.

In a qualitative sense, all nonempty open sets in $\mathbf{A}^n$ are "large" (think of the complement of a curve in $\mathbf{A}^2$ or of a surface in $\mathbf{A}^3$). Since a closed set $\mathscr{V}(I)$ is the intersection of the zero sets of the various $f(\mathsf{T}) \in I$, a typical nonempty open set can be written as the union of **principal open sets**—sets of nonzeros of individual polynomials. These therefore form a basis for the topology, but are still not very "small". For example, $\mathrm{GL}(n, \mathsf{K})$ is the principal open set in $\mathbf{A}^{n^2}$ defined by the nonvanishing of $\det (\mathsf{T}_{ij})$; $\mathrm{GL}(n, \mathsf{K})$ denotes here the group of all invertible $n \times n$ matrices over K.

1.3. Irreducible Components

In topology one often studies connectedness properties. But the union of two intersecting curves in $\mathbf{A}^n$ is connected, while at the same time capable of being analyzed further into "components." This suggests a different emphasis, based on a somewhat different topological property. For use later on, we formulate this in general terms.

Let X be a topological space. Then X is said to be **irreducible** if X cannot be written as the union of two proper, nonempty, closed subsets. A subspace Y of X is called irreducible if it is irreducible as a topological space (with the induced topology). Notice that X is irreducible if and only if any two nonempty open sets in X have nonempty intersection, or equivalently, any nonempty open set is dense. Evidently an irreducible space is connected, but not conversely.

> **Proposition A.** *Let X be a topological space.*
> *(a) A subspace Y of X is irreducible if and only if its closure $\bar{Y}$ is irreducible.*
> *(b) If $\varphi : X \to X'$ is a continuous map, and X is irreducible, then so is $\varphi(X)$.*

Proof. (a) In view of the preceding remarks, Y is irreducible if and only if the intersection of two open subsets of X, each meeting Y, also meets Y; and similarly for $\bar{Y}$. But an open set meets Y if and only if it meets $\bar{Y}$.

(b) If U, V are open sets in X' which meet $\varphi(X)$, we have to show that $U \cap V$ meets $\varphi(X)$ as well. But $\varphi^{-1}(U)$, $\varphi^{-1}(V)$ are (nonempty) open sets in X, so they have nonempty intersection (X being irreducible), whose image under φ lies in $U \cap V \cap \varphi(X)$. $\square$

Our intention is to decompose an affine variety into irreducible "components". Actually, an argument using Zorn's Lemma shows that any topological space can be written as the union of its maximal irreducible subspaces (which are necessarily closed, in view of part (a) of the proposition). To insure a *finite* decomposition of this sort, we exploit the fact that an affine variety has maximal condition on open subsets. This too can be formulated in general. Call a topological space **noetherian** if each nonempty collection of open sets has a maximal element (equivalently, if open sets satisfy ACC, or if closed sets satisfy the minimal condition, or if closed sets satisfy DCC).

Proposition B.   *Let X be a noetherian topological space. Then X has only finitely many maximal irreducible subspaces (necessarily closed, and having X as their union).*

Proof.   Consider the collection $\mathscr{A}$ of all finite unions of closed irreducible subsets of X (for example, $\emptyset \in \mathscr{A}$). If X itself does not belong to $\mathscr{A}$, use the noetherian property to find a closed subset Y of X which is *minimal* among the closed subsets (such as X) not belonging to $\mathscr{A}$. Evidently Y is neither empty nor irreducible; so $Y = Y_1 \cup Y_2$ (Y_i proper closed subsets of Y). The minimality of Y forces both Y_1 and Y_2 to lie in $\mathscr{A}$. But then Y also lies in $\mathscr{A}$, which is absurd. This proves that $X \in \mathscr{A}$.

Write $X = X_1 \cup \cdots \cup X_n$, where the X_i are irreducible closed subsets. If Y is any maximal irreducible subset of X, then since $Y = \bigcup_{i=1}^{n} (Y \cap X_i)$, we must have $Y \cap X_i = Y$ for some i. Thus $Y = X_i$ (by maximality). $\square$

The proposition allows us to write a noetherian space as the union of its finitely many maximal irreducible subspaces; these are called the **irreducible components** of X.

Let us return now to affine n-space. Which of its closed subsets $\mathscr{V}(I)$ are irreducible?

Proposition C.   *A closed set X in $\mathbf{A}^n$ is irreducible if and only if its ideal $\mathscr{I}(X)$ is prime. In particular, $\mathbf{A}^n$ itself is irreducible.*

Proof.   Write $I = \mathscr{I}(X)$. Suppose X is irreducible. To show that I is prime, let $f_1(\mathsf{T})f_2(\mathsf{T}) \in I$. Then each $x \in X$ is a zero of $f_1(\mathsf{T})$ or of $f_2(\mathsf{T})$, i.e., X is covered by $\mathscr{V}(I_1) \cup \mathscr{V}(I_2)$, I_i the ideal generated by $f_i(\mathsf{T})$. Since X is irreducible, it must lie wholly within one of these two sets, i.e., $f_1(\mathsf{T}) \in I$ or $f_2(\mathsf{T}) \in I$, and I is prime.

In the other direction, suppose I is prime, but $X = X_1 \cup X_2$ (X_i closed in X). If neither X_i covers X, we can find $f_i(\mathsf{T}) \in \mathscr{I}(X_i)$, with $f_i(\mathsf{T}) \notin I$. But $f_1(\mathsf{T})f_2(\mathsf{T})$ vanishes on X, so $f_1(\mathsf{T})f_2(\mathsf{T}) \in I$, contradicting primeness. $\square$

As remarked in (1.1), a prime ideal is always a radical ideal; so the result just proved fits neatly into the $1-1$ correspondence established in (1.1) between radical ideals in $\mathsf{K}[\mathsf{T}_1, \ldots, \mathsf{T}_n]$ and closed sets in $\mathbf{A}^n$.

1.4. Products of Affine Varieties

The cartesian product of two (or more) topological spaces can be topologized in a fairly straightforward way, so as to yield a "product" in the category of topological spaces (where the morphisms are continuous maps). Since we have not yet introduced morphisms of affine varieties, it would be premature to look for an analogous categorical product here. But it is reasonable to ask that the product of two affine varieties $X \subset \mathbf{A}^n$, $Y \subset \mathbf{A}^m$, should look set-theoretically like the cartesian product $X \times Y \subset \mathbf{A}^{n+m}$. In particular, this obliges us to *define* $\mathbf{A}^n \times \mathbf{A}^m$ to be $\mathbf{A}^{n+m}$, and suggests that we impose on $X \times Y$ its induced topology as a subspace of $\mathbf{A}^{n+m}$. The question remains: Is $X \times Y$ an affine variety, i.e., is it *closed*? The answer is yes: If X is the zero set of polynomials $f_i(\mathsf{T}_1, \ldots, \mathsf{T}_n)$ and Y is the zero set of polynomials $g_i(\mathsf{U}_1, \ldots, \mathsf{U}_m)$, then $X \times Y$ is defined by the vanishing of all $f_i(\mathsf{T})g_j(\mathsf{U})$. (However, it is not immediately clear how to describe $\mathscr{I}(X \times Y)$ in terms of $\mathscr{I}(X)$ and $\mathscr{I}(Y)$. The obvious guess does turn out to be correct.)

It must be emphasized that the induced topology on $X \times Y$ is *not* what we would get by taking the usual product topology. For example, very few sets are closed in the product topology of $\mathbf{A}^1$ times itself, as contrasted with $\mathbf{A}^2$. The topology on $\mathbf{A}^n \times \mathbf{A}^m$ which identifies this set with $\mathbf{A}^{n+m}$ may be called the **Zariski product topology**.

> **Proposition.** *Let $X \subset \mathbf{A}^n$, $Y \subset \mathbf{A}^m$, be closed irreducible sets. Then $X \times Y$ is closed and irreducible in $\mathbf{A}^{n+m}$.*

Proof. Only the irreducibility remains to be checked. Suppose $X \times Y$ is the union of two closed subsets Z_1, Z_2. We have to show that it coincides with one of them. If $x \in X$, $\{x\} \times Y$ is closed (since $\{x\}$ is closed). It is also irreducible: any decomposition as a union of closed subsets would imply a similar decomposition of Y, since a closed subset of $\{x\} \times Y$ clearly has to be of the form $\{x\} \times Z$ for some closed subset Z of Y. Therefore the intersections of $\{x\} \times Y$ with Z_1, Z_2 cannot both be proper. So $X = X_1 \cup X_2$, where $X_i = \{x \in X | \{x\} \times Y \subset Z_i\}$.

Next we observe that each X_i is *closed* in X: For each $y \in Y$, $X \times \{y\}$ is closed, so that $(X \times \{y\}) \cap Z_i$ is closed, which implies in turn that the set $X_y^{(i)}$ of first coordinates is closed in X. But $X_i = \bigcap_{y \in Y} X_y^{(i)}$.

From the irreducibility of X we conclude that either $X = X_1$ or $X = X_2$, i.e., either $X \times Y = Z_1$ or $X \times Y = Z_2$. $\square$

1.5. Affine Algebras and Morphisms

Every category needs morphisms. Since affine varieties are defined by polynomial equations, it is only natural to turn to **polynomial functions**. If X is closed in $\mathbf{A}^n$, each polynomial $f(\mathsf{T}) \in \mathsf{K}[\mathsf{T}]$ defines a K-valued function on X by the rule $x \mapsto f(x)$. But other polynomials may define the same function; indeed, a moment's consideration should convince the reader that the distinct polynomial functions on X are in 1–1 correspondence with the elements of the residue class ring $\mathsf{K}[\mathsf{T}]/\mathscr{I}(X)$. We denote this ring $\mathsf{K}[X]$ and call it the **affine algebra** of X (or the **algebra of polynomial functions** on X). It is a finitely generated algebra over K, which is **reduced** (i.e., has no nonzero nilpotent elements), in view of the fact that $\mathscr{I}(X)$ is its own radical. When X is irreducible, i.e., when $\mathscr{I}(X)$ is a prime ideal (Proposition 1.3C), $\mathsf{K}[X]$ is an integral domain. So we may form its field of fractions, denoted $\mathsf{K}(X)$ and called the **field of rational functions** on X. This is a finitely generated field extension of K. Although we are sometimes compelled to work with reducible varieties, we shall often be able to base our arguments on the irreducible case, where the function field is an indispensable tool.

The affine algebra $\mathsf{K}[X]$ stands in the same relation to X as $\mathsf{K}[\mathsf{T}]$ does to $\mathbf{A}^n$. With its aid we can begin to formulate a more intrinsic notion of "affine variety", thereby liberating X from the ambient space $\mathbf{A}^n$. To begin with, X is a noetherian topological space (in the Zariski topology), with basis consisting of **principal open subsets** $X_f = \{x \in X \,|\, f(x) \neq 0\}$ for $f \in \mathsf{K}[X]$. It is easy to see that the closed subsets of X correspond 1–1 with the radical ideals of $\mathsf{K}[X]$ (by adapting the Nullstellensatz from $\mathsf{K}[\mathsf{T}]$ to $\mathsf{K}[\mathsf{T}]/\mathscr{I}(X)$), the irreducible ones belonging to prime ideals. In particular, we find that the points of X are in 1–1 correspondence with the maximal ideals of $\mathsf{K}[X]$, or with the K-algebra homomorphisms $\mathsf{K}[X] \to K$. So X is in a sense recoverable from $\mathsf{K}[X]$.

Indeed, let R be an arbitrary reduced, finitely generated commutative algebra over K, say $R = \mathsf{K}[t_1, \ldots, t_n]$ (the number n and this choice of generators being nonunique). Then R is a homomorphic image of $\mathsf{K}[\mathsf{T}_1, \ldots, \mathsf{T}_n]$, which is "universal" among the commutative, associative K-algebras on n generators. Moreover, the fact that R is reduced just says that the kernel of the epimorphism sending T_i to t_i is a radical ideal I. So R is isomorphic to the affine algebra of the variety $X \subset \mathbf{A}^n$ defined by I. This points the way to an equivalence of categories, to which we shall return shortly. One advantage of this approach is that it enables us to give to any principal open subset X_f of an irreducible affine variety X its own structure of affine variety (in an affine space of higher dimension): Define R to be the subring of $\mathsf{K}(X)$ generated by $\mathsf{K}[X]$ along with $1/f$, and notice that R is automatically a (reduced) finitely generated K-algebra. Moreover, the maximal ideals of R correspond 1–1 with their intersections with $\mathsf{K}[X]$, which are just the maximal ideals excluding f. In turn, the points of the affine variety defined by R correspond naturally to the points of X_f. What we have done, in effect, is to identify points of $X_f \subset X \subset \mathbf{A}^n$ with points $(x_1, \ldots, x_n, 1/f(x))$ in $\mathbf{A}^{n+1}$.

Next let $X \subset \mathbf{A}^n$, $Y \subset \mathbf{A}^m$, be arbitrary affine varieties. By a **morphism** $\varphi : X \to Y$ we mean a mapping of the form $\varphi(x_1, \ldots, x_n) = (\psi_1(x), \ldots, \psi_m(x))$, where $\psi_i \in \mathsf{K}[X]$. Notice that a morphism $X \to Y$ is always induced by a morphism $\mathbf{A}^n \to \mathbf{A}^m$ (use any pre-images of the ψ_i in $\mathsf{K}[\mathbf{A}^n] = \mathsf{K}[\mathsf{T}]$), and that a morphism $X \to \mathbf{A}^1$ is the same thing as a polynomial function on X. A morphism $\varphi : X \to Y$ is continuous for the Zariski topologies involved.

Indeed, if $Z \subset Y$ is the set of zeros of polynomial functions f_i on Y, then $\varphi^{-1}(Z)$ is the set of zeros of the polynomial functions $f_i \circ \varphi$ on X.

With a morphism $\varphi : X \to Y$ is associated its **comorphism** $\varphi^* : \mathsf{K}[Y] \to \mathsf{K}[X]$ defined by $\varphi^*(f) = f \circ \varphi$. It is obvious that the image of φ^* does lie in $\mathsf{K}[X]$, that φ^* is a homomorphism of K-algebras, and that the usual functorial properties hold: $1^* = \text{identity}, (\varphi \circ \psi)^* = \varphi^* \circ \psi^*$. Moreover, knowledge of φ^* is tantamount to knowledge of φ: $\mathsf{K}[Y]$ is generated (as K-algebra) by the restrictions to Y of the coordinate functions $\mathsf{T}_1, \ldots, \mathsf{T}_m$ on $\mathbf{A}^m$, call them t_i, and $\varphi^*(t_i)$ is just the function ψ_i used above to define φ. This shows that every K-algebra homomorphism $\mathsf{K}[Y] \to \mathsf{K}[X]$ arises as the comorphism of some morphism $X \to Y$.

The preceding discussion establishes, in effect, a (contravariant) equivalence between the category of affine K-algebras (with the K-algebra homomorphisms as morphisms) and the category of affine varieties (with morphisms as defined above). This more intrinsic way to view affine varieties, cut loose from specific embeddings in affine space, will be explored further in §2. The "product" introduced in (1.4) turns out to be a categorical product, and corresponds in fact to the *tensor product* of K-algebras (which is known to be the "coproduct" in the category of commutative rings).

Suppose $\varphi : X \to Y$ is a morphism for which $\varphi(X)$ is dense in Y. Then φ^* is *injective* (cf. Exercise 11 or (2.5) below). In particular, if X and Y are irreducible, φ^* induces an embedding of $\mathsf{K}(Y)$ into $\mathsf{K}(X)$.

1.6. Projective Varieties

Geometers have long recognized the advantages of working in "projective space", where the behavior of loci at infinity can be put on an equal footing with the behavior elsewhere. From the algebraic viewpoint, the theory of projective varieties runs parallel to that of affine varieties, with homogeneous polynomials taking the place of arbitrary polynomials. We shall give only a brief introduction here, adequate for the later applications. In §2 the affine and projective theories will be subsumed under an abstract theory of "varieties", while in §6 the "completeness" of projective varieties (analogous to compactness) will be discussed systematically.

Projective n-space $\mathbf{P}^n$ may be defined to be the set of equivalence classes of $\mathsf{K}^{n+1} - \{(0, 0, \ldots, 0)\}$ relative to the equivalence relation:

$$(x_0, x_1, \ldots, x_n) \sim (y_0, y_1, \ldots, y_n)$$

if and only if there exists $a \in \mathsf{K}^*$ such that $y_i = ax_i$ for all i. Intuitively, $\mathbf{P}^n$ is just the collection of all lines through the origin in K^{n+1}. Sometimes it is convenient, when working with a vector space V of dimension $n + 1$, to identify the set of all 1-dimensional subspaces of V with $\mathbf{P}^n$; we write $\mathbf{P}(V)$ for $\mathbf{P}^n$ in this case.

Each point in $\mathbf{P}^n$ can be described by **homogeneous coordinates** x_0, $x_1, \ldots, x_n$, which are not unique but may be multiplied by any nonzero scalar. If a locus in $\mathbf{P}^n$ is to be described by polynomial equations (in indeterminates $X_0, X_1, \ldots, X_n$), this nonuniqueness forces us to require that the polynomials be homogeneous. Recall that $f(X_0, \ldots, X_n)$ is **homogeneous** of degree d if it is a linear combination of monomials $X_0^{i_0} X_1^{i_1} \cdots X_n^{i_n}$ with $\sum i_j = d$. Such a polynomial satisfies $f(ax_0, \ldots, ax_n) = a^d f(x_0, \ldots, x_n)$; in particular, if it takes the value 0 for one set of homogeneous coordinates of a point in $\mathbf{P}^n$, it takes the value 0 for any other choice.

Now we can topologize $\mathbf{P}^n$ by taking a **closed** set to be the common zeros of a collection of homogeneous polynomials, or equally well of the ideal they generate. Notice that the ideal generated by some homogeneous polynomials is a **homogeneous ideal** (i.e., contains the homogeneous parts of all its elements). It is a straightforward matter to define operators $\mathscr{V}$, $\mathscr{I}$, as in the affine case, thereby setting up an inclusion-reversing correspondence between **projective varieties** (closed subsets of $\mathbf{P}^n$) and homogeneous ideals. As in the affine case, ideals of the form $\mathscr{I}(X)$ are radical ideals. There is a version here of the Nullstellensatz, which requires only a minor adjustment. Namely, the ideal I_0 generated by $X_0, \ldots, X_n$ is proper, but clearly has no common zero in $\mathbf{P}^n$ (since the origin of K^{n+1} has been discarded). So we are led to the following formulation, which the reader can easily verify using the affine Nullstellensatz (1.1):

Proposition. *The operators $\mathscr{V}$, $\mathscr{I}$ set up a $1{-}1$ inclusion-reversing correspondence between the closed subsets of $\mathbf{P}^n$ and the homogeneous radical ideals of $K[X_0, \ldots, X_n]$ other than I_0.* $\square$

The discussion of irreducible components in (1.3) applies here as well. In particular, the irreducible projective varieties belong to the homogeneous prime ideals (other than I_0).

As in the affine case, the principal open sets form a basis for the Zariski topology on $\mathbf{P}^n$. Certain of these are especially useful, because they are naturally isomorphic to affine n-space. (This provides a suggestive link with the affine case, to be exploited in the general discussion of "varieties" in §2.) Let U_i be the set of points in $\mathbf{P}^n$ having i^{th} homogeneous coordinate nonzero. Then U_i corresponds $1{-}1$ with the points of $\mathbf{A}^n$, via $(x_0, \ldots, x_n) \mapsto \left(\dfrac{x_0}{x_i}, \ldots, \dfrac{x_{i-1}}{x_i}, \dfrac{x_{i+1}}{x_i}, \ldots, \dfrac{x_n}{x_i}\right)$. These quotients of homogeneous coordinates are called **affine coordinates** on U_i ($0 \leqslant i \leqslant n$), Notice that the U_i cover $\mathbf{P}^n$.

The correspondence between U_i and $\mathbf{A}^n$ is not just set-theoretic: the Zariski topologies also correspond. To see this, introduce indeterminates $T_1, \ldots, T_n$. To each polynomial $f(T_1, \ldots, T_n)$ we may associate a homogeneous polynomial $X_i^{\deg. f} f(X_0/X_i, \ldots, X_{i-1}/X_i, X_{i+1}/X_i, \ldots, X_n/X_i)$, where deg. f is the largest degree of any monomial occurring in $f(T)$. Then if $X \subset \mathbf{A}^n$ is the zero set of certain polynomials $f(T)$, the image of X in U_i is the intersection of U_i with the zero set in $\mathbf{P}^n$ of the corresponding "homogenized" polynomials. In the reverse direction, let $X \subset \mathbf{P}^n$ be the zero set of certain homogeneous polynomials $f(X_0, \ldots, X_n)$. For each i, consider $f(X_0/X_i, \ldots, X_{i-1}/X_i, 1, X_{i+1}/X_i, \ldots, X_n/X_i) = g(T_1, \ldots, T_n)$, $T_k = X_k/X_i$. It is clear that $X \cap U_i$ corresponds to the zero set in $\mathbf{A}^n$ of these polynomials $g(T)$.

The main point of the preceding discussion is that a subset of $\mathbf{P}^n$ is closed if and only if its intersections with the **affine open sets** U_i are all closed (U_i being identified canonically with $\mathbf{A}^n$). More generally, if X is closed in $\mathbf{P}^n$, a subset Y of X is closed in X (or in $\mathbf{P}^n$) if and only if all $Y \cap U_i$ are closed. This "affine criterion" will be put to good use immediately.

1.7. Products of Projective Varieties

Let $X \subset \mathbf{P}^n$, $Y \subset \mathbf{P}^m$ be two projective varieties. If there is to be a "product" of X and Y, its underlying set ought to be the Cartesian product. But this set cannot be straightforwardly identified with a subset of $\mathbf{P}^n \times \mathbf{P}^m$, due to the vagaries of homogeneous coordinates. Instead, we must resort to a more elaborate embedding. To this end, we map the Cartesian product $\mathbf{P}^n \times \mathbf{P}^m$ into $\mathbf{P}^q$, where $q = (n+1)(m+1) - 1$, by the recipe: $\varphi((x_0, \ldots, x_n), (y_0, \ldots, y_m)) = (x_0 y_0, \ldots, x_0 y_m, x_1 y_0, \ldots, x_1 y_m, \ldots, x_n y_0, \ldots, x_n y_m)$. Note that this is unambiguous.

We want to show that the image of φ is *closed* in $\mathbf{P}^q$, using the affine criterion developed in (1.6). Denote the homogeneous coordinates on $\mathbf{P}^n$ by X_i, on $\mathbf{P}^m$ by Y_j, and on $\mathbf{P}^q$ by Z_{ij} ($0 \leqslant i \leqslant n, 0 \leqslant j \leqslant m$). Let $\mathbf{P}_i^n, \mathbf{P}_j^m, \mathbf{P}_{ij}^q$ be the corresponding affine open subsets, with affine coordinates S_i, T_j, U_{ij}. Evidently φ maps $\mathbf{P}_i^n \times \mathbf{P}_j^m$ into $\mathbf{P}_{ij}^q$. For ease of notation, we treat just the (typical) case $i = j = 0$. In affine coordinates, φ sends $((s_1, \ldots, s_n), (t_1, \ldots, t_n))$ to $(\ldots, u_{k\ell}, \ldots)$, where $u_{k\ell} = s_k t_\ell$ ($k, \ell \geqslant 1$), $u_{k0} = s_k$, $u_{0\ell} = t_\ell$. So the image in $\mathbf{P}_{00}^q$ is just the locus of the equations $U_{k\ell} = U_{k0} U_{0\ell}$ ($k, \ell \geqslant 1$). This shows that the image of φ is closed, as asserted.

Moreover, it is easy to invert φ on each affine open set such as $\mathbf{P}_{00}^q$: Send $(\ldots, u_{k\ell}, \ldots)$ to $((u_{10}, \ldots, u_{n0}), (u_{01}, u_{02}, \ldots, u_{0m}))$. So φ actually induces isomorphisms of the affine products $\mathbf{P}_i^n \times \mathbf{P}_j^m$ onto their images. This allows us finally to deal with arbitrary closed sets $X \subset \mathbf{P}^n$, $Y \subset \mathbf{P}^m$. X is the union of its intersections X_i with the $\mathbf{P}_i^n$, and each X_i is closed in the affine space $\mathbf{P}_i^n$; similarly for Y. Thanks to (1.4), $X_i \times Y_j$ is closed in $\mathbf{P}_i^n \times \mathbf{P}_j^m$ and hence maps isomorphically onto a closed subset of the affine open set $\varphi(\mathbf{P}_i^n \times \mathbf{P}_j^m)$ in $\varphi(\mathbf{P}^n \times \mathbf{P}^m)$. It follows from the affine criterion (1.6) that $\varphi(X \times Y)$ is *closed* in $\varphi(\mathbf{P}^n \times \mathbf{P}^m)$, which in turn is closed in $\mathbf{P}^q$. To sum up:

Proposition. *The map* $\varphi: \mathbf{P}^n \times \mathbf{P}^m \to \mathbf{P}^{nm+n+m}$ *defined above is a bijection onto a closed subset. If X is closed in $\mathbf{P}^n$ and Y is closed in $\mathbf{P}^m$, then $\varphi(X \times Y)$ is closed in* $\mathbf{P}^{nm+n+m}$. $\quad\square$

Thus the Cartesian product of two projective varieties can be identified with another projective variety. Fortunately, the way in which this is done turns out to conform well with the categorical notion of "product" (2.4).

1.8. Flag Varieties

Some of the most interesting examples of projective varieties (from our point of view) result from the following construction, which goes back to Grassmann.

Let V be an n-dimensional vector space over K, with **exterior algebra** $\wedge V$ (the quotient of the tensor algebra on V by the ideal generated by all $v \otimes v$, $v \in V$). Recall that $\wedge V$ is a finite dimensional graded algebra over K, with $\wedge^0 V = K$, $\wedge^1 V = V$. If $v_1, \ldots, v_n$ is an ordered basis of V, then the $\binom{n}{d}$ wedge (or exterior) products $v_{i_1} \wedge \cdots \wedge v_{i_d}$ $(i_1 < i_2 < \cdots < i_d)$ form a basis of $\wedge^d V$. Notice that $\wedge^n V$ is 1-dimensional, i.e., the wedge product of an arbitrary basis of V is well-determined up to a nonzero scalar multiple. If W is a subspace of V, then $\wedge^d W$ may be identified canonically with a subspace of $\wedge^d V$.

The preceding remarks show that there is a map ψ from the collection $\mathfrak{G}_d(V)$ of all d-dimensional subspaces of V into $\mathbf{P}(\wedge^d V)$, defined by sending a subspace D to the point in projective space belonging to $\wedge^d D$ $(d \geqslant 1)$. We assert that ψ is *injective*. Indeed, let D, D' be two d-dimensional subspaces. Choose a basis of V so that $v_1, \ldots, v_d$ span D, while $v_r, \ldots, v_{r+d-1}$ span D'. Then $v_1 \wedge \cdots \wedge v_d$ cannot be proportional to $v_r \wedge \cdots \wedge v_{r+d-1}$ unless $r = 1$, i.e., unless $D = D'$.

In order to endow $\mathfrak{G}_d(V)$ with the structure of a projective variety, it now suffices to check that the image of ψ is *closed*. Thanks to the affine criterion (1.6), it is enough to do this on affine open sets which cover $\mathbf{P}(\wedge^d V)$. (Of course, the extreme cases $d = 1, d = n$, require no checking, since then $\mathfrak{G}_d(V)$ is respectively $\mathbf{P}(V)$ or a point.)

Fix an ordered basis $(v_1, \ldots, v_n)$ of V and the associated basis elements $v_{i_1} \wedge \cdots \wedge v_{i_d}$ of $\wedge^d V$. A typical affine open set U in $\mathbf{P}(\wedge^d V)$ then consists of points whose homogeneous coordinate relative to (say) $v_1 \wedge \cdots \wedge v_d$ is non-zero. Let us show that Im ψ intersects this U in a closed subset. Set $D_0 = $ span of $v_1, \ldots, v_d$. Clearly, $\psi(D)$ belongs to U if and only if the natural projection of V onto D_0 maps D isomorphically onto D_0. In this case, the inverse images of $v_1, \ldots, v_d$ comprise a basis of D having the form: $v_i + x_i(D)$, where $x_i(D) = \sum_{j>d} a_{ij} v_j$. (And this is the only basis of D having this form.) The wedge product looks like:

$$v_1 \wedge \cdots \wedge v_d + \sum_{1 \leqslant i \leqslant d} (v_1 \wedge \cdots \wedge x_i(D) \wedge \cdots \wedge v_d) + (*),$$

where $(*)$ involves basis vectors with two or more of $v_1, \ldots, v_d$ omitted. Here $v_1 \wedge \cdots \wedge x_i(D) \wedge \cdots \wedge v_d = \sum_{j>d} a_{ij}(v_1 \wedge \cdots \wedge v_j \wedge \cdots \wedge v_d)$, with v_j substituted for v_i. Thus $\pm a_{ij}$ $(1 \leqslant i \leqslant d, d + 1 \leqslant j \leqslant n)$ may be recovered as the coefficient of the basis element $v_1 \wedge \cdots \wedge \hat{v}_i \wedge \cdots \wedge v_d \wedge v_j$ (v_i omitted), in the wedge product of the above basis of D. Furthermore, the coefficients in $(*)$ are obviously *polynomial functions* of the a_{ij}, independent of D.

Conversely, if we prescribe the $d(n - d)$ scalars a_{ij} arbitrarily, it is clear that the resulting vectors $v_i + x_i(D)$ span a d-dimensional subspace of V

whose image under ψ lies in U. The upshot is that $\operatorname{Im} \psi \cap U$ consists of all points with (affine) coordinates $(\ldots a_{ij} \ldots, f_k(a_{ij}) \ldots)$, where the a_{ij} are arbitrary and the f_k are polynomial functions on $\mathbf{A}^{d(n-d)}$. This set can be viewed as the *graph* of a morphism from $\mathbf{A}^{d(n-d)}$ into another affine space. As such, it is closed in the Zariski product topology (cf. Exercise 8); and in turn $\operatorname{Im} \psi \cap U$ is closed in U (cf. (1.4)).

The **Grassmann varieties** $\mathfrak{G}_d(V)$ lead us to other projective varieties, as follows. A **flag** in V is, by definition, a chain $0 \subset V_1 \subset \cdots \subset V_k = V$ of subspaces of V, each properly included in the next. A **full flag** is one for which $k = \dim V$ (i.e., $\dim V_{i+1}/V_i = 1$). $\mathfrak{F}(V)$ denotes the collection of all full flags of V. We want to give it the structure of projective variety (to be called the **flag variety** of V).

Thanks to (1.7), it is possible to give the Cartesian product $\mathfrak{G}_1(V) \times \mathfrak{G}_2(V) \times \cdots \times \mathfrak{G}_n(V)$ the structure of a projective variety. $\mathfrak{F}(V)$ identifies in an obvious way with a subset, which we need only show to be *closed*. To avoid cumbersome notation, we just consider the product $\mathfrak{G}_d(V) \times \mathfrak{G}_{d+1}(V)$. Once it is proved that the set S of pairs (D, D') for which $D \subset D'$ is closed, the reader should have no difficulty in completing the argument.

As before, we may fix a basis $v_1, \ldots, v_n$ of V, and consider the various affine open subsets of $\mathbf{P}(\wedge^d V)$, $\mathbf{P}(\wedge^{d+1} V)$, whose products cover the product variety. We can limit our attention to pairs such as U, U', where U is defined as before relative to $v_1 \wedge \cdots \wedge v_d$, and U' consists of points in $\mathbf{P}(\wedge^{d+1} V)$ with nonzero coordinate relative to $v_1 \wedge \cdots \wedge v_{d+1}$. (The set S is already covered by products of the form $U \times U'$.) If D (resp. D') has image in U (resp. U'), we get (as before) canonical bases: $v_i + x_i(D)$, $1 \leq i \leq d$; $v_i + y_i(D')$, $1 \leq i \leq d + 1$. Here $x_i(D) = \sum_{j>d} a_{ij} v_j$, $y_i(D') = \sum_{j>d+1} b_{ij} v_j$. A quick computation with these bases shows that $D \subset D'$ if and only if $x_i(D) = y_i(D') + a_{i, d+1}(v_{d+1} + y_{d+1}(D'))$ for $1 \leq i \leq d$. This in turn translates into certain polynomial conditions on the a_{ij}, b_{ij}, whence S intersects $U \times U'$ in a closed set.

Exercises

1. If I, J are ideals in $\mathsf{K}[\mathsf{T}_1, \ldots, \mathsf{T}_n]$, recall that IJ is the ideal consisting of all sums of products $f(\mathsf{T})g(\mathsf{T})(f(\mathsf{T}) \in I, g(\mathsf{T}) \in J)$. Prove that $\mathscr{V}(IJ) = \mathscr{V}(I \cap J)$. Show by example that IJ may be included properly in $I \cap J$.
2. Each radical ideal in $\mathsf{K}[\mathsf{T}_1, \ldots, \mathsf{T}_n]$ is an intersection of prime ideals.
3. Any subspace of a noetherian topological space is also noetherian.
4. Let X be a noetherian topological space, Y a subspace having irreducible components $Y_1, \ldots, Y_n$. Prove that the $\bar{Y}_i$ are the irreducible components of $\bar{Y}$.
5. Find an open subset of $\mathbf{A}^2$ which (with its given Zariski topology) cannot be isomorphic to any affine variety. [Delete the point $(0, 0)$.]
6. Show that a map between affine varieties which is continuous for the Zariski topologies need not be a morphism. [Consider $\mathbf{A}^1 \to \mathbf{A}^1$.]

7. Prove that projection onto one of the coordinates defines a morphism $\mathbf{A}^n \to \mathbf{A}^1$, which in general fails to send closed sets to closed sets.

8. The graph of a morphism $X \to Y$ (X, Y affine varieties) is closed in $X \times Y$. What if X, Y are projective varieties?

9. Complete the proof in (1.8) that $\mathfrak{F}(V)$ is closed in the product

$$\mathfrak{G}_1(V) \times \cdots \times \mathfrak{G}_n(V).$$

10. Show that every automorphism of $\mathbf{A}^1$ ($=$ bijective morphism whose inverse is again a morphism) has the form: $x \mapsto ax + b$ ($a \in \mathsf{K}^*, b \in \mathsf{K}$).

11. If $\varphi : X \to Y$ is a morphism of affine varieties for which $\varphi(X)$ is dense in Y, then $\varphi^* : \mathsf{K}[Y] \to \mathsf{K}[X]$ is injective.

12. Let X be an irreducible affine variety, $f \in \mathsf{K}(X)$. The set of points $x \in X$ at which f is **defined** (i.e., f can be written as g/h, with $g, h \in \mathsf{K}[X]$ and $h(x) \neq 0$) is open.

Notes

Good references for the sort of algebraic geometry we require are Mumford [3, Chapter I] and Shafarevich [1], [2].

2. Varieties

The notion of "prevariety" is introduced here, as a common generalization of the notions of affine and projective variety. After defining morphisms and products, we discuss in (2.5) the additional assumption ("Hausdorff axiom") which characterizes "varieties".

2.1. Local Rings

A point on a projective variety has an open neighborhood which looks just like an affine variety. It is this "local" behavior which suggests the correct route to follow. There is an analogy with the theory of manifolds, where each point has a neighborhood indistinguishable from an open set in euclidean space. But the Zariski topology does not separate points in the ordinary way; so our construction will lead (in the irreducible case) to a covering by affine open sets which overlap a great deal.

To pinpoint the local behavior of an affine variety X, assume first that X is *irreducible*, with function field $\mathsf{K}(X)$. Consider the rational functions f which are defined at $x \in X$, i.e., for which there is an expression $f = g/h$ ($g, h \in \mathsf{K}[X]$) with $h(x) \neq 0$. One sees easily that these functions form a ring $\mathcal{O}_x$ including $\mathsf{K}[X]$, which we call the **local ring** of x on X. In fact, $\mathcal{O}_x$ results from the construction described in (0.10) and is a "local ring" in the technical sense: If $R = \mathsf{K}[X]$, $P = \mathscr{I}(x)$, then $R_P = \mathcal{O}_x$. The unique maximal ideal m_x of $\mathcal{O}_x$ consists of all rational functions representable as g/h ($g, h \in \mathsf{K}[X]$), where $g(x) = 0$, $h(x) \neq 0$.

The local rings of an irreducible affine variety X actually determine $\mathsf{K}[X]$ (hence determine X), as the following proposition shows.

Proposition. *Let X be an irreducible affine variety. Then* $\mathsf{K}[X] = \bigcap_{x \in X} \mathcal{O}_x$.

Proof. $\mathsf{K}[X]$ is evidently included in all $\mathcal{O}_x$. Conversely, let $f \in \mathsf{K}(X)$ be in all $\mathcal{O}_x$. This means that for a given x, $f = g/h$ for some $g, h \in \mathsf{K}[X]$ such that $h(x) \neq 0$. Of course, this representation of f is not unique. We consider the ideal I generated by all possible denominators h, as x ranges over X. If I were a proper ideal in $\mathsf{K}[X]$, it would have a common zero (by the analogue for X of the Nullstellensatz (1.1)), which is impossible. So $I = \mathsf{K}[X]$, allowing us to write $f = g/1$ $(g \in \mathsf{K}[X])$. $\square$

2.2. Prevarieties

Let X be an irreducible affine variety. To each (nonempty) open subset $U \subset X$, we may associate the subring of $\mathsf{K}(X)$ consisting of functions which are **regular** (or everywhere defined) on U:

$$\mathcal{O}_X(U) = \bigcap_{x \in U} \mathcal{O}_x.$$

For example, Proposition 2.1 shows that $\mathcal{O}_X(X) = \mathsf{K}[X]$ or, more generally, that $\mathcal{O}_X(X_f) = \mathsf{K}[X_f] = \mathsf{K}[X]_f$ (since the local rings of points on the affine variety X_f coincide with those on X).

$\mathcal{O}_X$ is an example of a **sheaf of functions** on X. For our purposes, a sheaf of functions on a topological space X is a function $\mathscr{S}$ which assigns to each open $U \subset X$ a K-algebra $\mathscr{S}(U)$ consisting of K-valued functions on U, subject to two further requirements:

(S1) If $U \subset V$ are two open sets, and $f \in \mathscr{S}(V)$, then $f|U \in \mathscr{S}(U)$.

(S2) Let U be an open set covered by open subsets U_i (i running over some index set I). Given $f_i \in \mathscr{S}(U_i)$, suppose that f_i agrees with f_j on $U_i \cap U_j$ for all $i, j \in I$. Then there exists $f \in \mathscr{S}(U)$ whose restriction to U_i is f_i $(i \in I)$.

It is clear that $\mathcal{O}_X$ satisfies (S1) and (S2). Moreover, in case $X = X_1 \cup \cdots \cup X_t$ is an arbitrary affine variety, with irreducible components X_i, there is only one reasonable way to define a sheaf on X which extends all the $\mathcal{O}_{X_i}$. Namely, write an open set $U \subset X$ as the union of the open sets $U_i = U \cap X_i$, and define $\mathcal{O}_X(U)$ to be the set of K-valued functions on U whose restriction to each U_i lies in $\mathcal{O}_{X_i}(U_i)$. In particular, $\mathcal{O}_X(X) = \mathsf{K}[X]$, as in the irreducible case.

In case X is an irreducible affine variety, we can recover the local rings $\mathcal{O}_x$ as **stalks** of the sheaf $\mathcal{O}_X$: The open sets containing a given point x form an inverse system, relative to inclusion, and it is immediate that $\mathcal{O}_x = \varinjlim_U \mathcal{O}_X(U)$ (direct limit over these U), since in this case the direct limit is just the union (in $\mathsf{K}(X)$). (This suggests defining $\mathcal{O}_x$ as such a direct limit when X is not irreducible.)

With the example $\mathbf{P}^n$ in mind, we next define an **irreducible prevariety** X to be an irreducible noetherian topological space, endowed with a sheaf $\mathcal{O}_X$ of K-valued functions, such that X is the union of finitely many open subsets U_i, each isomorphic to an affine variety when given the restricted sheaf of functions $\mathcal{O}_X|U_i$. (It is clear how $\mathcal{O}_X$ induces a sheaf of functions on any open subset of X.) By a **prevariety** we shall mean a noetherian topological space X, whose irreducible components X_i are irreducible prevarieties in such a way that $\mathcal{O}_{X_i}$ and $\mathcal{O}_{X_j}$ induce the same sheaf of functions on $X_i \cap X_j$ for all i, j. Then there is a unique sheaf $\mathcal{O}_X$ extending the $\mathcal{O}_{X_i}$, as in the affine case above. The elements of $\mathcal{O}_X(U)$ are called the **regular** functions on U. The open sets U_i above are called **affine open subsets** of X. More generally, we give this name to any open subset of X which, with its induced sheaf of functions, is isomorphic to an affine variety.

Let us see that $X = \mathbf{P}^n$ qualifies as an irreducible prevariety, given the Zariski topology and a covering by open subsets U_i (each corresponding to $\mathbf{A}^n$), as in (1.6). The sheaf $\mathcal{O}_X$ has to be defined so as to induce on U_i the sheaf canonically attached to $\mathbf{A}^n$. But this is easy enough. First attach to $x \in U_i$ its local ring $\mathcal{O}_x$ in $K(\mathbf{A}^n)$; note that this is independent of the choice of U_i containing x. Then define $\mathcal{O}_X(U) = \bigcap_{x \in U} \mathcal{O}_x$, to get the desired sheaf on X.

Arbitrary projective varieties $X \subset \mathbf{P}^n$ can be given an induced structure of prevariety. This is true more generally for *open* or *closed* subsets of a prevariety, as follows. Say $(X, \mathcal{O}_X)$ is an irreducible prevariety, $U \subset X$ a (nonempty) open set. Then $\mathcal{O}_X$ restricts (as above) to a sheaf of functions on U, making U also an irreducible prevariety. On the other hand, if $Z \subset X$ is closed (and irreducible), we can cover Z with finitely many closed subsets Z_i of the affine varieties U_i covering X. Each Z_i has a canonical sheaf $\mathcal{O}_{Z_i}$, and these agree on the intersections $Z_i \cap Z_j$; so they may be patched together as before to yield $\mathcal{O}_Z$, which is independent of the choices made.

Call a subset of a topological space **locally closed** if it is the intersection of an open set and a closed set. We call the locally closed subsets of a prevariety X, with their induced sheaves of functions described above, the **subprevarieties** of X. Actually, the cases of interest to us all turn out to be obtainable as open subsets of projective varieties: these are called **quasiprojective** varieties. But it is more natural to work in a slightly more general framework.

Notice that when X is an irreducible prevariety, covered by affine open sets U_i, the irreducibility forces $U_i \cap U_j$ to be nonempty. It follows that U_i, U_j must have the same function field, which we call the **function field** $K(X)$ of X.

2.3. Morphisms

A mapping $\varphi: X \to Y$ (X, Y prevarieties) should be called a **morphism** only if it respects the essential structure of X: its topology and its sheaf of functions. So we impose the following two conditions:

(M1) φ is continuous.

(M2) If $V \subset Y$ is open and $U = \varphi^{-1}(V)$, then $f \circ \varphi \in \mathcal{O}_X(U)$ whenever $f \in \mathcal{O}_Y(V)$.

It is easy to check that this definition is equivalent to the earlier one (1.5) when X, Y are affine varieties. Evidently the restriction of a morphism to a subprevariety is again a morphism. Note too that we get an obvious notion of **isomorphism** for prevarieties.

Let us take a closer look at condition (M2). The assignment $f \mapsto f \circ \varphi$ is a K-algebra homomorphism $\mathcal{O}_Y(V) \to \mathcal{O}_X(\varphi^{-1}(V))$, which we denote φ^* and call the **comorphism** of φ. (Strictly speaking, φ^* here ought to be denoted φ_V^* or the like.) In case X, Y are irreducible and $\varphi(X)$ is dense in Y, the co-morphism of φ can be thought of globally as a ring homomorphism $\mathsf{K}(Y) \to \mathsf{K}(X)$, whose restriction to $\mathcal{O}_Y(V)$ has image in $\mathcal{O}_X(\varphi^{-1}(V))$. Here φ^* is injective, enabling us to treat $\mathsf{K}(X)$ as a field extension of $\mathsf{K}(Y)$ (cf. the affine case (1.5)).

What effect does a morphism $\varphi : X \to Y$ have on *local rings*? Say X, Y are irreducible with $\varphi(X)$ dense in Y, $\varphi^* : \mathsf{K}(Y) \to \mathsf{K}(X)$. Since $\mathcal{O}_x$ ($x \in X$) is just the union ($=$ direct limit in this case) of all $\mathcal{O}_X(U)$ (U an open neighborhood of x), and similarly for $\mathcal{O}_y$ ($y \in Y$), it is clear that φ^* maps $\mathcal{O}_{\varphi(x)}$ into $\mathcal{O}_x$ (sending $\mathfrak{m}_{\varphi(x)}$ into $\mathfrak{m}_x$). Conversely, this condition (at least in the irreducible case) could be used in place of (M2), since $\mathcal{O}_X(U) = \bigcap_{x \in U} \mathcal{O}_x$.

It is important to be able to recognize when a mapping of prevarieties is a morphism. For this we develop an **affine criterion**.

Proposition. *Let $\varphi : X \to Y$ be a mapping (X, Y prevarieties). Suppose there is a covering of Y by affine open sets V_i ($i \in I$, I a finite index set) and a covering of X by open sets U_i, such that:*
(a) $\varphi(U_i) \subset V_i (i \in I)$;
(b) $f \circ \varphi \in \mathcal{O}_X(U_i)$ whenever $f \in \mathcal{O}_Y(V_i)$.

Then φ is a morphism.

Proof. First we reduce to the case in which all U_i are also *affine*: If U is an affine open subset of U_i, then (b) shows that composing with φ sends $\mathcal{O}_Y(V_i) = \mathsf{K}[V_i]$ into $\mathcal{O}_X(U) = \mathsf{K}[U]$. So it does no harm to replace U_i by an affine open covering (thereby enlarging the index set I).

Now the hypotheses insure that the restriction of φ to U_i is a morphism of affine varieties $\varphi_i : U_i \to V_i$, since φ_i is completely determined by the K-algebra homomorphism $\varphi_i^* : \mathsf{K}[V_i] \to \mathsf{K}[U_i]$ (cf. (1.5)). In particular, φ_i is *continuous*. This makes it obvious that φ is continuous.

It remains to verify (M2). Take an open set $V \subset Y$, and let $U = \varphi^{-1}(V)$. If $f \in \mathcal{O}_Y(V)$, then (b) implies that $f \circ \varphi \in \mathcal{O}_X(\varphi^{-1}(V \cap V_i))$. But $\varphi^{-1}(V \cap V_i) \supset U \cap U_i$, so $f \circ \varphi \in \mathcal{O}_X(U \cap U_i)$, for all $i \in I$. In turn, since U is the union of the $U \cap U_i$, and since $\mathcal{O}_X$ is a sheaf, $f \circ \varphi \in \mathcal{O}_X(U)$. $\quad\square$

For the rest of this subsection we concentrate on *irreducible* prevarieties. It is clear that a regular function $f \in \mathcal{O}_X(X)$ defines a morphism $X \to \mathbf{A}^1$, but of course a rational function need not be regular (cf. projective varieties!). Nonetheless, given $f \in \mathsf{K}(X)$, and given an affine covering $\{U_i\}$ of X, the subset of U_i where f is defined is open (Exercise 1.12), so the subset U of X where f is defined is also open. Thus f induces a morphism $U \to \mathbf{A}^1$. In turn, the subset of U on which $f \neq 0$ is open and may be denoted X_f. Similarly, we can define $\mathscr{V}(f) = \{x \in X \mid f(x) = 0\}$ for $f \in \mathcal{O}_X(X)$, as in the affine case.

Two irreducible prevarieties X, Y may have function fields related by a monomorphism $\sigma: \mathsf{K}(Y) \to \mathsf{K}(X)$. We claim that σ induces a "partial morphism", i.e., a morphism from a (nonempty) open subset of X into Y whose comorphism is essentially σ. Indeed, we may first replace X, Y by affine open subsets; this has no effect on the function fields. Thus $\mathsf{K}(Y)$ is of the form $\mathsf{K}(f_1, \ldots, f_n)$, where $\mathsf{K}[Y] = \mathsf{K}[f_1, \ldots, f_n]$. Set $g_i = \sigma(f_i) \in \mathsf{K}(X)$. Cut down as above, to an open subset of X on which all g_i are defined, then further to an affine open set U for which all $g_i \in \mathsf{K}[U]$. Now σ takes $\mathsf{K}[Y]$ into $\mathsf{K}[U]$, so there is a unique morphism $U \to Y$ having this as comorphism.

Finally, we introduce the notion of **birational morphism**: $\varphi: X \to Y$ is birational if φ^* is an isomorphism of $\mathsf{K}(Y)$ onto $\mathsf{K}(X)$. Irreducible prevarieties with isomorphic function fields are called **birationally equivalent**; they need not be isomorphic (cf. $\mathbf{A}^n$ and $\mathbf{P}^n$).

2.4. Products

For pairs of affine or projective varieties, we were able to give the same type of structure to the cartesian product set (cf. (1.4), (1.7)). For arbitrary prevarieties, the categorical notion of "product" is our surest guide. Given objects X, Y, a **product** of X and Y consists of an object Z, together with morphisms $\pi_1: Z \to X$, $\pi_2: Z \to Y$ (**projections**), satisfying the universal mapping property: For any object W and any morphisms $\varphi_1: W \to X$, $\varphi_2: W \to Y$, there exists a unique morphism $\psi: W \to Z$ such that $\pi_i \psi = \varphi_i$ ($i = 1, 2$). The definition is constructed so as to insure the *uniqueness* of the product, if it exists, but the *existence* has to be settled by a specific construction.

For prevarieties X, Y, the underlying set of a product prevariety would have to be the cartesian product: apply the universal property to morphisms $W \to \{x\}$, $W \to \{y\}$, where W is a prevariety consisting of a single point, to conclude that points of Z correspond bijectively to pairs (x, y). The construction in (1.7) suggests that we give $X \times Y$ the structure of a prevariety by patching together products of various affine open subsets of X, Y. So we begin by examining more closely the affine situation.

Proposition. *Let $X \subset \mathbf{A}^n$, $Y \subset \mathbf{A}^m$ be affine varieties, with $R = \mathsf{K}[X]$, $S = \mathsf{K}[Y]$. Endow the cartesian product $X \times Y$ with the Zariski product topology (1.4). Then:*

(a) $X \times Y$, *with the projections* $pr_1: X \times Y \to X$ *and* $pr_2: X \times Y \to Y$, *is a product* (*in the categorical sense*) *of the prevarieties* X, Y, *and* $K[X \times Y] \cong R \otimes_K S$.

(b) *If* $(x, y) \in X \times Y$, $\mathcal{O}_{(x, y)}$ *is the localization of* $\mathcal{O}_x \otimes_K \mathcal{O}_y$ *at the* (*maximal*) *ideal* $m_x \otimes \mathcal{O}_y + \mathcal{O}_x \otimes m_y$.

Proof.   (a) First we pin down the affine algebra of $X \times Y$, which by its construction is a closed subset of $\mathbf{A}^{n+m}$. Via the projections, polynomial functions on X, Y induce polynomial functions on $X \times Y$. Assign to a pair $(g, h) \in R \times S$ the polynomial function $f(x, y) = g(x)h(y)$ on $X \times Y$. This assignment is bilinear in each variable g, h, so it induces a K-algebra homomorphism $\sigma: R \otimes_K S \to K[X \times Y]$. It is clear that each polynomial in $m + n$ indeterminates $T_1, \ldots, T_n, U_1, \ldots, U_m$ can be expressed as a finite sum of products $g(T)h(U)$. This shows that σ is *surjective* (polynomial functions on $X \times Y$ being the restrictions of polynomial functions on $\mathbf{A}^{m+n}$).

To show that σ is *injective*, let $f = \sum_{i=1}^{r} g_i \otimes h_i$ be sent to 0. We may assume that f is written with r minimal. In case $f \neq 0$, we claim that $r = 1$. Indeed, not all h_i are 0 in this case, so we can fix some $y \in Y$ for which not all $h_i(y) = 0$. Since $\sum g_i(x)h_i(y) = 0$ for all $x \in X$, we get $\sum h_i(y)g_i = 0$ in R, i.e., the g_i are linearly dependent over K. If $r > 1$, we could reduce by one the number of g_i and get a contradiction to the minimality of r. So $r = 1$. Now the argument shows that $g_1 = 0$, so $f = 0$.

It remains to verify the universal mapping property for $X \times Y$. Given a prevariety W and morphisms $\varphi_1: W \to X$, $\varphi_2: W \to Y$, we have to construct a suitable morphism $\psi: W \to X \times Y$. There is a unique such mapping of sets which makes $\varphi_i = pr_i \circ \psi$. To check that it is a morphism, we use the affine criterion (2.3). $X \times Y$ is affine, so it just has to be seen that ψ pulls back polynomial functions on $X \times Y$ to regular functions on W. $K[X \times Y]$ being generated by the pullbacks of $K[X]$, $K[Y]$ under pr_i, and the φ_i being morphisms by assumption, the conclusion follows.

(b) If X, Y are irreducible, so is $X \times Y$ (1.4); part (a) shows that $R \otimes_K S$ is an integral domain, with fraction field isomorphic to $K(X \times Y)$. Now we have inclusions $R \otimes S \subset \mathcal{O}_x \otimes \mathcal{O}_y \subset \mathcal{O}_{(x, y)}$. Since $\mathcal{O}_{(x, y)}$ is the localization of $R \otimes S$ at the ideal $m_{(x, y)}$, it is equally well the localization of $\mathcal{O}_x \otimes \mathcal{O}_y$ at its ideal m vanishing at (x, y). Evidently $m_x \otimes \mathcal{O}_y + \mathcal{O}_x \otimes m_y \subset m$. Conversely, let $f = \sum g_i \otimes h_i \in m$, with $g_i \in \mathcal{O}_x$, $h_i \in \mathcal{O}_y$. If $g_i(x) = a_i$, $h_i(y) = b_i$, then $f - \sum a_i b_i = \sum (g_i - a_i) \otimes h_i + \sum a_i \otimes (h_i - b_i) \in m_x \otimes \mathcal{O}_y + \mathcal{O}_x \otimes m_y$. This forces $\sum a_i b_i = 0$ and concludes the proof. $\square$

The proposition shows that $X \times Y$ has intrinsic meaning in the category of prevarieties, when X and Y are affine, independent of any particular embeddings in affine space.

In order to treat the arbitrary prevarieties X, Y, we concentrate first on the *irreducible* ones. To endow the cartesian product $X \times Y$ with the structure of prevariety, we have to specify a topology and a covering by affine open sets. For all affine open sets $U \subset X$, $V \subset Y$, and all finite sets of polynomial functions f_i on U, g_i on V, we decree that the principal open sets $(U \times V)_{\Sigma f_i g_i}$ should be basic open sets in $X \times Y$. Notice that these sets do form a basis for a topology, since the intersection of two of them is another of the same type. Moreover, the description of the affine algebra of $U \times V$ in part (a) of the proposition shows that the topology induced on $U \times V$ coincides with the Zariski product topology there.

The *function field* of $X \times Y$ will have to be that of $U \times V$, where $U \subset X$, $V \subset Y$ are affine open sets. Since $\mathsf{K}[U \times V] = \mathsf{K}[U] \otimes \mathsf{K}[V]$, it is evident that $\mathsf{K}(U \times V)$ can be described as the field of fractions of (the integral domain!) $\mathsf{K}(U) \otimes \mathsf{K}(V)$. Call this field F. Part (b) of the proposition forces us to define the *local ring* of $(x, y) \in X \times Y$ to be the localization of $\mathcal{O}_x \otimes \mathcal{O}_y$ at $m_x \otimes \mathcal{O}_y + \mathcal{O}_x \otimes m_y$. In turn, we get a sheaf of functions on $X \times Y$ by assigning to each open set U the intersection of all $\mathcal{O}_{(x, y)}$, $(x, y) \in U$. (This agrees on each product of affine open sets with the affine product already defined.) It is clear that $X \times Y$ thus acquires the structure of prevariety. Moreover, the set-theoretic projections onto X, Y are morphisms: use the affine criterion (2.3).

To check the universal mapping property, let W be a prevariety, with morphisms $\varphi_1 : W \to X$, $\varphi_2 : W \to Y$. As before, there is a unique map of sets $\psi : W \to X \times Y$ for which $\varphi_i = \mathrm{pr}_i \circ \psi$. We appeal to the affine criterion (2.3) to prove that ψ is a morphism: By construction, products $U \times V$ of affine open sets in X, Y are affine open sets which cover $X \times Y$. Open sets of the form $W' = \varphi_1^{-1}(U) \cap \varphi_2^{-1}(V)$ cover W, and the universal property of $U \times V$ shows that the restriction of ψ to W' is a morphism.

This takes care of the irreducible case. For arbitrary prevarieties X, Y, having irreducible components X_i, Y_j, we form the prevarieties $X_i \times Y_j$ as above. We then topologize $X \times Y$ by declaring that a set is open if and only if its intersection with each $X_i \times X_j$ is open. Finally, we endow $X \times Y$ with a sheaf of functions as in (2.2). It is then a routine matter to verify that $X \times Y$ is a categorical product. Therefore:

Theorem.   *Products exist in the category of prevarieties.* $\square$

The reader should check that the construction of products of projective varieties (1.7) is duplicated abstractly by the foregoing process. However, the embedding in projective space specified in (1.7) is needed in order to see that the resulting product is again *projective*.

2.5. Hausdorff Axiom

It is possible to concoct examples of prevarieties which are geometrically pathological. For instance, let X be covered by two copies U, V of $\mathbf{A}^1$, with $x \in U$ equal to $x \in V$ except when $x = 0$ ("the affine line with a point doubled").

A prevariety X is called a **variety** if it satisfies the **Hausdorff axiom**: The **diagonal** $\Delta(X) = \{(x, x)|x \in X\}$ is closed in $X \times X$. (In the category of topological spaces, with $X \times X$ given the ordinary product topology, this condition is equivalent to the usual Hausdorff separation axiom.) An equivalent condition is this: (*) For morphisms $\varphi, \psi: Y \to X$, Y any prevariety, $\{y \in Y|\varphi(y) = \psi(y)\}$ is closed in Y. Indeed, by applying (*) to the situation $X \times X \overset{\mathrm{pr_1}}{\underset{\mathrm{pr_2}}{\rightrightarrows}} X$, we get $\Delta(X)$ closed in $X \times X$; in the other direction, use the set-up $Y \overset{\varphi \times \psi}{\to} X \times X \overset{\mathrm{pr_1}}{\underset{\mathrm{pr_2}}{\rightrightarrows}} X$, the inverse image of $\Delta(X)$ being $\{y \in Y|\varphi(y) = \psi(y)\}$.

The example above fails to pass the test (*), if we take the two maps $\mathbf{A}^1 \to U \subset X, \mathbf{A}^1 \to V \subset X$, since $\mathbf{A}^1 - \{0\}$ is not closed in $\mathbf{A}^1$. On the other hand, varieties do abound.

Examples: (1) An affine variety is a variety. (The diagonal is clearly given by polynomial conditions.)

(2) Subprevarieties of a variety are again varieties. These are therefore called **subvarieties**.

(3) If X, Y are varieties, so is $X \times Y$.

(4) A projective variety is a variety. (This results from the following nice criterion.)

Lemma. *Let X be a prevariety, and assume that each pair $x, y \in X$ lie in some affine open subset of X. Then X is a variety.*

Proof. Given a prevariety Y and morphisms $\varphi, \psi: Y \to X$, let $Z = \{y \in Y|\varphi(y) = \psi(y)\}$. We have to show that Z is closed. If $z \in \bar{Z}$, set $x = \varphi(z)$, $y = \psi(z)$. By hypothesis, x and y lie in some affine open set V. Then $U = \varphi^{-1}(V) \cap \psi^{-1}(V)$ is an open neighborhood of z, which must meet Z. But $Z \cap U = \{y \in U|\varphi'(y) = \psi'(y)\}$, where $\varphi', \psi': U \to V$ are the restrictions. Since V is a variety, $Z \cap U$ is *closed* in U. This means that $U - (Z \cap U)$ is an *open* set not meeting Z, so in particular it cannot contain z. We conclude that $z \in Z$. $\square$

The following proposition shows why it is better to deal with varieties than with prevarieties:

Proposition. *Let Y be a variety, X any prevariety.*

(a) *If $\varphi: X \to Y$ is a morphism, the* **graph** *$\Gamma_\varphi = \{(x, \varphi(x))|x \in X\}$ is closed in $X \times Y$.*

(b) *If $\varphi, \psi: X \to Y$ are morphisms which agree on a dense subset of X, then $\varphi = \psi$.*

Proof. (a) Γ_φ is the inverse image of $\Delta(Y)$ under the morphism $X \times Y \to Y \times Y$ which sends (x, y) to $(\varphi(x), y)$).

(b) The set $\{x \in X \mid \varphi(x) = \psi(x)\}$ is closed in X since Y is a variety. It is dense by assumption, so it coincides with X. $\quad\square$

In practice, we shall deal only with varieties in what follows: affine and projective varieties, their subvarieties, their products.

Exercises

1. Show that the definition of "morphism" (2.3) agrees with that given for affine varieties in (1.5).
2. If $\varphi : \mathbf{A}^1 \to \mathbf{A}^1$ is a birational morphism, then φ is necessarily an isomorphism.
3. Let char $\mathsf{K} = p > 0$. If X is an irreducible affine variety, let $\mathsf{K}(X)^p = \{ f^p \mid f \in \mathsf{K}(X) \}$. Prove that the inclusion $\mathsf{K}(X)^p \to \mathsf{K}(X)$ is the comorphism of a morphism $\varphi : X \to X$ (called the **Frobenius map**). Describe φ explicitly when $X \subset \mathbf{A}^n$.
4. Let X, Y be prevarieties. Prove that the projections $X \times Y \to X$, $X \times Y \to Y$ are open maps (i.e., send open sets to open sets). Must they send closed sets to closed sets?
5. If X, Y are prevarieties, and W is open (resp. closed) in X, then $W \times Y$ is open (resp. closed) in $X \times Y$.
6. Prove that a topological space X is T_2 if and only if $\{(x, x) \mid x \in X\}$ is closed in $X \times X$ (given the ordinary product topology).
7. Let $\varphi, \psi : Y \to X$ be morphisms (X, Y prevarieties). Prove that $\{y \in Y \mid \varphi(y) = \psi(y)\}$ is locally closed in Y.
8. If Y is a prevariety for which Proposition 2.5 (a) (resp. (b)) holds for all prevarieties X, then Y is a variety.
9. Let $\varphi : X \to Y$ be a morphism of varieties. Prove that pr_1 induces an isomorphism of the graph $\Gamma_\varphi \subset X \times Y$ onto X.

Notes

We have followed Mumford [3, I §6]. The current terminology is "variety" (resp. "separated variety") in place of our "prevariety" (resp. "variety"); but we prefer to follow the older usage, since all prevarieties we shall encounter are in fact varieties.

3. Dimension

Intuitively, a point is 0-dimensional, a curve 1-dimensional, a surface 2-dimensional. The object of this section is to give precise algebraic meaning to the geometric notion of dimension. This in turn will allow us to describe the behavior of varieties under morphisms, in the following section.

3.1. Dimension of a Variety

With an irreducible variety X is associated its field $\mathsf{K}(X)$ of rational functions. As a finitely generated field extension of K, $\mathsf{K}(X)$ has finite transcendence degree over K (0.4), abbreviated tr. deg.$_K$ $K(X)$. This number is called the **dimension** of X, written dim X. For example, dim $\mathbf{A}^n$ = dim $\mathbf{P}^n$ = n. So the dimension measures the maximum number of algebraically independent functions on X (or the number of "parameters" required to describe X). In case X has more than one irreducible component, say $X = X_1 \cup \cdots \cup X_t$, it is reasonable to define dim X as max (dim X_i).

Let X be irreducible. Since $K(X) = K(U)$ for any affine open subset U, dim X = dim U. Similarly, dim X = dim X_f for any $f \in K(X)$. For example, $GL(n, K)$ is the principal open subset of $\mathbf{A}^{n^2}$ defined by nonvanishing of the determinant, so its dimension is n^2.

Proposition. *Let X, Y be irreducible varieties of respective dimension m, n. Then dim $X \times Y = m + n$.*

Proof. In view of the preceding remarks, we may as well assume that X, Y are affine, with $X \subset \mathbf{A}^p$, $Y \subset \mathbf{A}^q$. (The reader who prefers to avoid embedding X and Y in affine spaces can proceed more intrinsically by using the identification $K[X] \otimes K[Y] = K[X \times Y]$ (2.4).) If $S_1, \ldots, S_p$ (resp. $T_1, \ldots, T_q$) are the coordinates on $\mathbf{A}^p$ (resp. $\mathbf{A}^q$), their restrictions s_i (resp. t_i) generate $K(X)$ (resp. $K(Y)$). From these generating sets we can extract transcendence bases (0.4), say $s_1, \ldots, s_m$ and $t_1, \ldots, t_n$. It is clear that $K(X \times Y) = K(s_1, \ldots, s_p, t_1, \ldots, t_q)$ and that $K(X \times Y)$ is algebraic over the subfield $K(s_1, \ldots, s_m, t_1, \ldots, t_n)$. So it suffices to show that the latter field is purely transcendental over K. Suppose there is a polynomial relation $f(s_1, \ldots, s_m, t_1, \ldots, t_n) = 0$. Then for each fixed $x = (x_1, \ldots, x_p) = (s_1(x), \ldots, s_p(x)) \in X$, the polynomial function $f(x_1, \ldots, x_m, t_1, \ldots, t_n)$ vanishes on Y. The algebraic independence of the t_i forces all coefficients $g(x_1, \ldots, x_m)$ of the polynomial $f(x_1, \ldots, x_m, T_1, \ldots, T_n)$ to be zero. In turn, each $g(s_1, \ldots, s_m) = 0$ (since $x \in X$ was arbitrary), so the algebraic independence of the s_i forces all $g(S_1, \ldots, S_m) = 0$. Finally, $f(S_1, \ldots, S_m, T_1, \ldots, T_n) = 0$. $\square$

3.2. Dimension of a Subvariety

The following is a first step toward understanding how dimension varies inside a given variety.

Proposition. *Let X be an irreducible variety, Y a proper, closed, irreducible subset. Then dim Y < dim X.*

Proof. It is harmless to assume that X is affine, say of dimension d. Let $R = K[X]$, $\bar{R} = K[Y] \cong R/P$ (where P is a *nonzero* prime ideal of R). It is clear that transcendence bases for $K(X)$, $K(Y)$ can already be found in R, $\bar{R}$. Suppose dim $Y \geqslant d$, and select algebraically independent elements $\bar{x}_1, \ldots, \bar{x}_d \in \bar{R}$ (images of $x_1, \ldots, x_d \in R$, which are clearly algebraically independent as well). Let $f \in P$ be nonzero. Since dim $X = d$, there must be a nontrivial polynomial relation $g(f, x_1, \ldots, x_d) = 0$, where $g(\mathsf{T}_0, \mathsf{T}_1, \ldots, \mathsf{T}_d) \in \mathsf{K}[\mathsf{T}]$. Because $f \neq 0$, we may assume that T_0 does not divide all monomials in $g(\mathsf{T}_0, \mathsf{T}_1, \ldots, \mathsf{T}_d)$, i.e., $h(\mathsf{T}_1, \ldots, \mathsf{T}_d) = g(0, \mathsf{T}_1, \ldots, \mathsf{T}_d)$ is nonzero. Now $h(\bar{x}_1, \ldots, \bar{x}_d) = 0$, contradicting the independence of the $\bar{x}_i$. $\quad\square$

Define the **codimension** $\mathrm{codim}_X Y$ of a subvariety Y of X to be $\dim X - \dim Y$.

Corollary. *Let X be an irreducible affine variety, Y a closed irreducible subset of codimension 1. Then Y is a component of $\mathscr{V}(f)$ for some $f \in \mathsf{K}[X]$.*

Proof. By assumption, $Y \neq X$, so there is a nonzero $f \in \mathsf{K}[X]$ vanishing on Y. Then $Y \subset \mathscr{V}(f) \subsetneq X$. Let Z be an irreducible component of $\mathscr{V}(f)$ containing Y. The proposition says that $\dim Z < \dim X$, while $\dim Y \leqslant \dim Z$, with equality only if $Y = Z$. Since $\mathrm{codim}_X Y = 1$, equality must hold. $\quad\square$

In the situation of the corollary, it is *not* usually possible to arrange that Y be precisely $\mathscr{V}(f)$. However, this can be done when Y has codimension 1 in some affine space $\mathbf{A}^n$, or more generally, when $\mathsf{K}[X]$ is a unique factorization domain (Exercise 6).

If subvarieties of codimension 1 (and hence, by induction, subvarieties of all possible dimensions) are to exist, the corollary indicates the shape they must have. We aim next for a converse to the corollary.

3.3. Dimension Theorem

The zero set in $\mathbf{A}^n$ of a single nonscalar polynomial $f(\mathsf{T}_1, \ldots, \mathsf{T}_n)$ is called a **hypersurface**; its irreducible components are just the hypersurfaces defined by the various irreducible factors of $f(\mathsf{T})$. More generally, when X is an affine variety, a nonzero nonunit $f \in \mathsf{K}[X]$ defines a hypersurface in X (whose components are not so easy to characterize unless $\mathsf{K}[X]$ happens to be a unique factorization domain). For example, $SL(n, \mathsf{K})$ is a hypersurface in $GL(n, \mathsf{K})$, or in $\mathbf{A}^{n^2}$, defined by $\det(\mathsf{T}_{ij}) = 1$.

Proposition. *All irreducible components of a hypersurface in $\mathbf{A}^n$ have codimension 1.*

Proof. It suffices to look at the zero set X of an irreducible polynomial $p(\mathsf{T})$. We can assume that (say) T_n actually occurs in $p(\mathsf{T})$ (which is nonscalar by assumption). Let t_i be the restriction of T_i to X, so $\mathsf{K}(X) = \mathsf{K}(t_1, \ldots, t_n)$. We claim that $t_1, \ldots, t_{n-1}$ are algebraically independent over K. Otherwise there exists a nontrivial polynomial relation $g(t_1, \ldots, t_{n-1}) = 0$, whence $g(\mathsf{T}_1, \ldots, \mathsf{T}_{n-1})$ vanishes on X; but $\mathscr{I}(X) = (p(\mathsf{T}))$, forcing $g(\mathsf{T})$ to be a multiple of $p(\mathsf{T})$. This is impossible, since T_n occurs in $p(\mathsf{T})$ but not in $g(\mathsf{T})$. We conclude that $\dim X \geqslant n-1$, which must be an equality in view of Proposition 3.2. $\quad\square$

We want to generalize this result to arbitrary affine varieties. Two technical tools are needed: Noether's Normalization Lemma (0.7) and the norm map $N_{E/F} : E \to F$ for a finite field extension E/F (0.5).

Theorem.   *Let X be an irreducible affine variety, $0 \neq f \in K[X]$ a nonunit, Y an irreducible component of $\mathscr{V}(f)$. Then Y has codimension 1 in X.*

Proof.   Let $P = \mathscr{I}(Y) \subset K[X]$, and let $Y_1, \ldots, Y_t$ be the components of $\mathscr{V}(f)$ other than Y, $P_i = \mathscr{I}(Y_i)$. The Nullstellensatz (1.1) implies that $\sqrt{(f)} = P \cap P_1 \cap \cdots \cap P_t$. Choose $g \in P_1 \cap \cdots \cap P_t - P$ (g exists, since Y is not contained in $Y_1 \cup \cdots \cup Y_t$). Then X_g is an (irreducible) affine variety having the same dimension as X, and $Y \cap X_g$ is precisely the zero set of f in X_g. Since $Y \cap X_g$ is a principal open set in Y, it suffices to prove that its codimension in X_g is 1. So we might as well assume at the outset that $Y = \mathscr{V}(f)$, $P = \sqrt{(f)}$.

Now we apply the Normalization Lemma (0.7) to the domain $R = K[X]$: R is integral over a subring S which is isomorphic to $K[T_1, \ldots, T_d]$, $d = \dim X$. Let $E = K(X)$, $F = $ field of fractions of S; so E/F is finite. Notice that $N_{E/F}$ takes elements of R into S. Indeed, if $h \in R$, then h satisfies a monic polynomial equation over S, and by the definition of "norm", $N_{E/F}(h)$ is a power of the constant term.

Set $f_0 = N_{E/F}(f) \in S$. We claim that $f_0 \in P = \sqrt{(f)}$. Say $f^k + a_1 f^{k-1} + \cdots + a_k = 0$ ($a_i \in S$), with $f_0 = (a_k)^m$ (by the preceding paragraph). Then $0 = (f^k + a_1 f^{k-1} + \cdots + a_k)a_k^{m-1} = f(f^{k-1}a_k^{m-1} + \cdots + a_{k-1}a_k^{m-1}) + f_0$, so f_0 is an R-multiple of f.

If $\sqrt{(f_0)}$ denotes the radical of (f_0) in S, what we have just shown implies that $\sqrt{(f_0)} \subset P \cap S$. The reverse inclusion also holds: Let $g \in P \cap S$, so in particular, $g \in \sqrt{(f)}$ and $g^l = fh$ for some $l \in \mathbb{Z}^+$, $h \in R$. Take norms to get $g^{l[E:F]} = N_{E/F}(g^l) = N_{E/F}(f)N_{E/F}(h)$, where the first equality holds because $g \in S$ and the second because $N_{E/F}$ is multiplicative. As remarked earlier, $N_{E/F}(h) \in S$, so we conclude that a power of g is an S-multiple of f_0, as claimed.

Now we have replaced the prime ideal $P = \sqrt{(f)}$ in R by the prime ideal $P \cap S = \sqrt{(f_0)}$ in S. The advantage of this is that S is a *unique factorization domain*. In particular, since the radical of (f_0) is prime, it is very easy to see that (up to a unit multiple) f_0 is just a power of an irreducible polynomial p, whence $P \cap S$ is just the principal ideal (p). It is clear that p is nonscalar.

If S is viewed as the affine algebra of $\mathbf{A}^n$, then $P \cap S$ defines a hypersurface in $\mathbf{A}^n$, which has codimension 1 (by the preceding proposition). This means that the fraction field of $S/(P \cap S)$ has transcendence degree $n - 1$ over K. On the other hand, R integral over S clearly implies that R/P is integral over $S/(P \cap S)$, so the two fraction fields have equal transcendence degree. But the fraction field of R/P is $K(Y)$; therefore, $\dim Y = n - 1$. $\square$

The theorem can easily be reformulated as an assertion about an arbitrary irreducible variety X: If U is open in X, and $0 \neq f \in \mathcal{O}_X(U)$ is a nonunit, then each irreducible component of the zero set of f in U has codimension 1 in X.

(Deduce this from the theorem by cutting down to an affine open subset of U which meets the irreducible component in question.)

3.4. Consequences

Theorem 3.3 guarantees that an n-dimensional variety X has irreducible closed subsets of all dimensions less than n. More precisely:

Corollary A.   *Let X be an irreducible variety, Y a closed irreducible subset of codimension $r \geqslant 1$. Then there exist closed irreducible subsets Y_i of codimension $1 \leqslant i \leqslant r$, such that $Y_1 \supset Y_2 \supset \cdots \supset Y_r = Y$.*

Proof.   It suffices to prove this for an affine open subset of X which meets Y, so we may as well let X be affine. For $r = 1$, there is nothing to prove. Proceed inductively. Since $Y \neq X$, there exists $f \neq 0$ in $\mathscr{I}(Y)$, and Y lies in some irreducible component Y_1 of $\mathscr{V}(f)$. But Theorem 3.3 says that $\operatorname{codim}_X Y_1 = 1$. By induction, Y_1 has closed irreducible subsets of the required type, going down to Y. $\square$

Another obvious induction yields:

Corollary B.   *Let X be an irreducible variety, $f_1, \ldots, f_r \in \mathcal{O}_X(X)$. Then each irreducible component of $\mathscr{V}(f_1, \ldots, f_r)$ has codimension at most r.* $\square$

The inequality cannot be improved (e.g., let $f_1 = f_2 = \cdots = f_r$). But we can say something useful in the other direction, which accords with the geometric idea that a surface (resp. curve, point) in 3-space should be definable by one (resp. two, three) polynomial equations.

Corollary C.   *Let X be an irreducible affine variety, Y a closed irreducible subset of codimension $r \geqslant 1$. Then Y is a component of $\mathscr{V}(f_1, \ldots, f_r)$ for some choice of $f_i \in \mathsf{K}[X]$.*

Proof.   It is easier to prove a more general statement: Given closed irreducible subsets $Y_1 \supset Y_2 \supset \cdots \supset Y_r$, with $\operatorname{codim}_X Y_i = i$, there exist $f_i \in \mathsf{K}[X]$ such that all components of $\mathscr{V}(f_1, \ldots, f_q)$ have codimension q in X and such that Y_q is one of these components ($1 \leqslant q \leqslant r$). (This really is a more general statement, thanks to Corollary A.)

The proof goes by induction on q (r being fixed). For $q = 1$, we appeal first to Corollary 3.2 for the existence of f_1, and then to Theorem 3.3 for the fact that all components of $\mathscr{V}(f_1)$ have codimension 1.

Assume that $f_1, \ldots, f_{q-1}$ have been found. Let $Z_1 = Y_{q-1}, Z_2, \ldots, Z_m$ be the components of $\mathscr{V}(f_1, \ldots, f_{q-1})$. Since each has codimension $q - 1$, none can lie in Y_q; so $\mathscr{I}(Z_i)$ does not include $\mathscr{I}(Y_q)$ ($1 \leqslant i \leqslant m$). The ideals $\mathscr{I}(Z_i)$ being prime, it follows (0.9) that their union also does not include $\mathscr{I}(Y_q)$. Choose f_q vanishing on Y_q but not vanishing identically on any Z_i. If Z is any component of $\mathscr{V}(f_1, \ldots, f_q)$, then Z of course lies in one of the components Z_i of $\mathscr{V}(f_1, \ldots, f_{q-1})$, as well as in $\mathscr{V}(f_q)$. Theorem 3.3 implies that $\mathscr{V}(f_q) \cap Z_i$ has codimension 1 in Z_i (since f_q does not vanish on Z_i), hence codimension q in X. On the other hand, Corollary B says that $\operatorname{codim}_X Z \leqslant q$. So Z must have codimension precisely q. Since f_q vanishes on Y_q, and Y_q has codimension q, we also conclude that Y_q is one of the components of $\mathscr{V}(f_1, \ldots, f_q)$. $\square$

Exercises

1. A variety has dimension 0 if and only if it is a finite set of points.
2. Show that the dimension of the Grassmann variety $\mathfrak{G}_d(V)$ of d-dimensional subspaces of an n-dimensional vector space V (1.8) is $d(n - d)$. Try to determine the dimension of the flag variety $\mathfrak{F}(V)$.
3. Let X be an arbitrary variety, Y a closed subset. Prove that $\dim Y \leqslant \dim X$.
4. The dimension of an irreducible variety X is the largest d for which there exists a chain of (nonempty) closed irreducible subsets $X_0 \subset X_1 \subset X_2 \subset \cdots \subset X_d = X$ (all inclusions proper).
5. In a variety X, the closed irreducible sets satisfy ACC.
6. Let X be an irreducible affine variety for which $\mathsf{K}[X]$ is a unique factorization domain, e.g., $X = \mathbf{A}^n$. Prove that each closed subset Y of codimension 1 has the form $\mathscr{V}(f)$ for some $f \in \mathsf{K}[X]$. [Treat first the case: Y irreducible. Show that minimal prime ideals of $\mathsf{K}[X]$ are principal.]

Notes

The exposition here follows Mumford [3, I §7].

4. Morphisms

In the study of linear algebraic groups, two kinds of morphisms will be especially prominent: group homomorphisms (which are simultaneously required to be morphisms of varieties) and canonical maps $G \to G/H$ (where the coset space has to be given a suitable structure of variety, not necessarily affine). This section provides the essential tools for the study of these and other morphisms. The proofs depend heavily on considerations of dimension. As in §3, it is usually enough to treat irreducible varieties, which can often (though not always) be assumed to be affine.

4.1. Fibres of a Morphism

By definition, the **fibres** of a morphism $\varphi : X \to Y$ are the closed sets $\varphi^{-1}(y)$, $y \in Y$. (If φ were a group homomorphism, these would be cosets of

Ker φ.) We want to examine the dimensions of fibres and, in particular, to prove that they are not "too small". Of course, $\varphi^{-1}(y)$ is empty if $y \notin \text{Im } \varphi$, so care must be taken in the formulation below. More generally, we shall consider sets $\varphi^{-1}(W)$, where $W \subset Y$ is closed and irreducible.

It usually does no harm to replace Y by the closure of $\varphi(X)$; this facilitates dimension comparisons. When X is irreducible and $\varphi(X)$ is dense in Y, we say that φ is **dominant**. We could say the same when X is not irreducible, but instead we reserve the term "dominant" for the following more special situation: φ maps each component of X onto a dense subset of some component of Y, and $\varphi(X)$ is dense in Y. Even when φ is dominant, the restriction of φ to an irreducible component of $\varphi^{-1}(W)$ (W closed irreducible in Y) certainly need not be dominant, viewed as a morphism from this component to W. But when it is so, we say that the component in question **dominates** W. One other remark: When $\varphi : X \rightarrow Y$ is dominant, X and Y irreducible, then φ^* induces an injection of $\mathsf{K}(Y)$ into $\mathsf{K}(X)$; in particular, $\dim X \geqslant \dim Y$.

Theorem.   *Let $\varphi : X \rightarrow Y$ be a dominant morphism of irreducible varieties, and set $r = \dim X - \dim Y$. Let W be a closed irreducible subset of Y. If Z is an irreducible component of $\varphi^{-1}(W)$ which dominates W, then $\dim Z \geqslant \dim W + r$. In particular, if $y \in \varphi(X)$, each component of $\varphi^{-1}(y)$ has dimension at least r.*

Proof.   If U is an affine open subset of Y which meets W, then $U \cap W$ is dense in W. For comparison of dimensions, we could therefore replace Y by U and X by the open subvariety $\varphi^{-1}(U)$. So we might as well assume at the outset that Y is *affine*.

Let $s = \text{codim}_Y W$. According to Corollary C of (3.4), W is a component of $\mathscr{V}(f_1, \ldots, f_s)$ for suitable $f_i \in \mathsf{K}[Y]$. Setting $g_i = \varphi^*(f_i) \in \mathcal{O}_X(X)$, we conclude that Z lies in $\mathscr{V}(g_1, \ldots, g_s)$. Since Z is irreducible, it lies in some component Z_0 of this set. But $W = \overline{\varphi(Z)}$, by assumption, while $\overline{\varphi(Z)} \subset \overline{\varphi(Z_0)} \subset \mathscr{V}(f_1, \ldots, f_s)$. W being a component of $\mathscr{V}(f_1, \ldots, f_s)$, it follows that $\overline{\varphi(Z)} = \overline{\varphi(Z_0)} = W$, whence $Z_0 \subset \varphi^{-1}(W)$. But Z is a component of $\varphi^{-1}(W)$, so $Z = Z_0$. i.e., Z is a component of $\mathscr{V}(g_1, \ldots, g_s)$. Now Corollary B of (3.4) says that $\text{codim}_X Z \leqslant s$. The theorem follows at once. $\square$

In the situation of the theorem, we now know that (nonempty) fibres of φ are not too small. Next we have to ask whether they can be "too big.". A simple geometric example helps to illustrate the possibilities: Define $\varphi : \mathbf{A}^2 \rightarrow \mathbf{A}^2$ by $\varphi(x, y) = (xy, y)$. The image consists of the complement of the "x-axis" along with the origin $(0, 0)$. This is easily seen to be dense, so φ is dominant (though not surjective). If U is the principal open set in $\mathbf{A}^2$ defined by $y \neq 0$, then φ induces a bijection from $U = \varphi^{-1}(U)$ onto U, which is even an isomorphism of varieties. But the fibre $\varphi^{-1}((0, 0))$ is 1-dimensional, consisting of all points $(x, 0)$.

This example suggests that we look for an *open* set in the target variety above which all fibres have the "correct" dimension: dim X − dim Y. But if such an open set is to exist, it must lie in $\varphi(X)$. So the study of fibres of φ leads also to the study of the image of φ. In preparation for the main theorem (4.3), we look briefly at a special type of morphism which turns out to be surjective.

4.2. Finite Morphisms

Let $\varphi : X \to Y$ be a morphism of *affine* varieties. If $\mathsf{K}[X]$ is integral over the subring $\varphi^* \mathsf{K}[Y]$, we call φ **finite**. Notice that if X, Y are irreducible, and if φ is also dominant, then $\mathsf{K}(X)$ is a finite algebraic extension of $\varphi^* \mathsf{K}(Y)$; so dim X = dim Y.

As a *nonexample*, take $\varphi(x, y) = (xy, y)$, discussed at the end of (4.1). Here $\varphi^* : \mathsf{K}[\mathsf{S}_1, \mathsf{S}_2] \to \mathsf{K}[\mathsf{T}_1, \mathsf{T}_2]$ is injective and sends S_1 to $\mathsf{T}_1 \mathsf{T}_2$, S_2 to T_2. It is easy to see that, e.g., T_1 fails to be integral over $\mathsf{K}[\mathsf{T}_1 \mathsf{T}_2, \mathsf{T}_2]$. But this also follows from our next result.

Proposition. *Let $\varphi : X \to Y$ be a finite, dominant morphism of affine varieties.*

(a) If Z is closed in X, then $\varphi(Z)$ is closed in Y, and the restriction of φ to Z is finite. In particular, φ is surjective.

(b) If W is a closed irreducible subset of Y, and Z is any component of $\varphi^{-1}(W)$, then $\varphi(Z) = W$.

Proof. Let $R = \mathsf{K}[X]$, $S = \mathsf{K}[Y]$. Since φ is dominant, φ^* is injective; we abuse notation by viewing S as a subring of R (of which R is an integral extension by assumption). If I is an ideal of R, then R/I may be regarded naturally as an extension of $S/(I \cap S)$. This extension is clearly integral as well.

(a) Now let $Z = \mathscr{V}(I)$ be closed in X, where $I = \mathscr{I}(Z)$. Then φ maps Z into the zero set Z' of $I' = I \cap S$, which is a radical ideal of S (hence equal to $\mathscr{I}(Z')$). The corresponding affine algebras are R/I and $S/I \cap S$, so the preceding remark shows that $\varphi : Z \to Z'$ is again finite (and dominant). It suffices now to prove that any finite dominant morphism is surjective. If $y \in Y$, then to say that $\varphi(x) = y$ is just to say that φ^* sends the local ring of y into that of x, or that φ^* sends the maximal ideal M' of S vanishing at y into the maximal ideal M of R vanishing at x. To show that φ is surjective, we therefore have to show that M' lies in *some* maximal ideal M of R (S still being viewed as a subring of R). But this follows from the Going Up Theorem (0.8), since R is integral over S.

(b) According to part (a), the restriction of φ to Z is again finite, so $\varphi(Z)$ is closed, irreducible. Now it suffices to show that dim Z = dim W. If $I = \mathscr{I}(Z)$, $I' = \mathscr{I}(W)$, then I is one of the prime ideals of R for which

$I \cap S = I'$. But as before, R/I is integral over S/I', so the corresponding extension of fields of fractions is algebraic and the dimensions coincide. $\square$

4.3. Image of a Morphism

Noether's Normalization Lemma (0.7) says that a finitely generated domain over K can be built up in two steps: a purely transcendental extension followed by an integral extension. In order to deal with a pair of irreducible affine varieties related by a dominant morphism $\varphi: X \to Y$, we have to develop a relativized version of this.

Let $S \subset R$ be two finitely generated domains over K, with respective fields of fractions $E \subset F$. Denote by R' the localization of R with respect to the multiplicative system S^* of nonzero elements of S (0.10). Of course, R' still has F as field of fractions. On the other hand, R' includes E and can thus be viewed as a finitely generated E-algebra. The Normalization Lemma says that R' is integral over $E[x_1, \ldots, x_r]$ for some $x_i \in R'$ which are algebraically independent over E. It is clear that the x_i can be chosen to lie in R (since any denominators occurring are units in R'). It is also clear that $r = \mathrm{tr.\,deg._E}\, F$.

Now compare the integral extension $E[x_1, \ldots, x_r] \subset R'$ with the extension $S[x_1, \ldots, x_r] \subset R$. The latter need not be integral. But R is finitely generated over S, and each generator satisfies a monic polynomial over $E[x_1, \ldots, x_r]$. By choosing a suitable common denominator $f \in S$, we can therefore guarantee that R_f is integral over $S_f[x_1, \ldots, x_r]$ (R_f, S_f the rings of quotients gotten by allowing powers of f as denominators). This set-up will be used to prove the following key theorem.

Theorem. *Let $\varphi: X \to Y$ be a dominant morphism of irreducible varieties, $r = \dim X - \dim Y$. Then Y has a nonempty open set U such that:*

(a) *$U \subset \varphi(X)$;*

(b) *if $W \subset Y$ is an irreducible closed set which meets U, and if Z is a component of $\varphi^{-1}(W)$ which meets $\varphi^{-1}(U)$, then $\dim Z = \dim W + r$.*

Proof. As in the proof of Theorem 4.1, it is harmless to replace Y by an affine open subset. We can also assume that X is affine: If we have found suitable open sets $U_i \subset Y$ for the restrictions of φ to finitely many affine open sets X_i which cover X, then $U = \bigcap_i U_i$ clearly satisfies both (a) and (b).

Let $R = K[X]$, $S = K[Y]$; view S as a subring of R via φ^* and use the method described above to find $x_1, \ldots, x_r \in R, f \in S$ such that R_f is integral over $S_f[x_1, \ldots, x_r]$, the latter being isomorphic to a polynomial ring over S_f. Now R_f, S_f are the respective affine algebras of the principal open sets X_f, Y_f (1.5); so $S_f[x_1, \ldots, x_r]$ may be viewed as the affine algebra of $Y_f \times \mathbf{A}^r$. The restriction of φ to X_f can be factored as $X_f \xrightarrow{\psi} Y_f \times \mathbf{A}^r \xrightarrow{\mathrm{pr}_1} Y_f$, where ψ is a *finite* morphism (and dominant). Set $U = Y_f$, and notice that $\varphi^{-1}(U) = X_f$.

Thanks to Proposition 4.2, ψ is surjective, as is pr_1. Therefore U does lie in $\varphi(X)$, proving (a). For (b), we may as well let $X = X_f$, $U = Y = Y_f$, with the above factorization $\varphi = \mathrm{pr}_1 \circ \psi$ (ψ finite). If W is a closed irreducible subset of Y, Z any component of $\varphi^{-1}(W)$, then Z is a component of $\psi^{-1}(W \times \mathbf{A}^r)$ and therefore maps *onto* $W \times \mathbf{A}^r$, with $\dim Z = \dim \psi(Z) = \dim W + r$ (4.2). $\square$

For use in (21.1) we record explicitly an assertion which is contained in in the above proof when $r = 0$.

Corollary (of proof). *Let $\varphi : X \to Y$ be a bijective morphism of irreducible varieties. Then* $\dim X = \dim Y$ *and there exist affine open subsets* $U \subset X$, $V \subset Y$, *such that* $\varphi(U) \subset V$ *and* $\varphi|U$ *is a finite morphism.* $\square$

4.4. Constructible Sets

Part (a) of Theorem 4.3 can be used to characterize the image of an arbitrary morphism. Recall that a subset of a topological space X is said to be locally closed if it is the intersection of an open set with a closed set. Call a finite union of locally closed sets **constructible**. (The constructible sets comprise the smallest collection of subsets of X containing all open and closed sets, and closed under the boolean operations, cf. Exercise 3.) Note that a constructible set contains a dense open subset of its closure. The following very useful result is due to Chevalley.

Theorem. *Let $\varphi : X \to Y$ be a morphism of varieties. Then φ maps constructible sets to constructible sets; in particular, $\varphi(X)$ is constructible.*

Proof. A locally closed subset of X is a subvariety, so a constructible set also is. It therefore suffices to prove that $\varphi(X)$ is constructible. In turn, we may as well assume that X, Y are irreducible. Proceed by induction on $\dim Y$, there being nothing to prove when this is 0. By induction, we need only consider dominant φ.

Choose an open subset U of Y contained in $\varphi(X)$, using Theorem 4.3(a). Then the irreducible components $W_1, \ldots, W_t$ of $Y - U$ have smaller dimension than Y (Proposition 3.2). The restrictions of φ to the various components Z_{ij} of $\varphi^{-1}(W_i)$ then have images which are constructible in W_i (by induction), hence also constructible in Y. Therefore, $\varphi(X)$ is constructible, being the union of U and the finitely many $\varphi(Z_{ij})$. $\square$

This theorem will be used repeatedly. Another fact, based on a similar induction, will be handy in one later argument (5.2). It is called the "upper semicontinuity of dimension".

Proposition. *Let $\varphi : X \to Y$ be a dominant morphism of irreducible varieties. For $x \in X$, let $\varepsilon_\varphi(x)$ be the maximum dimension of any component of $\varphi^{-1}(\varphi(x))$ passing through x. Then for all $n \in \mathbb{Z}^+$, $\{x \in X \mid \varepsilon_\varphi(x) \geqslant n\}$ is closed in X.*

Proof. Use induction on dim Y. Choose $U \subset \varphi(X)$ as in Theorem 4.3, and set $r = \dim X - \dim Y$, $E_n(\varphi) = \{x \in X | \varepsilon_\varphi(x) \geq n\}$. Theorem 4.1 implies that $\varepsilon_\varphi(x) \geq r$, so that $E_n(\varphi) = X$ is closed whenever $n \leq r$. On the other hand, Theorem 4.3 insures that $E_n(\varphi) \subset X - \varphi^{-1}(U)$ whenever $n > r$. Let $W_1, \ldots, W_t$ be the irreducible components of $Y - U$, Z_{ij} the various components of $\varphi^{-1}(W_i)$, $\varphi_{ij} : Z_{ij} \to W_i$ the restriction of φ. Since dim $W_i < \dim Y$, the induction hypothesis says that $E_n(\varphi_{ij})$ is closed in Z_{ij} (hence in X). But for $n > r$, $E_n(\varphi) = \bigcup_{i,j} E_n(\varphi_{ij})$, a finite union of closed sets. $\square$

4.5. Open Morphisms

The example at the end of (4.1) shows that the image of an open set under a morphism $\varphi : X \to Y$ need not be open. This is traceable to the fact that not all components of $\varphi^{-1}(W)$ (W closed, irreducible) need be of the same dimension. Such behavior is essentially confined to the complement of an open set in Y, thanks to Theorem 4.3 (b); so in a sufficiently "homogeneous" situation, the hypothesis of the following theorem will be fulfilled.

Theorem. *Let $\varphi : X \to Y$ be a dominant morphism of irreducible varieties, $r = \dim X - \dim Y$. Assume that for each closed irreducible subset $W \subset Y$, all irreducible components of $\varphi^{-1}(W)$ have dimension $r + \dim W$. Then φ maps open sets to open sets.*

Proof. The hypothesis implies in particular that φ is surjective and that all irreducible components of $\varphi^{-1}(W)$ dominate W.

Let $x \in X$ and let U be any open neighborhood of x. We have to show that $\varphi(x) = y$ lies in the interior of $\varphi(U) = V$. Suppose not. Then y lies in the closure of $Y - V$. Because V is constructible (Theorem 4.4), $Y - V$ is constructible. Thus y lies in the closure C of some locally closed set $O \cap C$ contained in $Y - V$ (where O is open in Y and C can be assumed to be irreducible, so that $O \cap C$ is dense in C). By hypothesis, the irreducible components of $C' = \varphi^{-1}(C)$ have equal dimension, and each dominates C. So $O' = \varphi^{-1}(O)$ meets each such component; $O' \cap C'$ is therefore *dense* in C'. But $O' \cap C' = \varphi^{-1}(O \cap C)$ lies in the closed set $X - U$, forcing $C' \subset X - U$. Since $x \in C'$, this is absurd. $\square$

This theorem will be applied in §12 to the canonical map $G \to G/H$, where G is a linear algebraic group and H a closed subgroup (the homogeneous space G/H being given a suitable structure of variety).

4.6. Bijective Morphisms

When a morphism is bijective, there is no mystery about either its fibres or its image. But its topological behavior and the effect of its comorphism on functions can be quite subtle. Here we are concerned mainly with the latter. The Frobenius map $\mathbf{A}^1 \to \mathbf{A}^1$, $x \mapsto x^p$ (where char $\mathsf{K} = p > 0$), illustrates how inseparability can complicate the whole question. (But even in characteristic 0 one cannot assert that a bijective morphism is an isomorphism, cf. Exercise 5.) The following theorem will be of great technical value in the study of homogeneous spaces in Chapter IV.

Theorem. *Let $\varphi: X \to Y$ be a dominant, injective morphism of irreducible varieties. Then $\mathsf{K}(X)$ is a finite, purely inseparable extension of $\varphi^*\mathsf{K}(Y)$.*

Proof. Since φ is dominant, φ^* is injective, so we may identify $\mathsf{K}(Y)$ with $\varphi^*\mathsf{K}(Y)$. Theorem 4.3 implies that dim X = dim Y. Therefore, $\mathsf{K}(X)$ is a finitely generated algebraic extension of $\mathsf{K}(Y)$, i.e., a *finite* extension.

The function fields do not change if we replace X, Y by nonempty open subsets. As a first reduction, let F be the separable closure of $\mathsf{K}(Y)$ in $\mathsf{K}(X)$ (= the set of elements of $\mathsf{K}(X)$ which are separable over $\mathsf{K}(Y)$). As a finitely generated extension of K, F may be viewed as the function field of some variety Z. Then the remarks in (2.3) allow us to factor φ as the composite of two (injective) morphisms $X \to Z, Z \to Y$, after cutting X, Z, Y down to suitable open sets. It suffices then to prove that $Z \to Y$ is birational, i.e., it suffices to prove the theorem when $\mathsf{K}(X)/\mathsf{K}(Y)$ is *separable*.

Let $n = [\mathsf{K}(X):\mathsf{K}(Y)]$. To show that $n = 1$, we shall find some $y \in Y$ whose fibre $\varphi^{-1}(y)$ has cardinality at least n.

The theorem of the primitive element allows us to choose a single generator f of $\mathsf{K}(X)$ over $\mathsf{K}(Y)$, whose minimal polynomial has the form $p(\mathsf{T}) = \mathsf{T}^n + \sum_{i=0}^{n-1} g_i\mathsf{T}^i$ $(g_i \in \mathsf{K}(Y))$. The separability further implies that f is not a root of $p'(\mathsf{T})$. Now X and Y may be replaced by open subsets so that the rational functions f, g_i (i.e., φ^*g_i) are everywhere defined on X. We may also assume that Y is *affine*, with $g_i \in S = \mathsf{K}[Y]$. Then f is *integral* over S, so $R = S[f]$ is integral over S. If U is an affine open subset of X, φ^* induces inclusions $S \subset R \subset \mathsf{K}[U]$. The ring R is an affine algebra in its own right, so it belongs to some affine variety X', and the factorization of φ^* corresponds to $U \to X' \to Y$. But $\mathsf{K}(U) = \mathsf{K}(X)$, while R has fraction field $\mathsf{K}(Y)(f) = \mathsf{K}(X)$. So it suffices to prove that the *finite* morphism $\psi: X' \to Y$ is birational. Unlike φ, this morphism might not be injective. But its restriction to a nonempty open set must be so: The image of U in X' contains a dense open set U' (4.3), and after deleting from it its intersection with the proper closed subset $\psi^{-1}(\psi(X' - U'))$ we are left with fibres consisting of single points. In turn, the image of this open set in Y contains a dense open subset V.

Now consider the fibre $\psi^{-1}(y)$, $y \in Y$. The point y corresponds to a K-algebra homomorphism $\varepsilon_y: S \to \mathsf{K}$, while the various $x \in \psi^{-1}(y)$ correspond to homomorphisms $\varepsilon_x: R \to \mathsf{K}$ extending ε_y. Integrality of R over S already guarantees at least one extension of ε_y (cf. (4.2)), but we need to be more precise in order to get n distinct extensions.

Since f is a root of $p(\mathsf{T})$, $f(x)$ is a root of the polynomial $p_\varepsilon(\mathsf{T})$ gotten by applying $\varepsilon = \varepsilon_y$ to each coefficient. We claim that for suitable choice of y, $p_\varepsilon(\mathsf{T})$ has n *distinct* roots (i.e., is a separable polynomial over K). For this we look at derivatives. Separability of $p(\mathsf{T})$ over $\mathsf{K}(Y)$ implies that $p'(f) \neq 0$. Since f is integral over S, $p'(f)$ is also; so there is a monic polynomial $q(\mathsf{T}) = \mathsf{T}^m + \sum h_i \mathsf{T}^i$ ($h_i \in S$) of which $p'(f)$ is a root (and we can assume that $h_0 \neq 0$, since $p'(f) \neq 0$). Choose $y \in V$ for which $h_0(y) \neq 0$. If $\psi(x) = y$, the choice of y clearly implies that $p'(f)(x) \neq 0$; in turn, $f(x)$ cannot be a root of $p'_\varepsilon(\mathsf{T})$. This justifies our claim that $p_\varepsilon(\mathsf{T})$ is separable over K for $\varepsilon = \varepsilon_y$.

For each of the n roots a of $p_\varepsilon(\mathsf{T})$, we can now construct a point $x \in \psi^{-1}(y)$ such that $f(x) = a$, thus completing the proof. The problem is to extend ε to a homomorphism $\varepsilon' : R \to \mathsf{K}$ for which $\varepsilon'(f) = a$. By a standard extension theorem (0.8), this can be done provided we know that for each $h(\mathsf{T}) \in S[\mathsf{T}]$, the condition $h(f) = 0$ implies $h_\varepsilon(a) = 0$. But $p(\mathsf{T})$ is the minimal polynomial of f over $\mathsf{K}(Y)$, so $h(f) = 0$ yields $h(\mathsf{T}) = p(\mathsf{T})k(\mathsf{T})$ ($k(\mathsf{T})$ a polynomial over $\mathsf{K}(Y)$, hence over S since $p(\mathsf{T})$ is monic and both $h(\mathsf{T})$, $p(\mathsf{T}) \in S[\mathsf{T}]$). From $p_\varepsilon(a) = 0$ we get $h_\varepsilon(a) = 0$, as required. $\square$

4.7. Birational Morphisms

A birational morphism need not be an isomorphism (cf. $\mathbf{A}^1 \to \mathbf{P}^1$ or the example at the end of (4.1)); but it is not too far from being one.

Proposition. *Let $\varphi : X \to Y$ be a birational morphism (X, Y irreducible). Then there is a nonempty open set $U \subset Y$ such that φ induces an isomorphism of $\varphi^{-1}(U)$ onto U.*

Proof. It does no harm to assume that Y is affine; set $S = \mathsf{K}[Y]$. Let $V \subset X$ be any affine open set, $R = \mathsf{K}[V]$. Consider $W = \overline{\varphi(X - V)}$, all of whose irreducible components have lower dimension than Y (since the components of $X - V$ are of smaller dimension than X, and $\dim X = \dim Y$). W being a proper closed subset of Y, we can find $0 \neq f \in S$ such that f vanishes on W, whence $\varphi^{-1}(Y_f) = X_{\varphi^* f} \subset V$. Replacing Y by Y_f and X by $X_{\varphi^* f}$, we may therefore assume that both X and Y are affine, with respective affine algebras R and S. By assumption, φ^* maps the field of fractions of S isomorphically onto that of R. Let $f_1, \dots, f_n$ generate R over K, where $f_i = \varphi^* g_i / \varphi^* h$ ($g_i, h \in S$). It is clear that φ^* maps the ring of quotients S_h isomorphically onto $R_{\varphi^* h}$. So we choose $U = Y_h$. $\square$

Exercises

1. The fibres of a finite morphism $\varphi : X \to Y$ (X, Y affine) are all finite.
2. Exhibit a subset of $\mathbf{A}^2$ which is constructible, but not locally closed.
3. The constructible subsets of a topological space X form the boolean algebra generated by the open (or closed) subsets of X: the smallest

collection containing all open sets, and closed under finite unions and complements.

4. For the example in (4.1), verify Proposition 4.4 directly by finding the sets $E_n(\varphi)$.
5. Define $\varphi : \mathbf{A}^1 \to \mathbf{A}^2$ by $\varphi(x) = (x^2, x^3)$. Verify that $X = \mathrm{Im}\ \varphi$ is closed in $\mathbf{A}^2$, and that $\varphi : \mathbf{A}^1 \to X$ is bijective, bicontinuous and birational, but not an isomorphism.

Notes

(4.1)–(4.4) are based on Mumford [3, I §8]. The proof of Theorem 4.5 is adapted from Chevalley [10, V, V, Prop. 3]; a somewhat more special criterion for openness, adequate for our needs in §12, is developed in Steinberg [13, appendix to 2.11]. The proof of Theorem 4.6 is adapted from Chevalley [10, II, V, Prop. 1 and Corollary].

5. Tangent Spaces

The tangent line to a curve at a given point is a good local approximation, whenever it is unambiguously defined, i.e., whenever the point in question is not a double point, cusp point, or other singularity. In this section we develop an intrinsic algebraic notion of tangent space to a variety at a point, which in the case of an algebraic group will be seen in §9 to carry the additional structure of a Lie algebra. The idea is to linearize problems and thereby simplify them. For our purposes it will be enough to consider those points which lie on only one irreducible component of a variety. So we assume, unless otherwise noted, that all varieties are irreducible.

5.1. Zariski Tangent Space

First we try to formulate geometrically the idea of "tangent space to a variety X at a point x". If X were a curve in $\mathbf{A}^2$ defined by a single equation $f(\mathsf{T}_1, \mathsf{T}_2) = 0$, we would describe the tangent (or tangents) at $x = (x_1, x_2)$ as the locus of the linear polynomial $\dfrac{\partial f}{\partial \mathsf{T}_1}(x)(\mathsf{T}_1 - x_1) + \dfrac{\partial f}{\partial \mathsf{T}_2}(x)(\mathsf{T}_2 - x_2)$. This locus consists of a straight line through x, unless both partial derivatives vanish at x.

By analogy, suppose $X \subset \mathbf{A}^n$ is defined by polynomials $f(\mathsf{T}_1, \ldots, \mathsf{T}_n)$. Set $d_x f = \sum\limits_{i=1}^{n} \dfrac{\partial f}{\partial \mathsf{T}_i}(x)(\mathsf{T}_i - x_i)$. Then write $\mathrm{Tan}(X)_x$ for the linear variety in $\mathbf{A}^n$ defined by the vanishing of all $d_x f$ as $f(\mathsf{T})$ ranges over $\mathscr{I}(X)$. It is easy to see that for any finite set of generators of $\mathscr{I}(X)$, the corresponding $d_x f$ generate the ideal of $\mathrm{Tan}(X)_x$ (Exercise 1); so this geometric tangent space may be computed explicitly in some cases. Notice that the tangent space to a linear variety (such as $\mathbf{A}^n$ itself) is just the variety.

If X is an arbitrary variety (not necessarily affine), we could choose an embedding in some $\mathbf{A}^n$ for an affine open neighborhood of $x \in X$ and thus define $\mathrm{Tan}(X)_x$ as above. But this procedure is hardly intrinsic, since it depends on the choices made. Instead, we look for an algebraic description of the tangent space in terms of the local ring $\mathcal{O}_x$.

For the moment, let $X \subset \mathbf{A}^n$ be affine, and let $M = \mathscr{I}(x)$ be the maximal ideal of $R = \mathsf{K}[X]$ vanishing at x. Since R/M can be identified with K, the R/M-module M/M^2 is a vector space over K (finite dimensional, since M is a finitely generated R-module). Now $d_x f$, for arbitrary $f(\mathsf{T}) \in \mathsf{K}[\mathsf{T}]$, can be viewed as a linear function on $\mathbf{A}^n$ (x being the "origin"), hence as a linear function on the vector subspace $\mathrm{Tan}(X)_x$ of $\mathbf{A}^n$. Since all $d_x f$ $(f(\mathsf{T}) \in \mathscr{I}(X))$ vanish on $\mathrm{Tan}(X)_x$ by definition, $d_x f$ is determined by the image of $f(\mathsf{T})$ in $R = \mathsf{K}[\mathsf{T}]/\mathscr{I}(X)$. We can therefore write $d_x f$ for $f \in R$. It is evident that d_x becomes in this way a K-linear map from R to the dual space of $\mathrm{Tan}(X)_x$. It is surjective, because a linear function g on $\mathrm{Tan}(X)_x$ is the restriction of a linear function on $\mathbf{A}^n$ (origin at x), given by a linear polynomial $f(\mathsf{T})$ whose $d_x f$ is the given g.

Since $R = \mathsf{K} + M$ (vector space direct sum), and since d_x (constant) $= 0$, we may as well view d_x as a map from M onto the dual space of $\mathrm{Tan}(X)_x$. We claim that $\mathrm{Ker}\, d_x = M^2$. Suppose $d_x f$ $(f \in M)$ vanishes on $\mathrm{Tan}(X)_x$, f the image of some nonconstant $f(\mathsf{T}) \in \mathsf{K}[\mathsf{T}]$. By construction, $d_x f = \sum a_i d_x f_i$ for some $a_i \in \mathsf{K}$, $f_i(\mathsf{T}) \in \mathscr{I}(X)$. Setting $g(\mathsf{T}) = f(\mathsf{T}) - \sum a_i f_i(\mathsf{T})$, we see that $d_x g$ vanishes on all of $\mathbf{A}^n$, i.e., is identically 0. Since $f(\mathsf{T})$ was nonconstant, we may assume that $g(\mathsf{T})$ is also. Then $g(\mathsf{T})$ must contain no linear term, i.e., $g(\mathsf{T})$ belongs to the square of the ideal $(\mathsf{T}_1, \ldots, \mathsf{T}_n)$. The image of this ideal in R is M, and $g(\mathsf{T})$ has the same image f in R as $f(\mathsf{T})$, so we conclude that $f \in M^2$, as asserted.

This identification of $\mathrm{Tan}(X)_x$ with the dual space $(M/M^2)^*$ falls short of giving an intrinsic notion of tangent space. But it is easy to pass to the local ring $(\mathcal{O}_x, \mathfrak{m}_x)$: Since $\mathcal{O}_x = R_M$, $\mathfrak{m}_x = MR_M$, there is a canonical isomorphism from the R/M-module M/M^2 onto the $\mathcal{O}_x/\mathfrak{m}_x$-module $\mathfrak{m}_x/\mathfrak{m}_x^2$, induced by the inclusion $R \to R_M$ (0.10). We can therefore cut loose entirely from the embedding of X in $\mathbf{A}^n$ and define the **tangent space** $\mathscr{T}(X)_x$ of X at x to be the dual vector space $(\mathfrak{m}_x/\mathfrak{m}_x^2)^*$ over $\mathsf{K} = \mathcal{O}_x/\mathfrak{m}_x$. This definition makes sense, of course, when X is an arbitrary irreducible variety (and even when X is reducible, provided we take care with the definition of $\mathcal{O}_x$). As a matter of notation, we use letters $\mathsf{x}, \mathsf{y}, \mathsf{z} \ldots$ to denote elements of $\mathscr{T}(X)_x$. (On a blackboard, ordinary capital letters might be used.)

A slightly different view of the tangent space is useful in some contexts. Define a **point derivation** $\delta : \mathcal{O}_x \to \mathsf{K}$ to be a map that behaves like a K-derivation of $\mathcal{O}_x$ followed by evaluation at x, i.e., δ is to be a K-linear map satisfying

$$(*) \quad \delta(fg) = \delta(f) \cdot g(x) + f(x) \cdot \delta(g).$$

It is clear that the point derivations of $\mathcal{O}_x$ form a vector space over K, call it

$\mathscr{D}_x$. We claim that $\mathscr{D}_x$ is naturally isomorphic to $\mathscr{T}(X)_x$. Indeed, if $f \in \mathcal{O}_x$ is constant or belongs to m_x^2, (*) shows that $\delta(f) = 0$ for $\delta \in \mathscr{D}_x$. Therefore, δ is completely determined by its effect on m_x, or by its induced effect on m_x/m_x^2. This injects $\mathscr{D}_x$ into $\mathscr{T}(X)_x$. In the other direction, a K-linear map $m_x/m_x^2 \to \mathsf{K}$ defines by composition with $m_x \to m_x/m_x^2$ a K-linear map $m_x \to \mathsf{K}$, which can be extended to $\mathcal{O}_x = \mathsf{K} + m_x$ by sending constants to 0. Then (*) is easy to check.

So there is a certain amount of flexibility in the way tangent spaces are thought of. For example, the tangent space of $\mathbf{A}^n$ (or $\mathbf{P}^n$) at a point x is geometrically just $\mathbf{A}^n$, viewed as a vector space with origin x; algebraically, the tangent space is the set of point derivations $\sum a_i \dfrac{\partial f}{\partial \mathsf{T}_i}(x)$ of the local ring of x in $\mathsf{K}(\mathsf{T}_1, \ldots, \mathsf{T}_n)$.

When X is the disjoint union of irreducible components, or just when x lies on a unique component Y, we can define $\mathscr{T}(X)_x$ to be $\mathscr{T}(Y)_x$. (Or we can use the dual space of m_x/m_x^2, for a suitably defined local ring $\mathcal{O}_x$ in the general case.)

Tangent spaces behave as expected relative to the formation of products:

Proposition. *Let X, Y be (irreducible) varieties, $x \in X$, $y \in Y$. Then $\mathscr{T}(X \times Y)_{(x, y)} \cong \mathscr{T}(X)_x \oplus \mathscr{T}(Y)_y$.*

Proof. This is obvious if we use the geometric description of tangent spaces. In algebraic terms, the assertion follows from the fact that $\mathcal{O}_{(x, y)}$ is the localization of $\mathcal{O}_x \otimes \mathcal{O}_y$ at the maximal ideal $m_x \otimes \mathcal{O}_y + \mathcal{O}_x \otimes m_y$, cf. Proposition 2.4. $\square$

5.2. Existence of Simple Points

In the case of a curve in affine space, the space of tangents at a point has (vector space) dimension 1 unless the point is "singular", and then the dimension goes up. We shall see shortly that $\dim \mathscr{T}(X)_x \geqslant \dim X$ for any variety. If equality holds, x is called a **simple point** of X. If all points of X are simple, X is called **smooth** (or **nonsingular**).

It is clear from the definition that $\mathbf{A}^n$ and $\mathbf{P}^n$ are smooth varieties, and it follows from Proposition 5.1 that the product of smooth varieties is smooth (cf. Proposition 3.1). But it is not so clear that an arbitrary variety possesses any simple points at all. Consider the special case of an irreducible hypersurface X in $\mathbf{A}^n$, where $\mathscr{I}(X)$ is the ideal generated by a single irreducible polynomial $f(\mathsf{T}_1, \ldots, \mathsf{T}_n)$. If $x = (x_1, \ldots, x_n) \in X$, the calculation in (5.1) shows that $\mathrm{Tan}(X)_x$ is the zero set of the linear polynomial $\sum_{i=1}^{n} \dfrac{\partial f}{\partial \mathsf{T}_i}(x)(\mathsf{T}_i - x_i)$. Since $\dim X = n - 1$, it follows that x will be simple unless this polynomial vanishes on $\mathbf{A}^n$, i.e., unless all partial derivatives $\dfrac{\partial f}{\partial \mathsf{T}_i}$ vanish at x

(then dim $\text{Tan}(X)_x = n$). If char $\mathsf{K} = 0$, this last condition cannot occur for every $x \in X$, else $f(\mathsf{T})$ would be constant. The same is true when char $\mathsf{K} = p > 0$: simultaneous vanishing of all $\dfrac{\partial f}{\partial \mathsf{T}_i}$ on X would force all powers of T_i in $f(\mathsf{T})$ to be multiples of p, so that $f(\mathsf{T}) = g(\mathsf{T})^p$ for some $g(\mathsf{T})$, contrary to the assumed irreducibility. Therefore X does have some simple points. In fact, the argument shows that they form a (dense) open set. Using this we can prove:

Theorem.   *Let X be any (irreducible) variety. Then* dim $\mathscr{T}(X)_x \geqslant$ dim X *for all $x \in X$, with equality holding for x in some dense open subset.*

Proof. $\mathsf{K}(X)$ is a **separably generated** extension of K (0.14), i.e., $\mathsf{K}(X)$ is a separable algebraic extension of a subfield $\mathsf{L} = \mathsf{K}(t_1, \ldots, t_d)$, the latter being purely transcendental over K ($d = $ dim X). The theorem of the primitive element allows us to find a single generator t_0 of the extension $\mathsf{K}(X)/\mathsf{L}$. Let $f(\mathsf{T}_0) \in \mathsf{L}[\mathsf{T}_0]$ be its minimal polynomial. This defines a rational function $f(\mathsf{T}_0, \mathsf{T}_1, \ldots, \mathsf{T}_d) \in \mathsf{K}(\mathsf{T}_0, \mathsf{T}_1, \ldots, \mathsf{T}_d)$, defined on an affine open subset of $\mathbf{A}^{d+1}$, where its set of zeros Y is a hypersurface with function field $\mathsf{K}(Y)$ isomorphic to $\mathsf{K}(X)$. It follows from the remarks in (2.3) and Proposition 4.7 that some nonempty open sets in X and Y are isomorphic. The points $y \in Y$ where dim $\mathscr{T}(Y)_y = $ dim $Y = d$ form a dense open subset of Y, so in particular dim $\mathscr{T}(X)_x = $ dim $X = d$ for all x in some dense open subset of X.

Next we apply the "upper semicontinuity of dimension" (Proposition 4.4) to get information about dim $\mathscr{T}(X)_x$ for arbitrary $x \in X$. Here it is enough to let X be an affine open neighborhood of x. So view X as a closed subset of some $\mathbf{A}^n$ and view all tangent spaces as linear subvarieties of $\mathbf{A}^n$. The pairs $(x, y) \in X \times \mathbf{A}^n$ for which $y \in \text{Tan}(X)_x$ are easily seen to form a closed subset T of the product. Projection onto the first factor defines a morphism $\varphi: T \to X$, whose fibre $\varphi^{-1}(x)$ has the dimension of $\mathscr{T}(X)_x$. For each m, $X_m = \{x \in X \,|\, \dim \mathscr{T}(X)_x \geqslant m\}$ is *closed* in X (4.4). But X_d was seen above to be *dense* in X, so $X_d = X$. $\square$

5.3. Local Ring of a Simple Point

The inequality dim $\mathscr{T}(X)_x \geqslant$ dim X (5.2) may be interpreted as stating that no fewer than dim X "local parameters" are required to determine x. To make this idea precise, we need a general lemma.

Lemma.   *Let R be a noetherian local ring, with unique maximal ideal M. Then M is generated as R-module by $f_1, \ldots, f_n \Leftrightarrow M/M^2$ is generated as R/M-module by the images of $f_1, \ldots, f_n$.*

Proof.   This is obvious in one direction. So let the images of the f_i generate M/M^2, and let N be the R-submodule of M generated by $f_1, \ldots, f_n$.

The finitely generated R-module M/N then satisfies: $M(M/N) = M/N$. According to Nakayama's Lemma (0.11), $M/N = 0$, i.e., $M = N$. $\square$

In particular, the *minimal number of generators n* of the ideal m_x coincides with the vector space dimension of m_x/m_x^2, or of its dual space $\mathcal{T}(X)_x$. What is the significance of this when x is a simple point, so that $n = \dim X$?

For a (noetherian) local ring (R, M), the **Krull dimension** of R is defined to be the greatest length k of any chain $0 \subsetneq P_1 \subsetneq P_2 \subsetneq \cdots \subsetneq M = P_k$ of prime ideals. In the case of $\mathcal{O}_x$, we observe that the Krull dimension is just $\dim X$. Indeed, we may assume that X is affine, so that $\mathcal{O}_x = \mathsf{K}[X]_{\mathcal{I}(x)}$, $\mathcal{I}(x)$ being a maximal ideal of $\mathsf{K}[X]$, it follows from the dimension theorem (cf. Exercise 3.4) that $\dim X$ is the length of a maximal chain of distinct prime ideals between 0 and $\mathcal{I}(x)$. But the prime ideals of $\mathsf{K}[X]$ contained in $\mathcal{I}(x)$ correspond 1–1 with the prime ideals of $\mathcal{O}_x$.

A local ring (R, M) is called **regular** if its Krull dimension coincides with the minimal number of generators of M ($= \dim_{R/M} M/M^2$, by the lemma). It is a general fact (0.12) that a regular local ring is an integral domain and is integrally closed (in its field of fractions). The discussion above therefore establishes:

Theorem A. *Let $x \in X$ be a simple point on the (irreducible) variety X. Then $\mathcal{O}_x$ is a regular local ring, hence is integrally closed (in its field of fractions $\mathsf{K}(X)$).* $\square$

With appreciably more labor, it can be shown that a regular local ring is even a UFD. We shall need to know this only when $\dim X = 1$, where it is easy to give a direct proof (Exercise 2).

The fact that the local ring of a simple point is integrally closed will enable us to apply the following result in the study of homogeneous spaces (12.3).

Theorem B. *Let X be an irreducible variety, $x \in X$ a point whose local ring $\mathcal{O}_x$ is integrally closed, $f \in \mathsf{K}(X)$ a function not in $\mathcal{O}_x$. Then there exists a subvariety Y of X containing x, such that $f' = 1/f \in \mathcal{O}_y$ for some $y \in Y$ and such that f' takes the value 0 on Y whenever it is defined.*

Proof. Let $R = \mathcal{O}_x$. Then $I = \{g \in R \,|\, gf \in R\}$ is a proper ideal of R (since $f \notin R$ by assumption), hence contained in m_x. Let $P = P_1, P_2, \ldots, P_t$ be the distinct minimal primes of I in R (these lie in m_x), cf. (0.13). For large enough n, $P^n P_2^n \cdots P_t^n \subset I$. For $i > 1$, it is clear that P_i generates the unit ideal in the local ring $R_P \subset \mathsf{K}(X)$, so that $P^n R_P \subset I R_P$. In particular, since $If \subset R$, $P^n f \subset (If)R_P \subset R_P$. Choose $k \geqslant 0$ as small as possible so that $P^k f \subset R_P$ (then $k > 0$), and let $g \in P^{k-1} f$, $g \notin R_P$ (then Pg lies in R_P).

R being integrally closed (by hypothesis), R_P is also (0.10). Now $g \notin R_P$, so g cannot be integral over R_P. It follows (0.6) that multiplication by g could not stabilize the finitely generated R_P-module PR_P, i.e., $PR_P g \not\subset PR_P$ (although $Pg \subset R_P$). In other words, Pg generates the unit ideal in R_P, hence contains a unit of R_P. Therefore $1/g \in PR_P$; in fact, $PR_P = (1/g)R_P$.

Now $h = f/g^k \in fP^kR_P \subset R_P$ (by choice of k). Observe that h is a *unit* in R_P: otherwise $h \in PR_P = (1/g)R_P$, or $f/g^{k-1} \in R_P$ (contradicting the choice of k). So $1/f = h^{-1}(1/g^k) \in PR_P$. If Y is the zero set of P (i.e., of the pullback of P in the affine algebra of some affine open subset of X containing x), this means that $1/f$ defines a rational function on Y and vanishes whenever defined. Moreover, $x \in Y$. $\quad\square$

5.4. Differential of a Morphism

Let $\varphi: X \to Y$ be a morphism of (irreducible) varieties. If $x \in X$, $y = \varphi(x)$, then φ^* maps $(\mathcal{O}_y, m_y)$ into $(\mathcal{O}_x, m_x)$. By composition with φ^*, a linear function x on m_x/m_x^2 therefore induces a linear function $d\varphi_x(\mathsf{x})$ on m_y/m_y^2. The resulting map $d\varphi_x: \mathcal{T}(X)_x \to \mathcal{T}(Y)_y$ is evidently K-linear. We call it the **differential** of φ at x.

Differentiation has the expected functorial properties: If $\varphi: X \to X$ is the identity map, so is $d\varphi_x$. If $\varphi: X \to Y$ and $\psi: Y \to Z$ are morphisms, with $x \in X$, $\varphi(x) = y \in Y$, $\psi(y) = z \in Z$, then $d(\psi \circ \varphi)_x = d\psi_y \circ d\varphi_x: \mathcal{T}(X)_x \to \mathcal{T}(Z)_z$.

For computational purposes, an explicit recipe is sometimes useful. Say $X \subset \mathbf{A}^n$, $Y \subset \mathbf{A}^m$, so φ is given by m coordinate functions $\varphi_i(\mathsf{T}_1, \ldots, \mathsf{T}_n)$. Let $x \in X$, $y = \varphi(x)$, and identify the respective tangent spaces with subspaces of K^n, K^m. This just identifies $a = (a_1, \ldots, a_n) \in \mathsf{K}^n$ with the point derivation $\mathcal{O}_x \to \mathsf{K}$ induced by $\sum a_i \dfrac{\partial}{\partial \mathsf{T}_i}$ (followed by evaluation at x). Then $d\varphi_x(a) = (b_1, \ldots, b_m)$, where $b_k = \sum_i \dfrac{\partial \varphi_k}{\partial T_i}(x)a_i$ (Exercise 3).

One important example is provided by the determinant, where det: $\mathrm{GL}(n, \mathsf{K}) \to \mathrm{GL}(1, \mathsf{K}) \subset \mathbf{A}^1$ is given by a single polynomial in n^2 indeterminates T_{ij} $(1 \leqslant i, j \leqslant n)$. Since $\mathrm{GL}(n, \mathsf{K})$ is an affine open subset of $\mathbf{A}^{n^2}$, its tangent space at each point is just K^{n^2}. It is convenient to view this vector space as the set $\mathrm{M}(n, \mathsf{K})$ of all $n \times n$ matrices. Then a quick calculation with the above formula for $d\varphi_e$ yields the result: $d(\det)_e(a) = a_{11} + a_{22} + \cdots + a_{nn} \in \mathrm{M}(1, \mathsf{K}) = \mathsf{K}$. In other words, *the differential of the determinant is the trace.*

A couple of other computations will be referred to later:

(1)   If $\varphi: \mathbf{A}^n \to \mathbf{A}^n$ is a linear map, then $d\varphi_x$ may be identified with φ, if $\mathcal{T}(\mathbf{A}^n)_x$ is identified with $\mathbf{A}^n$.

(2)   Let V be an n-dimensional vector space over K, and let $\varphi: V - \{0\} \to \mathbf{P}(V)$ be the canonical map, $V - \{0\}$ being viewed as a principal open subset of $V = \mathbf{A}^n$. Let $0 \neq v \in V$, and consider $d\varphi_v$, where $\mathcal{T}(V - \{0\})_v$ is identified with V. If v is taken as a first basis vector, the map φ can be described in corresponding affine coordinates by $\varphi(x_1, \ldots, x_n) = (x_2/x_1, x_3/x_1, \ldots, x_n/x_1)$. It follows readily that the kernel of the linear map $d\varphi_v$ is precisely the subspace $\mathsf{K}v$ (and therefore $d\varphi_v$ is *surjective*, by comparison of dimensions).

(3)   If Y is a subvariety of X, $y \in Y$, then the inclusion map $i: Y \to X$

induces a monomorphism $di_y \colon \mathcal{T}(Y)_y \to \mathcal{T}(X)_y$. So we view the former as a subspace of the latter.

5.5. Differential Criterion for Separability

A field extension E/F is said to be **separable** if either char F $= 0$, or else char F $= p$ and the pth powers of elements $x_1, \ldots, x_r \in$ E linearly independent over F are again linearly independent over F. We shall need to know various facts (0.14): (1) This is equivalent to the usual notion of separability when E/F is finite. (2) If F $\subset$ L $\subset$ E, E/F separable, then L/F is separable (but E/L need not be). (3) If F is perfect, E/F is always separable. (4) For finitely generated extensions of F, "separable" is equivalent to "separably generated".

For our purposes, the question of separability comes up in connection with dominant morphisms $\varphi \colon X \to Y$ of irreducible varieties (cf. Theorem 4.6). Here φ^* identifies $K(Y)$ with a subfield of $K(X)$, over which $K(X)$ is finitely generated but not always separable. For example, the Frobenius map $x \mapsto x^p$ is a bijective morphism $\mathbf{A}^1 \to \mathbf{A}^1$, whose comorphism maps $K(\mathbf{A}^1) = K(T)$ onto the subfield $K(T^p)$. If $K(X)/\varphi^*K(Y)$ is separable, we call the morphism φ **separable**. In characteristic 0, all morphisms are therefore separable. For the time being let char K $= p > 0$.

The separability of a finite field extension can be decided by testing the minimal polynomials of certain field elements for multiple roots; for this a derivative test is appropriate. Here we shall develop an analogous "differential criterion" for separability. Some further general facts are required (0.15): (5) Let E/F be separably generated, and let $\mathcal{D}$ be the vector space over L of F-derivations E $\to$ L (L some extension field of E). Then $\dim_L \mathcal{D} = $ tr. $\deg_F$ E. (6) An extension E/F is separable if (and only if) all derivations F $\to$ L extend to derivations E $\to$ L (L $\supset$ E any auxiliary field). (7) All derivations E $\to$ L are E^p-derivations; in particular, if E is perfect, all derivations of E are zero. This applies in particular to the algebraically closed field K.

Since K is perfect, any function field $K(X)$ is automatically separably generated (and separable) over K, in view of (3) and (4) above. So (5) implies that $\dim X = \dim_L$ Der $(K(X), L)$ for any extension L of K. (All derivations are K-derivations, so we henceforth omit the subscript.) In particular, consider a dominant morphism $\varphi \colon X \to Y$. Then $\dim X = \dim_{K(X)}$ Der $(K(X), K(X))$ and $\dim Y = \dim_{K(X)}$ Der $(\varphi^*K(Y), K(X))$. On the other hand, φ induces a $K(X)$-linear map, which we denote $\delta\varphi$, by restriction:

$$\delta\varphi \colon \mathrm{Der}\,(K(X), K(X)) \to \mathrm{Der}\,(\varphi^*K(Y), K(X)).$$

If we know that $\delta\varphi$ is *surjective*, then it will follow from (6) that $K(X)/\varphi^*K(Y)$ is separable (i.e., that φ is separable), and conversely.

To test the surjectivity of $\delta\varphi$, we make a comparison with $d\varphi_x \colon \mathcal{T}(X)_x \to \mathcal{T}(Y)_{\varphi(x)}$, where both x and $\varphi(x)$ are assumed to be *simple* points. Thus $d\varphi_x$

is a K-linear map from a vector space of dimension dim X into one of dimension dim Y. By formally extending the base field from K to $K(X)$, we get the following picture:

$$\text{Der}\,(K(X), K(X)) \xrightarrow{\;\delta\varphi\;} \text{Der}\,(\varphi^*K(Y), K(X))$$

$$\Lambda \qquad\qquad\qquad\qquad\qquad \Lambda$$

$$\vdots \qquad\qquad\qquad\qquad\qquad \vdots$$

$$V \qquad\qquad\qquad\qquad\qquad V$$

$$K(X) \otimes \mathcal{T}(X)_x \xrightarrow{\;1\otimes d\varphi_x\;} K(X) \otimes \mathcal{T}(Y)_{\varphi(x)}$$

The dotted map on the left can be supplied by identifying elements of $\mathcal{T}(X)_x$ (point derivations of $\mathcal{O}_x$) with derivations $\mathcal{O}_x \to \mathcal{O}_x$ and hence with derivations of the fraction field $K(X)$. (The quotient rule for differentiation shows that there is one and only one way to extend a derivation of $\mathcal{O}_x$ to $K(X)$.) Similarly, the dotted map on the right can be added. Note that $K(X)$ just plays the role of an auxiliary field (large enough to contain all fields involved).

It is evident that the diagram is commutative. Since the dimensions (over $K(X)$) agree on the left side and also agree on the right side, we conclude that $\delta\varphi$ will be surjective if and only if $d\varphi_x$ is surjective. This establishes the desired criterion for separability when char $K = p > 0$. On the other hand, field extensions are always separable when char $K = 0$; using (6) and the preceding argument, we see that K-derivations always extend, so the same criterion follows:

Theorem. *Let $\varphi: X \to Y$ be a dominant morphism of irreducible varieties. Suppose that x, $\varphi(x)$ are simple points. Then $d\varphi_x: \mathcal{T}(X)_x \to \mathcal{T}(Y)_{\varphi(x)}$ is surjective if and only if φ is separable.* $\quad\square$

An explicit calculation based on this criterion will be carried out in (6.4).

Exercises

1. Let X be an affine variety, $X \subset \mathbf{A}^n$. If $f_1(\mathrm{T}), \dots, f_r(\mathrm{T})$ generate $\mathscr{I}(X)$, prove that $d_x f_1, \dots, d_x f_r$ generate the ideal of $\mathrm{Tan}(X)_x$, for $x \in X$.
2. Let dim $X = 1$, $x \in X$ simple. Prove that $\mathcal{O}_x$ is a UFD. [Show that, up to units, a generator of m_x is the only irreducible element of $\mathcal{O}_x$, hence that $\mathcal{O}_x$ is a PID with unique nonzero prime ideal m_x.]
3. Verify the formula for $d\varphi_x$ in (5.4).
4. The canonical morphism $V - \{0\} \to \mathbf{P}(V)$ is separable.

Notes

For Theorem B of (5.3), see Chevalley [8, exposé 8, lemme 1].

6. Complete Varieties

The facts about completeness in (6.1) and (6.2) will play an important role from §21 on, in the study of homogeneous spaces G/H which are projective varieties. Some rather special observations about $\mathbf{P}^1$ are made in (6.3) and (6.4), to be applied at a crucial stage of the structure theory of reductive groups (25.3).

6.1. Basic Properties

A variety X is called **complete** if for all varieties Y, $\mathrm{pr}_2: X \times Y \to Y$ is a closed map (i.e., sends closed sets to closed sets). The geometric meaning of completeness is not made intuitively obvious by this definition, but a kind of "compactness" is intended. Indeed, for nice enough Hausdorff spaces, the criterion just stated (with $X \times Y$ given the ordinary product topology) is equivalent to compactness.

Evidently a single point, viewed as a variety, is complete. It is also clear that X is complete if and only if all its irreducible components are, and that the auxiliary varieties Y in the definition can be taken to be irreducible (and even affine) if a given variety X is to be tested for completeness. What is not so clear is that "interesting" complete varieties exist. It will be shown in (6.2) that projective varieties pass the test. On the other hand, not all varieties are complete: The locus of the equation $T_1 T_2 = 1$ in $\mathbf{A}^1 \times \mathbf{A}^1$ projects to a nonclosed subset of $\mathbf{A}^1$, so $\mathbf{A}^1$ cannot be complete.

We assemble here a list of elementary facts about completeness, some of which should remind the reader of properties enjoyed by compact Hausdorff spaces.

Proposition.   *Let X, Y be varieties.*
(a) *If X is complete and Y is closed in X, then Y is complete.*
(b) *If X and Y are complete, then $X \times Y$ is complete.*
(c) *If $\varphi: X \to Y$ is a morphism and X is complete, then $\varphi(X)$ is closed and complete.*
(d) *If Y is a complete subvariety of X, then Y is closed.*
(e) *If X is complete and affine, then* $\dim X = 0$.
(f) *A complete quasiprojective variety is projective.*

Proof.   (a) (b) These follow at once from the definitions.
(c) Since we are dealing with varieties, the graph of φ is closed in $X \times Y$ (2.5). Its image under pr_2 is $\varphi(X)$, which is closed because X is complete. To test $\varphi(X)$ for completeness, we can assume $Y = \varphi(X)$. For any variety Z, consider $\mathrm{pr}_2: X \times Z \to Z$, $\mathrm{pr}_2': Y \times Z \to Z$. If W is closed in $Y \times Z$, then $\mathrm{pr}_2'(W) = \mathrm{pr}_2(\varphi \times 1)^{-1}(W)$ is closed in Z because X is complete.

(d) Apply (c) to the inclusion morphism $Y \to X$.

(e) As remarked above, $\mathbf{A}^1$ is not complete. In view of (c), the only morphisms $X \to \mathbf{A}^1$ from an irreducible complete variety X are the constant maps. So a complete irreducible affine variety X satisfies $\mathsf{K}[X] = \mathsf{K}$, forcing X to be a point.

(f) This follows from (d). $\square$

6.2. Completeness of Projective Varieties

Theorem.   *Any projective variety is complete.*

Proof.   Thanks to Proposition 6.1(a), it is enough to show that $\mathrm{pr}_2 \colon \mathbf{P}^n \times Y \to Y$ is closed, for any variety Y. We can even assume that Y is irreducible and affine, with affine algebra R. The affine open sets $U_i = \mathbf{P}_i^n \times Y$ cover the product. If $\mathsf{X}_0, \ldots, \mathsf{X}_n$ are homogeneous coordinates on $\mathbf{P}^n$, then the affine algebra of U_i can be described as $R_i = R[\mathsf{X}_0/\mathsf{X}_i, \ldots, \mathsf{X}_n/\mathsf{X}_i]$. ($\mathsf{X}_0/\mathsf{X}_i, \ldots, \mathsf{X}_n/\mathsf{X}_i$ are affine coordinates on $\mathbf{P}_i^n$, as in (1.6).)

Take any closed set Z in $\mathbf{P}^n \times Y$, and any point $y \in Y - \mathrm{pr}_2(Z)$. We want to find a neighborhood of y in Y of the form Y_f which is disjoint from $\mathrm{pr}_2(Z)$. This amounts to finding $f \in R, f \notin M = \mathscr{I}(y)$, such that f vanishes on $\mathrm{pr}_2(Z)$, i.e., such that the pullback of f in R_i belongs to $\mathscr{I}(Z_i)$ for all i, $Z_i = Z \cap U_i$. The existence of such f will follow from a version of Nakayama's Lemma, applied to a suitable R-module, which we now proceed to construct.

First, consider the polynomial ring $S = R[\mathsf{X}_0, \ldots, \mathsf{X}_n]$, with natural grading $S = \sum S_m$. We construct a homogeneous ideal $I \subset S$ by letting I_m consist of all $f(\mathsf{X}_0, \ldots, \mathsf{X}_n) \in S_m$ such that $f(\mathsf{X}_0/\mathsf{X}_i, \ldots, \mathsf{X}_n/\mathsf{X}_i) \in \mathscr{I}(Z_i)$ for each i.

Next fix i, and let $f \in \mathscr{I}(Z_i)$. We claim that multiplication by a sufficiently high power of X_i will take f into I. Indeed, if we view f as a polynomial in $\mathsf{X}_0/\mathsf{X}_i, \ldots, \mathsf{X}_n/\mathsf{X}_i$, then $\mathsf{X}_i^m f$ becomes a homogeneous polynomial (of degree m) in $\mathsf{X}_0, \ldots, \mathsf{X}_n$ for large m. In turn, $(\mathsf{X}_i^m/\mathsf{X}_j^m)f \in R_j$ vanishes on $Z_i \cap U_j = Z_j \cap U_i$, while $(\mathsf{X}_i^{m+1}/\mathsf{X}_j^{m+1})f$ vanishes at all points of Z_j not in U_i. Since j is arbitrary, we conclude that $\mathsf{X}_i^{m+1}f$ lies in I_{m+1}.

Now Z_i and $\mathbf{P}_i^n \times \{y\}$ are disjoint closed subsets of the affine variety U_i, so their ideals $\mathscr{I}(Z_i)$ and MR_i generate the unit ideal R_i. In particular, there exists an equation $1 = f_i + \sum_j m_{ij}g_{ij}$, where $f_i \in \mathscr{I}(Z_i)$, $m_{ij} \in M$, $g_{ij} \in R_i$. Thanks to the preceding paragraph, multiplication by a sufficiently high power of X_i takes f_i into I. We can choose this power large enough to work in these equations for all i, and to take all g_{ij} into S as well. So we obtain: $\mathsf{X}_i^m \in I_m + MS_m$ (for all i). Enlarging m even more, we can get all monomials of degree m in $\mathsf{X}_0, \ldots, \mathsf{X}_n$ to lie in $I_m + MS_m$. This implies that $S_m = I_m + MS_m$.

Now apply (0.11) to the finitely generated R-module S_m/I_m, which satisfies: $M(S_m/I_m) = S_m/I_m$. The conclusion is that there exists $f \in R, f \notin M$, such that f annihilates S_m/I_m. Thus $fS_m \subset I_m$; in particular, $f\mathsf{X}_i^m \in I_m$, so that f vanishes on $\mathrm{pr}_2(Z)$. $\square$

6.3. Varieties Isomorphic to $\mathbf{P}^1$

The local ring $\mathcal{O}_x$ of a simple point on an irreducible variety X is an integrally closed domain (Theorem 5.3 A). This can easily be strengthened to the statement that $\mathcal{O}_x$ is a UFD when dim $X = 1$ (Exercise 5.2). In this case the maximal ideal m_x is *principal*, therefore generated by an irreducible element q. Since $\mathcal{O}_x - m_x$ consists of units, it is clear that q is (essentially) the only irreducible element of $\mathcal{O}_x$. This implies that $\mathcal{O}_x$ is a **valuation ring**, i.e., for each $f \in \mathsf{K}(X)$, either $f \in \mathcal{O}_x$ or else $1/f \in \mathcal{O}_x$. Indeed, if $f = g/h$ ($g, h \in \mathcal{O}_x$), then we may assume that not both g, h are divisible by q; so either g or h is a unit in $\mathcal{O}_x$.

Now specialize to the case in which $\mathsf{K}(X) \cong \mathsf{K}(T)$, the function field of $\mathbf{A}^1$ or $\mathbf{P}^1$, and assume that all points of X are simple. The preceding discussion shows that either T or $1/T$ belongs to each local ring $\mathcal{O}_x$. It is an elementary exercise to determine all the valuation rings in $\mathsf{K}(T)$ which include either $\mathsf{K}[T]$ or $\mathsf{K}[1/T]$: they are (aside from $\mathsf{K}(T)$) precisely the local rings belonging to the points of $\mathbf{P}^1$ (Exercise 2). Notice too that there cannot be any proper inclusions between pairs of these valuation rings.

Theorem. *Let X be a smooth variety of dimension 1, and let $\varphi : \mathbf{P}^1 \to X$ be a dominant morphism. Then X is isomorphic to $\mathbf{P}^1$ (although φ need not be an isomorphism).*

Proof. Thanks to Proposition 6.1 (c), φ is surjective, and X is complete (and irreducible). φ^* identifies $\mathsf{K}(X)$ with a subfield of $\mathsf{K}(\mathbf{P}^1) = \mathsf{K}(T)$ having transcendence degree 1 over K. According to Lüroth's Theorem (0.4), every such subfield of $\mathsf{K}(T)$ is isomorphic to $\mathsf{K}(T)$. So we identify $\mathsf{K}(X)$ with $\mathsf{K}(T)$ and show that this, coupled with the completeness of X, forces X to be isomorphic to $\mathbf{P}^1$.

Let $f, g \in \mathsf{K}(X)$ correspond respectively to T, $1/T$. As remarked above, the fact that $\mathcal{O}_x$ ($x \in X$) is a valuation ring implies that either $f \in \mathcal{O}_x$ or $g \in \mathcal{O}_x$; when both $f, g \in \mathcal{O}_x$, we have $f(x)g(x) = 1$. The (open) subsets U, V of X on which f, g are defined cover X. We can define a morphism $U \to \mathbf{P}^1$ (resp. $V \to \mathbf{P}^1$) by sending x to the point whose homogeneous coordinates are $(f(x), 1)$ (resp. $(1, g(x))$). Since $f(x)g(x) = 1$ for $x \in U \cap V$, these patch together to yield a morphism $\psi : X \to \mathbf{P}^1 (2.3)$.

Because X is complete, ψ is surjective (Proposition 6.1 (c)). By construction, ψ^* is an isomorphism of function fields. To conclude that ψ is an isomorphism, it remains to show that ψ^* maps local rings of $\mathbf{P}^1$ isomorphically *onto* local rings of X. But, as remarked above, there are no proper inclusions among the valuation rings of $\mathsf{K}(T)$ in question. $\quad\square$

6.4. Automorphisms of $\mathbf{P}^1$

For use in (25.3), we have to determine the group Aut $\mathbf{P}^1$ of automorphisms of the projective line. First we exhibit some particular automorphisms, induced by the natural action of GL(2, K) on the nonzero vectors in K^2. We may view this action as a permutation of the lines (through the origin) in K^2, which is doubly transitive: Any pair of distinct lines may be sent to any other pair of distinct lines, since GL(2, K) acts transitively on bases of K^2. The scalar matrices $\begin{pmatrix} a & 0 \\ 0 & a \end{pmatrix}$, $a \in K^*$, leave lines stable, so we obtain an action of PGL(2, K) = GL(2, K)/K*. Denote this group by G, and write the image of $\begin{pmatrix} a & b \\ c & d \end{pmatrix}$ as $\begin{bmatrix} a & b \\ c & d \end{bmatrix}$.

Now $\mathbf{P}^1$ can be thought of as the set of all lines through the origin in K^2, a typical point of $\mathbf{P}^1$ being written as $\begin{bmatrix} x \\ y \end{bmatrix}$ ($x, y \in K$, not both 0), with $\begin{bmatrix} x \\ y \end{bmatrix} = \begin{bmatrix} ax \\ ay \end{bmatrix}$ for all $a \in K^*$. So G acts on $\mathbf{P}^1$ as a group of permutations: $\begin{bmatrix} a & b \\ c & d \end{bmatrix}\begin{bmatrix} x \\ y \end{bmatrix} = \begin{bmatrix} ax + by \\ cx + dy \end{bmatrix}$. Moreover, it is clear that each element of G defines an automorphism of $\mathbf{P}^1$ as *variety*.

Theorem.   PGL(2, K) $\cong$ Aut $\mathbf{P}^1$.

Proof.   The idea is to characterize an automorphism of $\mathbf{P}^1$ by its effect on the triple $(0, 1, \infty)$, where $0 = \begin{bmatrix} 0 \\ 1 \end{bmatrix}$, $1 = \begin{bmatrix} 1 \\ 1 \end{bmatrix}$, $\infty = \begin{bmatrix} 1 \\ 0 \end{bmatrix}$. (Here $\mathbf{P}^1 = \mathbf{A}^1 \cup \{\infty\}$, the points $\begin{bmatrix} x \\ 1 \end{bmatrix}$ corresponding to the points of the affine line.)

Define a map $\varphi: G \mapsto \mathbf{P}^1 \times \mathbf{P}^1 \times \mathbf{P}^1$ by $\varphi(g) = (g(0), g(1), g(\infty))$, i.e., $\begin{bmatrix} a & b \\ c & d \end{bmatrix} \mapsto \left(\begin{bmatrix} b \\ d \end{bmatrix}, \begin{bmatrix} a + b \\ c + d \end{bmatrix}, \begin{bmatrix} a \\ c \end{bmatrix} \right)$. If $\varphi(g) = (0, 1, \infty)$, then $g = \begin{bmatrix} a & 0 \\ 0 & a \end{bmatrix} = \begin{bmatrix} 1 & 0 \\ 0 & 1 \end{bmatrix}$, which implies that φ is injective. On the other hand, if (x, y, z) is a triple of distinct points of $\mathbf{P}^1$, we can find as follows an element of G sending $(0, 1, \infty)$ to (x, y, z). First use the double transitivity of G to send (x, y, z) into $\left(0, \begin{bmatrix} u \\ v \end{bmatrix}, \infty \right)$, where $\begin{bmatrix} u \\ v \end{bmatrix} \neq 0, \infty$ (and thus $uv \neq 0$). In turn, $\begin{bmatrix} u^{-1} & 0 \\ 0 & v^{-1} \end{bmatrix}$ sends $\left(0, \begin{bmatrix} u \\ v \end{bmatrix}, \infty \right)$ to $(0, 1, \infty)$.

Now let σ be an arbitrary automorphism of $\mathbf{P}^1$. Then $(\sigma(0), \sigma(1), \sigma(\infty))$ is a triple of distinct points, which by the preceding argument has the form $(g(0), g(1), g(\infty))$ for some $g \in G$. The automorphism $\tau = \sigma^{-1}g$ therefore fixes 0, 1, ∞. In particular, τ restricts to an automorphism of $\mathbf{A}^1 \subset \mathbf{P}^1$. Such an automorphism is easily seen to have the form: $a \mapsto ra + s$ ($r \in K^*$, $s \in K$), cf. Exercise 1.10. Since 0, 1 are fixed, we conclude that $\tau = 1$ and $\sigma = g \in G$. $\square$

The map φ introduced in the preceding proof turns out, under closer scrutiny, to be a morphism (even an isomorphism) of affine varieties. GL(2, K) can of course be regarded as a principal open subset of $\mathbf{A}^4$, and the orbit map $\psi: GL(2, K) \to Y \subset \mathbf{P}^1 \times \mathbf{P}^1 \times \mathbf{P}^1$ defined by $\psi(g) = (g(0), g(1), g(\infty))$ is clearly a morphism ($Y = \text{Im } \varphi$). In Chapter IV we shall see how to regard $PGL(2, K)$ as a 3-dimensional affine variety, so that the canonical map $\pi: GL(2, K) \to PGL(2, K)$ is a morphism (and so that φ is a morphism). The factorization $\psi = \varphi\pi$ will then imply that φ is *separable*, provided we know that ψ is. (This in turn will be seen in (12.4) to imply that $\varphi: PGL(2, K) \to Y$ is an isomorphism of varieties.)

Let us use the *differential criterion* (5.5) to verify that ψ is separable. Evidently $GL(2, K)$ is smooth (either as a principal open subset of $\mathbf{A}^4$, or as an "algebraic group" (7.1)). Since Y has a transitive group of automorphisms, the fact (5.2) that some points are simple forces all points to be simple. So we need only verify the surjectivity of (say) $d\psi_e$, where $e = \begin{pmatrix} 1 & 0 \\ 0 & 1 \end{pmatrix}$. For this we identify the tangent space at e with $M(2, K)$ (i.e., $\mathbf{A}^4$) and apply the formula of (5.4).

Recall (1.7) how $\mathbf{P}^1 \times \mathbf{P}^1 \times \mathbf{P}^1$ is embedded in $\mathbf{P}^7$. ψ sends $\begin{pmatrix} a & b \\ c & d \end{pmatrix}$ to $\left(\begin{bmatrix} b \\ d \end{bmatrix}, \begin{bmatrix} a+b \\ c+d \end{bmatrix}, \begin{bmatrix} a \\ c \end{bmatrix} \right)$, which has homogeneous coordinates ($ab(a+b)$, $ab(c+d)$, $bc(a+b)$, $bc(c+d)$, $ad(a+b)$, $ad(c+d)$, $cd(a+b)$, $cd(c+d)$). Note that $\psi(e)$ belongs to the affine open set in $\mathbf{P}^7$ specified by nonvanishing of the 6th coordinate. So in affine coordinates, ψ is given in a neighborhood of e by seven coordinate functions, e.g., $\psi_1(T_1, T_2, T_3, T_4) = T_2(T_1 + T_2)/T_4(T_3 + T_4)$. The partial derivatives must then be evaluated at e. Set $x = \begin{pmatrix} 1 & 0 \\ 0 & 0 \end{pmatrix}$, $y = \begin{pmatrix} 0 & 1 \\ 0 & 0 \end{pmatrix}$, $z = \begin{pmatrix} 0 & 0 \\ 0 & 1 \end{pmatrix}$. A routine calculation (Exercise 3) shows that $d\psi_e$ sends x to $(0, 0, 0, 0, 1, 0, 0)$, y to $(1, 1, 0, 0, 1, 0, 0)$, z to $(0, 0, 0, 0, -1, 1, 1)$. The image of $d\psi_e$ is therefore a 3-dimensional subspace of the 3-dimensional space $\mathcal{T}(Y)_{\psi(e)}$. Conclusion : ψ is separable.

Once it is established in (12.4) that φ is an isomorphism, it will follow that for any algebraic group H (7.1) acting on $\mathbf{P}^1$, there is a morphism $H \to PGL(2, K)$ giving the action. Indeed, the effect of an automorphism of $\mathbf{P}^1$ on the triple $(0, 1, \infty)$ completely determines the automorphism, as shown above; so we have only to compose the orbit map $H \to Y$ with the inverse of φ.

Exercises

1. Must every dominant morphism $\mathbf{P}^1 \to \mathbf{P}^1$ be an isomorphism?
2. Prove that a valuation ring R in $K(T)$ which includes $K[T]$ and is distinct from $K(T)$ must consist of all $f(T)/g(T)$, where $f(T), g(T) \in K[T]$ are relatively prime and where some fixed linear polynomial $T - a$ does not

divide $g(T)$. Describe similarly the valuation rings which include $K[1/T]$. Show that no proper inclusions exist between pairs of such valuation rings.

3. Verify the differential calculation at the end of (6.4).

Notes

The proof (due to Grothendieck) of Theorem 6.2 is reproduced from Mumford [3, I §9]. For some facts about curves which include Theorem 6.3, consult Mumford [3, III §8, Theorem 5]. The discussion in (6.4) is taken from Borel [4, 10.8]. Theorem 6.4 is a special case of the Fundamental Theorem of Projective Geometry.

Chapter II

Affine Algebraic Groups

7. Basic Concepts and Examples

7.1. The Notion of Algebraic Group

Let G be a variety (irreducible or not) endowed with the structure of a group. If the two maps $\mu: G \times G \to G$, where $\mu(x, y) = xy$, and $\iota: G \to G$, where $\iota(x) = x^{-1}$, are morphisms of varieties, we call G an **algebraic group**. The reader who is familiar with the concept of "analytic group" will see here an obvious parallel. But there is a subtle difference: $G \times G$ is here given the Zariski topology rather than the product topology, so an algebraic group is *not* a topological group (except in dimension 0). Indeed, G is T_1 without being T_2 (except in dimension 0), while a T_1 topological group is automatically T_2.

Translation by an element $y \in G$ $(x \mapsto xy)$ is clearly an isomorphism of varieties $G \to G$, and therefore all geometric properties at one point of G can be transferred to any other point, by suitable choice of y. For example, since G has simple points (5.2), all points must be simple: G is *smooth*.

There is an obvious notion of **isomorphism**: algebraic groups G and G' are called isomorphic if there exists an isomorphism of varieties $\varphi: G \to G'$ which is simultaneously an isomorphism of groups. An **automorphism** of G is an isomorphism of G onto G. An algebraic group whose underlying variety is complete (6.1) is called an **abelian variety**. We shall not attempt to study these here (see Notes below). Instead *we always reserve the term "algebraic group" for those groups whose underlying varieties are affine*, unless the contrary is expressly stated. With this convention in mind, we proceed to list some examples.

The **additive group G_a** is the affine line $\mathbf{A}^1$ with group law $\mu(x, y) = x + y$ (so $\iota(x) = -x$, $e = 0$). The **multiplicative group G_m** is the affine open subset $\mathsf{K}^* \subset \mathbf{A}^1$ with group law $\mu(x, y) = xy$ (so $\iota(x) = x^{-1}$, $e = 1$). Each of these groups is irreducible (as a variety) and 1-dimensional; after a substantial amount of preparation, we shall eventually be able to prove that (up to isomorphism) they are the only algebraic groups with these two properties (§20). Generalizing the additive group, we see that affine n-space $\mathbf{A}^n$ has a natural (additive) structure of algebraic group. In each of these examples the underlying group is commutative.

Denote by $GL(n, \mathsf{K})$ the set of all $n \times n$ invertible matrices with entries in K; this is a group under matrix multiplication, called the **general linear group**. The set $M(n, \mathsf{K})$ of all $n \times n$ matrices over K may be identified with $\mathbf{A}^{n^2}$, and $GL(n, \mathsf{K})$ with the principal open subset defined by the nonvanishing of the polynomial det. Viewed thus as an affine variety, $GL(n, \mathsf{K})$ has its algebra

51

of polynomial functions generated by (the restrictions of) the n^2 coordinate functions T_{ij} along with $1/\det(\mathsf{T}_{ij})$. The formulas for matrix multiplication and inversion make it clear that $GL(n, \mathsf{K})$ is an algebraic group. Notice that $\mathbf{G}_m$ is the same thing as $GL(1, \mathsf{K})$.

It is easy to construct further examples, based on the obvious fact that *a closed subgroup of an algebraic group is again an algebraic group*. For instance, the group $\mathsf{T}(n, \mathsf{K})$ of all upper triangular $n \times n$ matrices is the set of zeros in $GL(n, \mathsf{K})$ of the polynomials T_{ij} $(i > j)$, while the subgroup $D(n, \mathsf{K})$ (resp. $U(n, \mathsf{K})$) consisting of diagonal matrices (resp. upper triangular matrices with all diagonal entries 1) is closed for similar reasons. Notice that $U(2, \mathsf{K})$ is naturally isomorphic to $\mathbf{G}_a$.

In another direction, given a (not necessarily associative) finite dimensional algebra $\mathfrak{A}$, the group Aut $\mathfrak{A}$ of its algebra automorphisms may be regarded as a subgroup of $GL(n, \mathsf{K})$ by choosing a basis for $\mathfrak{A}$ ($n = \dim \mathfrak{A}$); in fact, it is readily seen to be a *closed* subgroup (Exercise 3).

We mention finally that the **direct product** of two (or more) algebraic groups, i.e., the usual direct product of groups endowed with the Zariski topology, is again an algebraic group. For example, $D(n, \mathsf{K})$ may be viewed as the direct product of n copies of $\mathbf{G}_m$, while affine n-space may be viewed as the direct product of n copies of $\mathbf{G}_a$.

7.2. Some Classical Groups

We introduce next some families of linear groups which play a central role in the theory to be developed in this book. The parameter ℓ is in each case the dimension of the (closed) subgroup of diagonal matrices in the group under discussion.

A_ℓ: The **special linear group** $SL(\ell + 1, \mathsf{K})$ consists of the matrices of determinant 1 in $GL(\ell + 1, \mathsf{K})$; it is clearly a group (because of the product rule for det) and is closed (being the set of zeros of $\det(\mathsf{T}_{ij}) - 1$). Since it is defined by a single polynomial, $SL(\ell + 1, \mathsf{K})$ is a hypersurface in $M(\ell + 1, \mathsf{K})$ (3.3), so its dimension is $(\ell + 1)^2 - 1 = \ell^2 + 2\ell$.

C_ℓ: The **symplectic group** $Sp(2\ell, \mathsf{K})$ consists of all $x \in GL(2\ell, \mathsf{K})$ satisfying ${}^t x \begin{pmatrix} 0 & J \\ -J & 0 \end{pmatrix} x = \begin{pmatrix} 0 & J \\ -J & 0 \end{pmatrix}$, where $J = \begin{pmatrix} & & 1 \\ & \cdot^{\cdot^{\cdot}} & \\ 1 & & \end{pmatrix}$, ${}^t x = $ transpose of x. It is immediate that $Sp(2\ell, \mathsf{K})$ is a group. That it is closed follows from the fact that the indicated equation imposes certain (mildly complicated) polynomial conditions on x. In this case the dimension is difficult to compute directly.

B_ℓ: This is the **special orthogonal group** $SO(2\ell + 1, \mathsf{K})$; if char $\mathsf{K} \neq 2$, it consists of all $x \in SL(2\ell + 1, \mathsf{K})$ which satisfy ${}^t x s x = s$, where $s = \begin{pmatrix} 1 & 0 & 0 \\ 0 & 0 & J \\ 0 & J & 0 \end{pmatrix}$. Again it is easy to check that this condition defines a closed subgroup of the general linear group.

D_ℓ: This is another **special orthogonal group** $SO(2\ell, \mathsf{K})$, defined by the same condition as B_ℓ, ${}^t x s x = s$, where s is now $\begin{pmatrix} 0 & J \\ J & 0 \end{pmatrix}$ (if char $K \neq 2$).

The symplectic and special orthogonal groups arise geometrically as groups of linear transformations preserving certain skew-symmetric (resp. symmetric) bilinear forms. When char $K = 2$, special care has to be taken with the definitions, cf. Dieudonné [13], Carter [1, Ch. 1].

7.3. Identity Component

Let G be an algebraic group. We assert that only one irreducible component of G can pass through e. Indeed, let $X_1, \ldots, X_m$ be the distinct components containing e. The image of the irreducible variety $X_1 \times \cdots \times X_m$ (1.4) under the product morphism is an irreducible subset $X_1 \cdots X_m$ of G, which again contains e. So $X_1 \cdots X_m$ lies in some X_i. On the other hand, each of the components $X_1, \ldots, X_m$ clearly lies in $X_1 \cdots X_m$. This forces $m = 1$. Denote by G° this unique irreducible component of e, and call it the **identity component** of G.

Proposition. *Let G be an algebraic group.*
(a) *G° is a normal subgroup of finite index in G, whose cosets are the connected as well as irreducible components of G.*
(b) *Each closed subgroup of finite index in G contains G°.*

Proof. (a) For each $x \in G^\circ$, $x^{-1} G^\circ$ is an irreducible component of G passing through e, so $x^{-1} G^\circ = G^\circ$. Therefore, $G^\circ = (G^\circ)^{-1}$, and further, $G^\circ G^\circ = G^\circ$, i.e., G° is a (closed) subgroup of G. For any $x \in G$, $xG^\circ x^{-1}$ is also an irreducible component of G containing e, so $xG^\circ x^{-1} = G^\circ$ and G° is normal. Its (left or right) cosets are translates of G°, hence must also be irreducible components of G; there can only be finitely many of them (G being a noetherian space). Since they are disjoint, these are also the connected components of G.

(b) If H is a closed subgroup of finite index in G, then each of its finitely many left cosets is also closed and so is the union of those distinct from H. As the complement of this closed set, H must be *open*. Therefore the left cosets of H partition G° into a finite union of open sets; since G° is connected and meets H, we get $G^\circ \subset H$. $\square$

Henceforth we shall refer to an algebraic group G as **connected** when $G = G^\circ$, because the term "irreducible" has an entirely different meaning in the context of linear groups or group representations.

Most of the groups encountered above are in fact connected, e.g., $\mathbf{G}_a$ and $\mathbf{G}_m$. That $GL(n, K)$ is connected follows from its being a principal open set in an affine space. However, the connectedness of $SL(n, K)$ and other classical groups is not apparent from the definition alone, so a more indirect method must be developed (7.5). While our main interest will be in connected algebraic groups, certain disconnected ones will be forced upon us (e.g., the group of "monomial" matrices in $GL(n, K)$, cf. Exercise 7).

7.4. Subgroups and Homomorphisms

The following lemma is trivial but useful.

Lemma. *Let U, V be two dense open subsets of an algebraic group G. Then $G = U \cdot V$.*

Proof. Since inversion is a homeomorphism, V^{-1} is again a dense open set. So is its translate xV^{-1} (for any given $x \in G$). Therefore, U must meet xV^{-1}, forcing $x \in U \cdot V$. $\square$

We have already pointed out that a *closed* subgroup of an algebraic group is again an algebraic group. What can be said about an arbitrary subgroup?

Proposition A. *Let H be a subgroup of the algebraic group G, $\bar{H}$ its closure.*

(a) $\bar{H}$ is a subgroup of G.

(b) If H is constructible, then $H = \bar{H}$.

Proof. (a) Inversion being a homeomorphism, it is clear that $\bar{H}^{-1} = \overline{H^{-1}} = \bar{H}$. Similarly, translation by $x \in H$ is a homeomorphism, so $x\bar{H} = \overline{xH} = \bar{H}$. In turn, if $x \in \bar{H}$, $Hx \subset H\bar{H}$, so $\bar{H}x = \overline{Hx} \subset \bar{H}$. This says that $\bar{H}$ is a group.

(b) If H is constructible, it contains a dense open subset U of $\bar{H}$. Since $\bar{H}$ is a group, by part (a), the lemma above shows that $\bar{H} = U \cdot U \subset H \cdot H = H$. $\square$

Corollary. *Let A, B be closed subgroups of an algebraic group G. If B normalizes A, then AB is a closed subgroup of G.*

Proof. Since $B \subset N_G(A)$, AB is a subgroup. As the image of $A \times B$ under the product morphism $G \times G \to G$, it is constructible (4.4), therefore closed, by part (b) of the proposition. $\square$

By definition, a **morphism of algebraic groups** is a group homomorphism $\varphi : G \to G'$ which is also a morphism of varieties.

Proposition B. *Let $\varphi : G \to G'$ be a morphism of algebraic groups. Then:*
(a) Ker φ is a closed subgroup of G.
(b) Im φ is a closed subgroup of G'.
(c) $\varphi(G^\circ) = \varphi(G)^\circ$.
(d) dim G = dim Ker φ + dim Im φ.

Proof. (a) φ is continuous and Ker φ is the inverse image of the closed set $\{e\}$.

(b) $\varphi(G)$ is a subgroup of G'. It is also a constructible subset of G' (4.4), so it is closed by part (b) of Proposition A.

(c) $\varphi(G^\circ)$ is closed (part (b)) and connected (= irreducible), hence lies in $\varphi(G)^\circ$. Being of finite index in $\varphi(G)$, it must equal $\varphi(G)^\circ$, thanks to Proposition 7.3(b).

(d) Theorem 4.3 implies that dim $G - \dim \varphi(G) = \dim \varphi^{-1}(x)$ for some (indeed, for "most") $x \in \varphi(G)$. But all fibres $\varphi^{-1}(x)$ have the dimension of Ker φ, so we are done. $\square$

A good example to keep in mind is $\det: \mathrm{GL}(n, \mathsf{K}) \to \mathrm{GL}(1, \mathsf{K}) = \mathbf{G}_m$, which is evidently a morphism of algebraic groups. The image is $\mathbf{G}_m$, the kernel $\mathrm{SL}(n, \mathsf{K})$. From part (d) of the corollary we recover the fact that dim $\mathrm{SL}(n, \mathsf{K}) = n^2 - 1$.

When the target group under a morphism is $\mathrm{GL}(n, \mathsf{K})$, we say that φ is a **rational representation**. In this connection, it is sometimes desirable to view $\mathrm{GL}(V)$ as an algebraic group ($V = n$-dimensional vector space over K). Since a change of basis in K^n corresponds to an inner automorphism $x \mapsto yxy^{-1}$ in $\mathrm{GL}(n, \mathsf{K})$, the Zariski topology on $\mathrm{GL}(V)$ can be specified unambiguously by an arbitrary choice of basis for V, identifying V with K^n.

7.5. Generation by Irreducible Subsets

It is essentially an exercise in linear algebra to show that $\mathrm{SL}(n, \mathsf{K})$ is generated by subgroups U_{ij} ($i \neq j$), where U_{ij} consists of all matrices with 1's on the diagonal, arbitrary entry in the (i, j) position, and 0's elsewhere. (In the same spirit, $\mathrm{U}(n, \mathsf{K})$ is generated by those U_{ij} for which $i < j$.) Evidently U_{ij} is isomorphic (as algebraic group) to $\mathbf{G}_a$, since multiplication of such matrices just involves addition of the (i, j) entries. Our next proposition, based on the use of constructible sets in (7.4), allows us to deduce from the connectedness of $\mathbf{G}_a$ that $\mathrm{SL}(n, \mathsf{K})$ is connected. The same technique can of course be applied to other groups.

First, a definition. Given an arbitrary subset M of an algebraic group G, denote by $\mathscr{A}(M)$ the intersection of all closed subgroups of G containing M. This is the smallest closed subgroup of G containing M; we call it the **group closure** of M.

Proposition. *Let G be an algebraic group, I an index set, $f_i: X_i \to G$ ($i \in I$) a family of morphisms from irreducible varieties X_i, such that $e \in Y_i = f_i(X_i)$ for each $i \in I$. Set $M = \bigcup_{i \in I} Y_i$. Then:*

(a) *$\mathscr{A}(M)$ is a connected subgroup of G.*

(b) *For some finite sequence $a = (a(1), \ldots, a(n))$ in I, $\mathscr{A}(M) = Y_{a(1)}^{e_1} \cdots Y_{a(n)}^{e_n}$* $(e_i = \pm 1)$.

Proof. It is harmless to enlarge I so as to insure that the morphisms $x \mapsto f_i(x)^{-1}$ from X_i to G also occur. For each finite sequence $a = (a(1), \ldots, a(n))$ in I, set $Y_a = Y_{a(1)} \cdots Y_{a(n)}$. As the image of the irreducible variety $X_{a(1)} \times \cdots \times X_{a(n)}$ under the morphism $f_{a(1)} \times \cdots \times f_{a(n)}$ composed with multiplication in G, Y_a is constructible (4.4), and $\bar{Y}_a$ is an irreducible variety (1.3) passing through e. Using the maximal condition on irreducible closed subsets of G°, we can therefore find a sequence a for which $\bar{Y}_a$ is maximal.

Given any two finite sequences b, c in I, we claim that (*) $\overline{Y}_b \overline{Y}_c \subset \overline{Y}_{(b,\,c)}$, where (b, c) is the longer sequence obtained by juxtaposition. The proof is in two steps. For $x \in Y_c$, the (continuous) map $y \mapsto yx$ sends Y_b into $Y_{(b,\,c)}$, hence $\overline{Y}_b$ into $\overline{Y}_{(b,\,c)}$, i.e., $\overline{Y}_b Y_c \subset \overline{Y}_{(b,\,c)}$. In turn, $x \in \overline{Y}_b$ sends Y_c into $\overline{Y}_{(b,\,c)}$, hence $\overline{Y}_c$ as well.

Because $\overline{Y}_a$ is maximal, and e lies in each $\overline{Y}_b$, (*) implies that

$$\overline{Y}_a \subset \overline{Y}_a \overline{Y}_b \subset \overline{Y}_{(a,\,b)} = \overline{Y}_a$$

for any b. Setting $b = a$, we have $\overline{Y}_a$ stable under multiplication. Choosing b such that $Y_b = Y_a^{-1}$ (cf. first sentence of proof), we also have $\overline{Y}_a$ stable under inversion. Conclusion: $\overline{Y}_a$ is a closed *subgroup* of G containing all Y_i ($i \in I$), so $\overline{Y}_a = \mathscr{A}(M)$, proving (a). Moreover, since Y_a is constructible, Lemma 7.4 shows that $\overline{Y}_a = Y_a \cdot Y_a = Y_{(a,\,a)}$, so the sequence (a, a) satisfies (b). $\square$

Corollary. *Let G be an algebraic group, Y_i ($i \in I$) a family of closed connected subgroups of G which generate G (as an abstract group). Then G is connected.* $\square$

The proposition itself (not just its corollary) will be of further use in (17.2) when we consider the closure and connectedness properties of commutator groups.

7.6. Hopf Algebras

An affine variety is completely determined by its affine algebra (1.5). Therefore it is interesting to reformulate the axioms for an (affine) algebraic group G as a set of conditions on the algebra $A = \mathsf{K}[G]$. If e is viewed as a morphism from a group of one element into G (K being the affine algebra!), then $e^*: A \to \mathsf{K}$ sends $f \mapsto f(e)$. From $\mu: G \times G \to G$ we obtain $\mu^*: A \to A \otimes_{\mathsf{K}} A$, sending $f \mapsto \sum g_i \otimes h_i$ if $f(xy) = \sum g_i(x)h_i(y)$. To $\iota: G \to G$ corresponds $\iota^*: A \to A$, where $(\iota^* f)(x) = f(x^{-1})$. It is also useful to write $p: G \to G$ for the constant morphism $p(x) = e$. Then $p^*: A \to A$ satisfies $(p^* f)(x) = f(e)$.

Now the group axioms for G (associative law, identity, inverses), as given by the commutativity of the lefthand diagrams, translate into corresponding conditions for $(A, e^*, \mu^*, \iota^*, p^*)$:

$$
\begin{array}{ccc}
G \times G \times G & \xrightarrow{\ \mu \times 1\ } & G \times G \\
{\scriptstyle 1 \times \mu}\downarrow & & \downarrow{\scriptstyle \mu} \\
G \times G & \xrightarrow{\ \ \mu\ \ } & G
\end{array}
\qquad\qquad
\begin{array}{ccc}
A \otimes A \otimes A & \xleftarrow{\ \mu^* \otimes 1\ } & A \otimes A \\
{\scriptstyle 1}\quad{\scriptstyle \mu^*}\uparrow & & \uparrow{\scriptstyle \mu^*} \\
A \otimes A & \xleftarrow{\ \ \mu^*\ \ } & A
\end{array}
$$

$$
\begin{array}{ccc}
G & \xrightarrow{\ p \times 1\ } & G \times G \\
{\scriptstyle 1 \times p}\downarrow & \searrow{\scriptstyle 1} & \downarrow{\scriptstyle \mu} \\
G \times G & \xrightarrow{\ \ \mu\ \ } & G
\end{array}
\qquad\qquad
\begin{array}{ccc}
A & \xleftarrow{\ p^* \otimes 1\ } & A \otimes A \\
{\scriptstyle 1 \otimes p^*}\uparrow & \nwarrow{\scriptstyle 1} & \uparrow{\scriptstyle \mu^*} \\
A \otimes A & \xleftarrow{\ \ \mu^*\ \ } & A
\end{array}
$$

$$
\begin{array}{ccc}
G & \xrightarrow{\ \iota \times 1\ } & G \times G \\
{\scriptstyle 1 \times \iota}\downarrow & \searrow{\scriptstyle p} & \downarrow{\scriptstyle \mu} \\
G \times G & \xrightarrow{\ \ \mu\ \ } & G
\end{array}
\qquad\qquad
\begin{array}{ccc}
A & \xleftarrow{\ \iota^* \otimes 1\ } & A \otimes A \\
{\scriptstyle 1 \otimes \iota^*}\uparrow & \nwarrow{\scriptstyle p^*} & \uparrow{\scriptstyle \mu^*} \\
A \otimes A & \xleftarrow{\ \ \mu^*\ \ } & A
\end{array}
$$

Notice that p^* is the composite of $e^*: A \to \mathsf{K}$ and the natural inclusion $\mathsf{K} \to A$. Therefore, *to define a structure of algebraic group on the variety corresponding to A, we need only specify maps e^*, μ^*, ι^* making the three diagrams commute.* (Then A becomes an example of a **Hopf algebra** with identity.)

It is instructive to write down the group law for some of our earlier examples. $\mathbf{G}_a$ has $\mathsf{K}[\mathsf{T}]$ as affine algebra, so it is enough to say what the various maps do to $\mathsf{T}: e^*(\mathsf{T}) = 0$, $\mu^*(\mathsf{T}) = (\mathsf{T} \otimes 1) + (1 \otimes \mathsf{T})$, $\iota^*(\mathsf{T}) = -\mathsf{T}$. $\mathbf{G}_m$ has $\mathsf{K}[\mathsf{T}, \mathsf{T}^{-1}]$ as its affine algebra, and $e^*(\mathsf{T}) = 1$, $\mu^*(\mathsf{T}) = \mathsf{T} \otimes \mathsf{T}$, $\iota^*(\mathsf{T}) = \mathsf{T}^{-1}$. For $\mathrm{GL}(n, \mathsf{K})$ we have $A = \mathsf{K}[\mathsf{T}_{11}, \mathsf{T}_{12}, \ldots, \mathsf{T}_{nn}, d^{-1}]$, $d = \det(\mathsf{T}_{ij})$. Here $e^*(\mathsf{T}_{ij}) = \delta_{ij}$, $\mu^*(\mathsf{T}_{ij}) = \sum_h \mathsf{T}_{ih} \otimes \mathsf{T}_{hj}$, $\iota^*(\mathsf{T}_{ij}) = (-1)^{i+j} d^{-1} \det(\mathsf{T}_{rs})_{r \neq j,\, s \neq i}$. (The reader should verify these formulas.)

Exercises

1. Prove that $\mathbf{G}_a$ and $\mathbf{G}_m$ are not isomorphic.
2. Show that $\mathrm{T}(n, \mathsf{K})$, $\mathrm{D}(n, \mathsf{K})$, $\mathrm{U}(n, \mathsf{K})$ have respective dimensions $n(n + 1)/2$, n, $n(n - 1)/2$.
3. Let $\mathfrak{A}$ be a finite dimensional K-algebra. Prove that $\mathrm{Aut}\ \mathfrak{A}$ is a closed subgroup of $\mathrm{GL}(\mathfrak{A})$.
4. The only automorphisms of $\mathbf{G}_m$ (as algebraic group) are $x \mapsto x$, $x \mapsto x^{-1}$, while $\mathrm{Aut}\ \mathbf{G}_a \cong \mathsf{K}^*$.
5. A closed subset of an algebraic group which contains e and is closed under taking products is a subgroup of G.
6. G° is a *characteristic* subgroup of G, i.e., stable under all automorphisms of G.
7. Let $N \subset \mathrm{GL}(n, \mathsf{K})$ be the group of **monomial** matrices, i.e., matrices having precisely one nonzero entry in each row and each column. Prove that N is a closed subgroup of $\mathrm{GL}(n, \mathsf{K})$, with $N^\circ = \mathrm{D}(n, \mathsf{K})$ and $[N : N^\circ] = n!$ [Note that the finite group N/N° is isomorphic to the symmetric group S_n.]
8. Prove that $\mathrm{T}(n, \mathsf{K})$, $\mathrm{D}(n, \mathsf{K})$, $\mathrm{U}(n, \mathsf{K})$ are all connected.
9. In Proposition 7.5 (b), show that n can be taken to be $\leqslant 2 \dim G$.
10. Show by example that the subgroup of an algebraic group generated by two non-irreducible closed subsets need not be closed. [Use the cyclic subgroups of $\mathrm{GL}(2, \mathbb{C})$ generated by $\begin{pmatrix} 1 & 0 \\ 0 & -1 \end{pmatrix}$ and $\begin{pmatrix} 1 & 1 \\ 0 & -1 \end{pmatrix}$.]
11. Let G be a connected algebraic group. Prove that any finite normal subgroup H lies in the **center** $Z(G) = \{x \in G \mid xy = yx \text{ for all } y \in G\}$. [Consider the morphism $G \to H$ defined by $x \mapsto xyx^{-1}$ $y \in H$.]
12. Define a morphism $G \times G \times G \times G \to G \times G$ by $\mu \times \mu$, and let X be be the inverse image of the diagonal $\{(x, x) \mid x \in G\}$. If G is connected, prove that X is a closed irreducible subset of $G \times G \times G \times G$. [Note that $(w, x, y, z) \mapsto (w, x, y)$ induces an isomorphism of X onto $G \times G \times G$.]

Notes

The idea of (7.5) goes back to Chevalley [4, §7]. For the theory of abelian varieties consult Lang [2], Mumford [2], Weil [1].

8. Actions of Algebraic Groups on Varieties

In studying algebraic groups we shall very often exploit their actions on themselves or on other naturally associated varieties. The present section introduces most of the relevant machinery, and then uses some of it to prove that every (affine) algebraic group is isomorphic to a closed subgroup of some $GL(n, K)$.

8.1. Group Actions

If G is an abstract group, X a set, we say that G **acts** on X if there is a map $\varphi: G \times X \to X$, denoted for brevity by $\varphi(x, y) = x \cdot y$, such that:

> (A1) $x_1 \cdot (x_2 \cdot y) = (x_1 x_2) \cdot y$ for $x_i \in G, y \in X$;
> (A2) $e \cdot y = y$ for all $y \in X$.

These two conditions can also be construed as the requirement that φ induce a group homomorphism from G to the symmetric group on X. The triple (G, X, φ) is sometimes called an **algebraic transformation space**.

Let G act on X. We say that G acts **transitively** if $G \cdot y = X$ for arbitrary $y \in X$. In any case, the set $G \cdot y$ is called the **orbit** of y; evidently the distinct orbits under G form a partition of X, and G acts (transitively) on each orbit. X^G denotes the set of **fixed points** of G (i.e., the set of those $y \in X$ whose G-orbit consists of y alone). In general, if $y \in X$, we define the **isotropy group** (or **stabilizer**) of y to be $G_y = \{x \in G | x \cdot y = y\}$. It is clear that G_y is a group. Moreover, the **orbit map** $G \to G \cdot y$ defined by $x \mapsto x \cdot y$ induces a bijection $G/G_y \to G \cdot y$. In case $z = x \cdot y$ for some $x \in G$, it is easy to check that the isotropy groups G_y and G_z are conjugate: $xG_yx^{-1} = G_z$. (Conversely, if G_y and G_z are conjugate, then y and z lie in the same G-orbit.) If H is any subgroup of G, there is a natural transitive action of G on the left coset space G/H ($yH \mapsto xyH$), H being the isotropy group of the coset H. In view of the preceding remarks, every transitive action of G has essentially this form.

Consider now some natural actions of G on itself. For $x \in G$, $y \mapsto xyx^{-1} = $ Int $x(y)$ defines an action by **inner automorphisms**, with corresponding homomorphism $G \to$ Aut G. (Here the kernel is $Z(G)$, the **center** of G, and the image Int G is easily seen to be a normal subgroup of Aut G.) The orbit of y is its **conjugacy class**, the isotropy group its **centralizer** $C_G(y)$. The set of fixed points is just $Z(G)$.

G also acts on itself as a group of **left** (resp. **right**) **translations**, via $y \mapsto xy$ (resp. $y \mapsto yx^{-1}$). These actions are evidently transitive and the isotropy groups trivial.

8.2. Actions of Algebraic Groups

Let G be an algebraic group, X a variety. If we are given a morphism $\varphi : G \times X \to X$ such that axioms (A1), (A2) of (8.1) are satisfied, we say that G acts **morphically** on X (or just "acts", if no confusion is possible). For example, the actions of G on itself described above are clearly of this type. As in (8.1), we have the notions of orbit, isotropy group, ... Another useful notion is that of **transporter**: $\mathrm{Tran}_G(Y, Z) = \{x \in G \mid x \cdot Y \subset Z\}$, where Y and Z are subsets of X. This is clearly a submonoid of G. Finally, write $C_G(Y) = \bigcap_{y \in Y} G_y$; this is the **centralizer** of Y in G.

Proposition. *Let the algebraic group G act morphically on the variety X. Let Y, Z be subsets of X, with Z closed.*

(a) $\mathrm{Tran}_G(Y, Z)$ *is a closed subset of G.*

(b) *For each $y \in X$, G_y is a closed subgroup of G; in particular, $C_G(Y)$ is closed.*

(c) *The fixed point set of $x \in G$ is closed in X; in particular, X^G is closed.*

(d) *If G is connected, G stabilizes each connected component of X, hence acts trivially on X in case X is finite.*

Proof. For each $y \in X$, the orbit map $\varphi_y : G \to X$ ($x \mapsto x \cdot y$) is the composite of $x \mapsto (x, y)$ and φ; therefore it is a morphism. As y runs over Y, the various inverse images $\varphi_y^{-1}(Z)$ are closed in G (because Z is closed in X) and $\mathrm{Tran}_G(Y, Z)$ is their intersection, whence (a). Since $G_y = \mathrm{Tran}_G(\{y\}, \{y\})$, G_y is a closed subgroup of G, thanks to (a). In turn, $C_G(Y) = \bigcap_{y \in Y} G_y$, so (b) follows.

For (c), let $x \in G$, and consider the morphism $\psi : X \to X \times X$ defined by $y \mapsto (y, x \cdot y)$. The fixed point set X^x is precisely the inverse image under ψ of the diagonal, which is closed (X being a variety (2.5)); (c) follows.

Finally, let G be connected. The stabilizer H in G of a connected component of X is closed, thanks to part (a), and is a subgroup. Since G permutes the finitely many components of X, H has finite index in G. Therefore (Proposition 7.3 (b)) $H = G$. $\quad\square$

Corollary. *Let G be an algebraic group, H a closed subgroup. Then $N_G(H)$ and $C_G(H)$ are closed subgroups, as is $C_G(x)$ for $x \in G$. (Here $N_G(H) = \{x \in G \mid xHx^{-1} = H\}$, the **normalizer** of H in G.)*

Proof. For $C_G(H)$ or $C_G(x)$ this results from part (b) of the proposition, if we let G act on itself by inner automorphisms. For $N_G(H)$ it results from part (a) and the fact that $N_G(H) = \mathrm{Tran}_G(H, H)$. Indeed, the automorphism $\mathrm{Int}\, y$ ($y \in G$) maps H to a closed subgroup of G of the same dimension as H, whose identity component is of index $[H:H^\circ]$. So $\mathrm{Int}\, x$ maps H into H if and only if it maps H onto H. $\quad\square$

We shall frequently use the fact that normalizers, centralizers, and fixed point sets are closed. *Orbits, however, are often not closed* (conjugacy classes in G, for example), cf. (8.3) below. It may be well to emphasize also that the *connectedness* of normalizers and centralizers is not to be taken for granted. Some of the most subtle arguments in later chapters will involve precisely this question.

Suppose next that $\varphi : G \to \mathrm{GL}(V)$ is a (rational) representation of the algebraic group G. If we identify V with affine n-space ($n = \dim V$), it is clear that the recipe $x \cdot v = \varphi(x)(v)$ ($x \in G$, $v \in V$) defines an action of G on V. In this case we may call V a (rational) **G-module**. For later reference we record here some related notions and constructions.

Associated with φ is the **dual** or **contragredient** representation $G \to \mathrm{GL}(V^*)$, where V^* is the dual vector space. This is defined by the rule: $(x \cdot f)(v) = f(x^{-1} \cdot v)$, where $f \in V^*, v \in V, x \in G$. We have to write x^{-1} in order to insure that $y \cdot (x \cdot f) = (yx) \cdot f$. If dual bases for V, V^* are chosen, it becomes clear at once that the dual representation really is a rational representation of G.

Next, consider a pair of representations $\varphi : G \to \mathrm{GL}(V)$, $\psi : G \to \mathrm{GL}(W)$. The **tensor product** space $V \otimes W$ has as basis all $v_i \otimes w_j$ if v_i (resp. w_j) runs over a basis of V (resp. W). We can make $x \in G$ act on $V \otimes W$ by requiring that $x \cdot (v_i \otimes w_j) = (x \cdot v_i) \otimes (x \cdot w_j)$ and extending linearly. (Of course, there is no need to choose bases, if we appeal directly to the universal property of a tensor product.) It is clear that this defines a (rational) representation $G \to \mathrm{GL}(V \otimes W)$.

It is a standard fact that the vector space $V^* \otimes V$ identifies naturally with the vector space $\mathrm{End}\, V$, with $f \otimes v$ corresponding to the endomorphism $w \mapsto f(w)v$ of V. Therefore, a (rational) representation $G \to \mathrm{GL}(V)$ induces an action of G on $\mathrm{End}\, V$, which is in fact the action sending $t \in \mathrm{End}\, V$ to xtx^{-1} ($x \in G$). The reader can check that a fixed point of G in $\mathrm{End}\, V$ is just a G-module homomorphism $V \to V$.

When G acts on two varieties X, Y, a morphism $\varphi : X \to Y$ is called **G-equivariant** provided $\varphi(z \cdot x) = z \cdot \varphi(x)$ for all $z \in G$, $x \in X$. For example, a subgroup G of $\mathrm{GL}(V)$ acts on itself by left multiplication as well as on V. If $v \in V$, the resulting orbit map $G \to V$ (sending z to $z \cdot v$) is a G-equivariant morphism. (Cf. also Exercise 6.)

8.3. Closed Orbits

Proposition. *Let the algebraic group G act morphically on the (nonempty) variety X. Then each orbit is a smooth, locally closed subset of X, whose boundary is a union of orbits of strictly lower dimension. In particular, orbits of minimal dimension are closed (so closed orbits exist).*

Proof. Say $Y = G \cdot y$ is the orbit of $y \in X$. As the image of G under the orbit map, Y is constructible (4.4), hence contains an open dense subset of $\overline{Y}$. But G acts transitively on Y (leaving $\overline{Y}$ stable), so Y is smooth and contains a neighborhood in $\overline{Y}$ of each of its points, i.e., Y is open in $\overline{Y}$. Therefore, $\overline{Y} - Y$ is closed and of strictly lower dimension than $\overline{Y}$. Being G-stable, this boundary is just the union of other G-orbits. $\square$

This proposition will be of great use later on. It comes into play when X is not known in advance to have any points fixed by G.

8.4. Semidirect Products

An action of one group on another (as group automorphisms) permits construction of a larger group. First recall the (abstract) group-theoretic construction: When G acts (as a group of automorphisms) on N, the cartesian product $N \times G$ becomes a group if we define $(x_1, y_1)(x_2, y_2) = (x_1(y_1 \cdot x_2), y_1 y_2)$. This is called a **semidirect product** and is denoted $N \rtimes G$. Of course, the direct product $N \times G$ arises as a special case of this construction (when G acts trivially on N). N is embedded as a normal subgroup via $x \mapsto (x, e)$, and G as a subgroup via $y \mapsto (e, y)$, so that each element of $N \rtimes G$ may be written uniquely in the form xy $(x \in N, y \in G)$. The action of G on N is now realized by inner automorphisms. This construction may be summarized by an exact sequence (in which σ is the canonical "section" just described, π the projection onto the second factor):

$$e \to N \overset{i}{\to} N \rtimes G \underset{\underset{\sigma}{\curvearrowleft}}{\overset{\pi}{\to}} G \to e$$

How can we recognize a semidirect product? Given a group G' with subgroups N and G (N being normal), G acts on N by inner automorphisms, allowing us to construct a semidirect product $N \rtimes G$. It is easy to check that the homomorphism $N \rtimes G \to G'$ defined by $(x, y) \mapsto xy$ is an isomorphism precisely when $G' = NG$ and $N \cap G = e$.

This entire discussion can be carried over almost word-for-word to the case of algebraic groups. If G, N are algebraic groups and G acts morphically on N, then the variety $N \times G$ becomes an algebraic group $N \rtimes G$ under the above recipe, and the maps i, π, σ are morphisms of algebraic groups.

In the reverse direction, a given algebraic group G', with closed subgroups G and N (N being normal), will be identifiable in the above way with $N \rtimes G$ provided the natural morphism $N \rtimes G \to G'$ is an isomorphism of algebraic groups. (For this it is *not* always enough that the morphism be an isomorphism of *abstract* groups.) *Example*: $T(n, K)$ is easily seen to be the semidirect product (as algebraic group) of its subgroups $D(n, K)$ and $U(n, K)$, the latter being normal (Exercise 2).

8.5. Translation of Functions

When an algebraic group G acts on an *affine* variety X (e.g., on itself), we also obtain interesting linear actions of G on the affine algebra $K[X]$ and certain of its finite dimensional subspaces. Namely, denote by τ_x the comorphism attached to the morphism $y \mapsto x^{-1} \cdot y$ $(x \in G, y \in X)$. So if $f \in K[X]$, $y \in X$, $(\tau_x f)(y) = f(x^{-1} \cdot y)$. The inverse appears here in order to insure that $\tau : G \to \mathrm{GL}(K[X])$, where $\tau(x) = \tau_x$, is a group homomorphism. We call τ_x **translation of functions** by x. Notice that τ_x is actually a K-algebra automorphism of $K[X]$.

For example, when $X = G$, G acts on itself by left (resp. right) translations (8.1): $y \mapsto xy$ (resp. $y \mapsto yx^{-1}$). The associated morphism introduced above is then $y \mapsto x^{-1}y$ (resp. $y \mapsto yx$), and its comorphism λ_x (resp. ρ_x) is called **left** (resp. **right**) **translation of functions** by x:

$$(\lambda_x f)(y) = f(x^{-1}y),$$
$$(\rho_x f)(y) = f(yx).$$

As above, $\lambda : G \to \mathrm{GL}(\mathsf{K}[G])$ and $\rho : G \to \mathrm{GL}(\mathsf{K}[G])$, where $\lambda(x) = \lambda_x$ and $\rho(x) = \rho_x$, are both group homomorphisms. Moreover, λ_x and ρ_y obviously commute (for each pair $x, y \in G$). These operators will prove to be extremely useful in what follows. As a sample, we use right translation to characterize membership in a closed subgroup:

Lemma. *Let H be a closed subgroup of an algebraic group G, I the ideal of $\mathsf{K}[G]$ vanishing on H. Then $H = \{x \in G \,|\, \rho_x(I) \subset I\}$.*

Proof. In one direction, let $x \in H$. If $f \in I$, $(\rho_x f)(y) = f(yx) = 0$ for all $y \in H$ (since $yx \in H$), so $\rho_x f \in I$. In the reverse direction, let $\rho_x(I) \subset I$. In particular, if $f \in I$, then $\rho_x f$ vanishes at $e \in H$, i.e., $f(x) = 0$. So $x \in H$. $\quad\square$

8.6. Linearization of Affine Groups

It was observed in (7.1) that any closed subgroup of $\mathrm{GL}(n, \mathsf{K})$ is an (affine) algebraic group. The converse is also true, as we shall see by constructing a finite dimensional subspace of $\mathsf{K}[G]$ on which G acts by translations. First we lay the groundwork.

Proposition. *Let the algebraic group G act morphically on an affine variety X, and let F be a finite dimensional subspace of $\mathsf{K}[X]$.*

(a) There exists a finite dimensional subspace E of $\mathsf{K}[X]$ including F which is stable under all translations τ_x $(x \in G)$.

(b) F itself is stable under all τ_x $(x \in G)$ if and only if $\varphi^* F \subset \mathsf{K}[G] \otimes_{\mathsf{K}} F$, where $\varphi : G \times X \to X$ gives the action of G on X.

Proof. (a) We may assume that F is the span of a single $f \in \mathsf{K}[X]$ (and "add up" the resulting spaces E afterward). Write (non-uniquely) $\varphi^* f = \sum f_i \otimes g_i \in \mathsf{K}[G] \otimes \mathsf{K}[X]$. For each $x \in G$, $y \in X$, $(\tau_x f)(y) = f(x^{-1} \cdot y) = \sum f_i(x^{-1}) g_i(y)$, whence $\tau_x f = \sum f_i(x^{-1}) g_i$. The functions g_i therefore span a *finite dimensional* subspace of $\mathsf{K}[X]$ which contains all translates of f. So the space E spanned by all $\tau_x f$ does the trick.

(b) If $\varphi^* F \subset \mathsf{K}[G] \otimes F$, then the proof of part (a) shows that the functions g_i can be taken to lie in F, i.e., F is stable under all τ_x $(x \in G)$. Conversely, let F be stable under translations, and extend a vector space basis $\{f_i\}$ of F to a basis $\{f_i\} \cup \{g_j\}$ of $\mathsf{K}[X]$. If $\varphi^* f = \sum r_i \otimes f_i + \sum s_j \otimes g_j$, we have $\tau_x f = \sum r_i(x^{-1}) f_i + \sum s_j(x^{-1}) g_j$. Since this belongs to F, the functions s_j must vanish identically on G (and hence be 0), i.e., $\varphi^* F \subset \mathsf{K}[G] \otimes F$. $\quad\square$

Theorem. *Let G be an (affine) algebraic group. Then G is isomorphic to a closed subgroup of some $GL(n, K)$.*

Proof. Choose generators $f_1, .., f_n$ for the affine algebra $K[G]$. By applying part (a) of the preceding proposition to the span F of the f_i, we can find a larger finite dimensional subspace E of $K[G]$ stable under all right translations ρ_x ($x \in G$). Changing notation, we may assume that $f_1, .., f_n$ is a vector space basis of E (and generates $K[G]$). If $\varphi: G \times E \to E$ denotes the given action, use part (b) of the proposition to write $\varphi^* f_i = \sum_j m_{ij} \otimes f_j$, where $m_{ij} \in K[G]$. Then $(\rho_x f_i)(y) = f_i(yx) = \sum m_{ij}(x) f_j(y)$, whence $\rho_x f_i = \sum_j m_{ij}(x) f_j$. In other words, the matrix of $\rho_x | E$ (relative to the basis $f_1, .., f_n$) is $(m_{ij}(x))$. This shows that $\psi: G \to GL(n, K)$, where $\psi(x) = (m_{ij}(x))$, is a morphism of algebraic groups.

Notice that $f_i(x) = f_i(ex) = \sum m_{ij}(x) f_j(e)$, or $f_i = \sum f_j(e) m_{ij}$. This shows that the m_{ij} also generate $K[G]$; in particular, ψ is *injective*. Moreover, the image group $G' = \psi(G)$ is closed in $GL(n, K)$ (part (b) of Proposition 7.4 B). To complete the proof, we therefore need only show that ψ is an isomorphism of varieties. But the restrictions to G' of the coordinate functions T_{ij} are sent by ψ^* to the respective m_{ij}, which were just shown to generate $K[G]$. So ψ^* is surjective, hence it identifies $K[G']$ with $K[G]$. $\square$

Just as Cayley's theorem in group theory reduces the study of abstract groups to that of permutation groups, this theorem reduces the study of (affine) algebraic groups to that of linear groups. However, the "arbitrariness" of the concrete representation chosen makes us prefer to remain mostly in the general context. For our purposes, the theorem will be useful mainly as a technical aid in certain proofs: we embed a given group into a general linear group (as a closed subgroup) and then exploit the special properties of matrices (e.g., behavior of eigenvalues). This procedure is hardly very elegant, and could often be avoided, but it does make some of the proofs more transparent.

In a sense, the action of G on $K[G]$ by right translations already contains all information about the linear representations of G. We shall exploit heavily the passage from x to ρ_x in the next chapter.

Exercises

G denotes an arbitrary algebraic group.
1. Let $GL(n, K)$ act in the usual way on K^n (identified with affine n-space). Prove that there are precisely two orbits, one closed and one not.
2. Verify the assertion (8.4) that $T(n, K)$ is the semidirect product of $D(n, K)$ and $U(n, K)$.
3. $K[G]$ is the union of finite dimensional subspaces stable under right translations by G.
4. Each finite dimensional subspace of $K[G]$ lies in a finite dimensional subspace stable under all λ_x and ρ_x ($x \in G$).

5. Show that the following recipe defines on $\mathbf{A}^2$ the structure of (connected) algebraic group. Let char $\mathsf{K} = p > 0$, and let q, r be powers of p. Define $(a, b) \cdot (c, d) = (a + c, \; b + d + a^q c^r)$. In case $q \neq r$, notice that this group is not commutative. Try to find a closed subgroup of $\mathrm{U}(5, \mathsf{K})$ isomorphic to this group.

6. $\mathrm{GL}(V)$ acts morphically on $V - \{0\}$ (viewed as a principal open set in affine space) and on $\mathbf{P}(V)$. The canonical map $V - \{0\} \to \mathbf{P}(V)$ is a G-equivariant morphism for any closed subgroup G of $\mathrm{GL}(V)$.

7. Let V be a G-module which is isomorphic to the G-module V^* (8.2). Prove that there exists a nondegenerate bilinear form $\beta: V \times V \to \mathsf{K}$ which is invariant under G in the sense that $\beta(x \cdot v, x \cdot w) = \beta(v, w)$ for all $x \in G$, $v, w \in V$.

8. In $\mathrm{GL}(n, \mathsf{K})$, the normalizer of $\mathrm{D}(n, \mathsf{K})$ is the group N of monomial matrices (cf. Exercise 7.7).

Notes

For an intrinsic approach to affine algebraic groups, based on their Hopf algebras rather than on their realizations as linear groups, see Hochschild [8].

Chapter III

Lie Algebras

9. Lie Algebra of an Algebraic Group

9.1. Lie Algebras and Tangent Spaces

The object of this section is to attach to an algebraic group a Lie algebra (in a suitably functorial way). For our purpose, a **Lie algebra** over K is a subspace of an associative K-algebra which is closed under the **bracket** operation $[x, y] = xy - yx$. An important example is the **general linear algebra** $\mathfrak{gl}(n, K)$, which is the associative algebra $M(n, K)$ viewed as Lie algebra. (This will turn out to be essentially the Lie algebra of $GL(n, K)$.)

Let G be an algebraic group, $A = K[G]$. Recall (8.5) that G acts on A via left (resp. right) translation: $(\lambda_x f)(y) = f(x^{-1}y)$ (resp. $(\rho_x f)(y) = f(yx)$). It is routine to verify that the bracket of two *derivations* of A is again a derivation; therefore, Der A is a Lie algebra. So is the subspace $\mathscr{L}(G) = \{\delta \in \text{Der } A | \delta\lambda_x = \lambda_x\delta$ for all $x \in G\}$ = space of **left invariant derivations** of A, since the bracket of two derivations which commute with λ_x obviously does likewise. We call $\mathscr{L}(G)$ the **Lie algebra of G**.

Several natural questions arise at this point: (1) Is $\mathscr{L}(G)$ finite dimensional, and if so, what is its dimension? (2) How does a morphism of algebraic groups $\varphi: G \to G'$ relate the two Lie algebras $\mathscr{L}(G)$, $\mathscr{L}(G')$? (3) How does the structure of $\mathscr{L}(G)$ reflect the group structure of G? (4) If H is a closed subgroup of G, how is $\mathscr{L}(H)$ related to $\mathscr{L}(G)$?

Considerable light can be cast on these matters by comparing $\mathscr{L}(G)$ with the *tangent space* $\mathscr{T}(G)_e$ (5.1). Recall that $\mathscr{T}(G)_e$ is identified with $\mathscr{T}(G^{\circ})_e$; it has the structure of a vector space over K, of dimension equal to dim G (since e is a simple point (7.1)). We shall usually write $\mathfrak{g}$ for $\mathscr{T}(G)_e$. Given a morphism $\varphi: G \to G'$ which sends e to e (e.g., a morphism of algebraic groups), we obtain a linear map $d\varphi_e: \mathfrak{g} \to \mathfrak{g}'$. Differentiation has the functorial properties (5.4): $d(1_G)_e = 1_{\mathfrak{g}}, d(\psi \circ \varphi)_e = d\psi_e \circ d\varphi_e$. We also know that $\mathscr{T}(G)_e$ may be described algebraically as the space of point derivations from the local ring at e into K. But such derivations are already uniquely determined by their effect on the subring $A = K[G]$. This suggests that we might pass from $\mathscr{L}(G)$ to $\mathfrak{g}$ by evaluating functions at e. Formally, we define a K-linear map $\theta: \mathscr{L}(G) \to \mathfrak{g}$ by $(\theta\delta)(f) = (\delta f)(e)$ $(\delta \in \mathscr{L}(G), f \in A)$.

Theorem. *Let G be an algebraic group, $\mathfrak{g} = \mathscr{T}(G)_e$, $\mathscr{L}(G)$ as above. Then $\theta: \mathscr{L}(G) \to \mathfrak{g}$ is a vector space isomorphism. In case $\varphi: G \to G'$ is a morphism of algebraic groups, $d\varphi_e: \mathfrak{g} \to \mathfrak{g}'$ is a homomorphism of Lie algebras ($\mathfrak{g}$, $\mathfrak{g}'$ being given the bracket product of $\mathscr{L}(G)$, $\mathscr{L}(G')$).*

65

The proof will be given in (9.2). Since that proof is rather formal, it may be well to explain informally the underlying idea. If we follow any derivation $\delta : A \to A$ by evaluation at $x \in G$, we obtain a tangent vector at x. (So δ determines a "tangent vector field" on G, assigning to each point of G a tangent vector.) Left invariance of δ simply means that the tangent vector at e determined by δ is "sent" (via left translations) to the various tangent vectors at other points x determined by δ. Therefore, δ ought to be uniquely determined by the tangent vector $f \mapsto (\delta f)(e)$ (and any tangent vector at e should give rise to such a δ).

The theorem answers questions (1) and (2) above. (Questions (3) and (4) will also be dealt with shortly.) When $\varphi : G \to G'$ is a morphism of algebraic groups, we shall write $d\varphi$ in place of $d\varphi_e$. Consider, for example, the *inner automorphism* Int x $(y) = xyx^{-1}$ $(x, y \in G)$. Its differential $d(\text{Int } x)$ will be of great importance; we denote it by Ad x. According to the theorem, Ad x is an *automorphism* of the Lie algebra $\mathfrak{g}$: the invertibility follows from the fact that $(\text{Ad } x)(\text{Ad } x^{-1}) = d(\text{Int } x) \circ d(\text{Int } x^{-1}) = d(\text{Int } e) = 1_{\mathfrak{g}}$. Indeed, $(\text{Ad } x)(\text{Ad } y) = \text{Ad } xy$, so $\text{Ad} : G \to \text{Aut } \mathfrak{g} \subset \text{GL}(\mathfrak{g})$ is a homomorphism (of abstract groups), the **adjoint representation** of G.

It is interesting to compute explicitly the effect of Ad x on $\mathscr{L}(G)$ (via the identification θ). We claim that Ad $x(\delta) = \rho_x \delta \rho_x^{-1}$. (This is reasonable: since right and left translations commute, the right side is again in $\mathscr{L}(G)$.) If Ad $x(\delta) = \delta'$, then by definition, $(\delta' f)(e) = \delta(\varphi^* f)(e)$ for all $f \in K[G]$, where $\varphi = \text{Int } x$. This applies in particular to functions of the form $\rho_x f$. But $\varphi^*(\rho_x f)(y) = (\rho_x f)(xyx^{-1}) = f(xy) = (\lambda_{x^{-1}} f)(y)$, i.e., $\varphi_*(\rho_x f) = \lambda_{x^{-1}} f$. Now $(\delta' \rho_x)(f)(e) = \delta(\lambda_{x^{-1}} f)(e) = \lambda_{x^{-1}}(\delta f)(e)$ (δ is left invariant!) $= (\delta f)(x) = \rho_x \delta(f)(e)$. This shows that $\delta' = \rho_x \delta \rho_x^{-1}$, as claimed.

9.2. Convolution

In order to prove Theorem 9.1, we construct a backward map $\eta : \mathfrak{g} \to \mathscr{L}(G)$, sending a tangent vector x to a derivation $*\mathsf{x}$ (called **right convolution** by x):

$$(f * \mathsf{x})(x) = \mathsf{x}(\lambda_{x^{-1}} f) \qquad (x \in G, f \in A).$$

It has to be checked first that $*\mathsf{x}$ really is a left invariant derivation of A. If $x, y \in G$, $f, g \in A$, we have:

$$\begin{aligned}
(fg * \mathsf{x})(x) &= \mathsf{x}(\lambda_{x^{-1}}(fg)) \\
&= \mathsf{x}((\lambda_{x^{-1}} f)(\lambda_{x^{-1}} g)) \\
&= \mathsf{x}(\lambda_{x^{-1}} f)g(x) + f(x)\mathsf{x}(\lambda_{x^{-1}} g) \\
&= ((f * \mathsf{x})g + f(g * \mathsf{x}))(x)
\end{aligned}$$

(so $*\mathsf{x}$ is a derivation, K-linearity being obvious), and

$$\begin{aligned}
(\lambda_y(f * \mathsf{x}))(x) &= (f * \mathsf{x})(y^{-1} x) \\
&= \mathsf{x}(\lambda_{x^{-1}y} f) \\
&= \mathsf{x}(\lambda_{x^{-1}}(\lambda_y f)) \\
&= ((\lambda_y f) * \mathsf{x})(x)
\end{aligned}$$

(so $*\mathsf{x}$ is left invariant).

Now it is clear that η is a K-linear map. To see that it is inverse to θ, we must compute the two composites $\eta \circ \theta$ and $\theta \circ \eta$:

$$(f * \theta(\delta))(x) = \theta(\delta)(\lambda_{x^{-1}}f)$$
$$= \delta(\lambda_{x^{-1}}f)(e)$$
$$= \lambda_{x^{-1}}(\delta f)(e)$$
$$= (\delta f)(x)$$

(so for $\delta \in \mathscr{L}(G)$, $(\eta \circ \theta)(\delta) = \delta$), and

$$\theta(*x)(f) = (f * x)(e)$$
$$= x(\lambda_{e^{-1}}f)$$
$$= x(f)$$

(so for $x \in \mathfrak{g}$, $(\theta \circ \eta)(x) = x$).

It remains to show that if $\varphi: G \to G'$ is a morphism of algebraic groups, then $d\varphi: \mathfrak{g} \to \mathfrak{g}'$ preserves the bracket operation. For $x, y \in \mathfrak{g}$, let $x' = d\varphi(x)$, $y' = d\varphi(y)$. Let $f' \in K[G']$ and write $f = \varphi * f'$. By definition, $[x', y'](f') = (f' * y' * x')(e) - (f' * x' * y')(e) = x'(f' * y') - y'(f' * x') = x(\varphi^*(f' * y')) - y(\varphi^*(f' * x'))$. On the other hand, $d\varphi([x, y])$ sends f' to $(f * y * x)(e) - (f * x * y)(e) = x(f * y) - y(f * x)$. We claim that each term is equal to the corresponding term in the preceding equation; for this it suffices to prove that $f * x = \varphi^*(f' * x')$, i.e., that $(\varphi^* f') * x = \varphi^*(f' * d\varphi(x))$. Each side is a function on G, so we test the values at $x \in G$: $(\varphi^* f' * x)(x) = x(\lambda_{x^{-1}}\varphi^* f')$, while $(f' * d\varphi(x))(\varphi(x)) = d\varphi(x)(\lambda_{\varphi(x)^{-1}}f') = x(\varphi^*(\lambda_{\varphi(x)^{-1}}f'))$. Finally, it suffices to prove that $\lambda_{x^{-1}}\varphi^* f' = \varphi^*(\lambda_{\varphi(x)^{-1}}f')$, which follows immediately when we evaluate both sides at $y \in G$. $\square$

9.3. Examples

Let $G = \mathbf{G}_a$, so $K[G] = K[T]$. The Lie algebra $\mathscr{L}(G)$ is 1-dimensional, so its bracket operation is identically 0. We claim that the derivation $\delta = \dfrac{d}{dT}$ is left invariant (hence spans $\mathscr{L}(G)$). For this it suffices to test the polynomial T, letting $x \in G$ be arbitrary: $\lambda_{-x}\delta T = \lambda_{-x}1 = 1$, while $\delta\lambda_{-x}T = \delta(T + x) = \delta T = 1$.

Next let $G = \mathbf{G}_m$, $K[G] = K[T, T^{-1}]$ (the ring of Laurent polynomials). Again $\mathscr{L}(G)$ is 1-dimensional. If we specify $\delta T = T$, then δ extends in one and only one way to a derivation of $K[T, T^{-1}]$, which is in fact left invariant (since $\delta(aT) = a\delta T$ for $a \in K^*$).

These two examples are easy enough to work out by *ad hoc* methods. But for groups of larger dimension, a more systematic procedure is needed for calculation of $\mathscr{L}(G)$. Since the tangent space is often more readily accessible to us than the abstractly defined space of left invariant derivations, it is advantageous to introduce the bracket operation directly into $\mathscr{T}(G)_e$. This we shall do via an associative product $x \cdot y$ for tangent vectors at e, which of course must involve in some way the group operation. For $A = K[G]$, write $\mu^* f = \sum f_i \otimes g_i (f, f_i, g_i \in A)$, where $\mu : G \times G \to G$ is the given multiplication in G. If $x, y \in \mathcal{T}(G)_e$ are viewed as derivations $A \to K$, then it makes sense to define a K-linear map $x \otimes y : A \otimes A \to K$ by $(x \otimes y)(f \otimes g) = (xf)(yg)$ (check that this is unambiguous by starting with a bilinear map $A \times A \to K$). Then let $x \cdot y = (x \otimes y) \circ \mu^* : A \to K$.

Rather than verify directly that $x \cdot y$ is an associative operation and that $[x, y] = x \cdot y - y \cdot x$ corresponds to the Lie bracket in $\mathcal{L}(G)$, we just note that via $\theta : \mathcal{L}(G) \xrightarrow{\sim} \mathfrak{g}$, the ordinary associative product $(*x)(*y)$ goes over into $x \cdot y$. Observe first that if $\mu^* f = \sum f_i \otimes g_i$, then

$$f * x = \sum f_i x(g_i).$$

(Evaluate both sides at $x \in G$, and use the fact that $\lambda_{x^{-1}} f = \sum f_i(x) g_i$.) Now, by definition, $(x \cdot y)(f) = \sum x(f_i) y(g_i)$. On the other hand, using the observation just made,

$$\begin{aligned}
((f*y)*x)(e) = \left(\left(\sum f_i y(g_i)\right)*x\right)(e) &= \left(\sum (f_i * x) y(g_i)\right)(e) \\
&= \sum x(f_i) y(g_i).
\end{aligned}$$

We can now compute the Lie algebra of $G = \mathrm{GL}(n, K)$. Since G is an open set in affine n^2-space, its tangent space at e has as canonical basis the various partial differentiation operators $\partial / \partial T_{ij}$ followed by evaluation at the identity matrix. It is convenient therefore to specify a tangent vector x by the n^2 numbers $x_{ij} = x(T_{ij})$, arranged in a square matrix. With this convention, we have $(x \cdot y)(T_{ij}) = (x \otimes y)(\sum_h T_{ih} \otimes T_{hj}) = \sum_h x_{ih} y_{hj}$ (the usual matrix product). In other words, the K-linear map $x \mapsto (x_{ij})$ of $\mathfrak{g}$ into $\mathrm{M}(n, K)$ identifies $\mathfrak{g}$ as Lie algebra with $\mathfrak{gl}(n, K)$. (The map is injective, since only the 0 tangent vector kills all T_{ij}, hence surjective, since the dimensions are both n^2.) All we have done here is to view $\mathfrak{g}$ as the "geometric" tangent space to G at e (namely, affine n^2-space), with e translated to 0.

9.4. Subgroups and Lie Subalgebras

Let H be a closed subgroup of the algebraic group G. The inclusion $\eta : H \to G$ is an isomorphism onto a closed subgroup: η^* maps $K[G]$ onto $K[H] \cong K[G]/I$ (I the ideal vanishing on H). Therefore, $d\eta_e$ identifies $\mathcal{T}(H)_e$ with the subspace of $\mathcal{T}(G)_e$ consisting of those x for which $x(I) = 0$. But η is also a morphism of algebraic groups, so $d\eta : \mathfrak{h} \to \mathfrak{g}$ is a Lie algebra homomorphism, which allows us to view $\mathfrak{h}$ as a Lie subalgebra of $\mathfrak{g}$ (9.1). Recall (Lemma 8.5) that H may be characterized as the set of all $x \in G$ for which $\rho_x(I) \subset I$. There is a similar characterization of $\mathfrak{h}$, based on right convolution:

Lemma. *Let H be a closed subgroup of G, I the ideal of $K[G]$ vanishing on H. Then $\mathfrak{h} = \{x \in \mathfrak{g} \mid I * x \subset I\}$.*

Proof. In one direction, let $x \in \mathfrak{h}$. If $f \in I$, $x \in H$, then $(f*x)(x) = x(\lambda_{x^{-1}} f) = 0$ since $\lambda_{x^{-1}} f$ again belongs to I. So $f * x \in I$. In the reverse direction, let $x \in \mathfrak{g}$ satisfy $I*x \subset I$. If $f \in I$, then $(f*x)(e) = x(\lambda_{e^{-1}}f) = x(f) = 0$ (since $e \in H$), forcing $x \in \mathfrak{h}$. $\square$

This lemma is not of much practical use in computing $\mathfrak{h}$ (but will be of some theoretical use later on). Instead, one can frequently describe an algebraic group as a closed subgroup of some $GL(n, K)$ and then compute explicitly its Lie algebra as a subalgebra of $\mathfrak{gl}(n, K)$.

Take $T(n, K)$, for example. This may be viewed as the principal open set in $\mathbf{A}^{n(n+1)/2}$ (embedded in the obvious way in $\mathbf{A}^{n^2}$) defined by the nonvanishing of det, so the tangent space at e is this affine space. In other words, after translating the tangent space from e to 0, the Lie algebra of $T(n, K)$ is the set $\mathfrak{t}(n, K)$ of *all* upper triangular matrices. Similarly, the Lie algebra of $D(n, K)$ is the set $\mathfrak{d}(n, K)$ of all diagonal matrices. $U(n, K)$ is the closed subset of $M(n, K)$ defined by vanishing of the *linear* polynomials $T_{ii} - 1, T_{ij} \, (i > j)$, so according to (5.1) its tangent space at e (again translated to 0) consists of all matrices having zeros on or below the diagonal. This Lie algebra is (perversely) denoted $\mathfrak{n}(n, K)$. Notice that in each of these cases, the dimension of the group (or Lie algebra) is easy to compute; this is not always true, of course.

Consider also $SL(n, K)$, the zero set of $f(T) = \det(T_{ij}) - 1$. As noted in (5.4), $\sum_{i,j} \dfrac{\partial f}{\partial T_{ij}} (e)T_{ij} = \sum_i T_{ii}$, so according to (5.1), the Lie algebra of $SL(n, K)$ lies in the set $\mathfrak{sl}(n, K)$ of matrices of *trace* 0. But in each case the dimension is $n^2 - 1$ (cf. (7.2)), so equality must hold.

9.5. Dual Numbers

There is another way to view the Lie algebra, within a scheme-theoretic framework. We shall indicate this briefly, referring the reader to Borel [4] and Demazure, Gabriel [1] for a rigorous development. The present subsection is not needed elsewhere in the text.

Let G be an algebraic group, $A = K[G]$. The points of G correspond 1–1 to the maximal ideals of A (1.5), or to the K-algebra homomorphisms $A \to K$. We just write $\mathrm{Hom}(A, K)$ for the latter. To $x \in G$ corresponds the homomorphism ε_x, where $\varepsilon_x(f) = f(x)$. How does the multiplication μ in G carry over to $\mathrm{Hom}(A, K)$? Given $x, y \in G$, we just form the composite $A \xrightarrow{\mu^*} A \otimes A \xrightarrow{\varepsilon_x \otimes \varepsilon_y} K \otimes K \to K$, the last arrow being multiplication in K.

If K is replaced by an arbitrary (commutative) K-algebra B, the set of K-algebra homomorphisms $\mathrm{Hom}(A, B)$ acquires a group structure in a similar way. This is thought of as the "points of G in B". So G is now viewed as a *functor* which assigns to each K-algebra B a group $G(B) = \mathrm{Hom}(A, B)$. For example, in place of the single group $GL(n, K)$ we could study the functor $\mathbf{GL}_n$ which assigns to B the group $GL(n, B)$. Similarly, $\mathbf{G}_a$ and $\mathbf{G}_m$ could be viewed as functors.

The algebra of **dual numbers** is defined to be $K[T]/(T^2)$, or $K[\delta]$ (where $\delta^2 = 0$). A typical element of this algebra is therefore of the form $a + b\delta$

$(a, b \in \mathsf{K})$. If G is a functor of the type just described, we write $\mathscr{T}(G) = G(\mathsf{K}[\delta])$ and call this the **tangent bundle** of G. In effect, this will be an assemblage of all the tangent spaces at points of $G(\mathsf{K})$. The inclusion $\mathsf{K} \to \mathsf{K}[\delta]$ and the projection $\mathsf{K}[\delta] \to \mathsf{K}$ $(a + b\delta \mapsto a)$ induce group homomorphisms $G(\mathsf{K}) \to \mathscr{T}(G)$, $\mathscr{T}(G) \to G(\mathsf{K})$, leading in turn to a split exact sequence, in which the kernel is interpreted as the tangent space at e:

$$0 \to \mathfrak{g} \to \mathscr{T}(G) \to G(\mathsf{K}) \to e.$$

Indeed, a typical element of $\mathscr{T}(G)$ can be written in the form $\varepsilon_x + \delta\mathsf{x}$, where $x \in G$ and $\mathsf{x} \in \mathscr{T}(G)_x$; this homomorphism $A \to \mathsf{K}[\delta]$ sends $f \mapsto \varepsilon_x(f) + \delta\mathsf{x}(f)$. The product $(\varepsilon_x + \delta\mathsf{x})(\varepsilon_y + \delta\mathsf{y})$ equals $\varepsilon_{xy} + \delta(\varepsilon_x\mathsf{y} + \varepsilon_y\mathsf{x})$.

This set-up, once justified, makes certain computations to be encountered in §10 quite natural and transparent. For example, Ad x is just the restriction to $\mathfrak{g}$ of Int ε_x. More generally, a morphism of algebraic groups $\varphi : G \to G'$ induces a homomorphism $\mathscr{T}(G) \to \mathscr{T}(G')$, whose restriction to $\mathfrak{g}$ is just $d\varphi$. This makes it possible to compute certain differentials very directly.

Exercises

1. Let G be a closed subgroup of $\mathrm{GL}(n, \mathsf{K})$, J the ideal of all polynomials in $\mathsf{K}[\mathsf{T}_{ij}]$ vanishing on G. Prove that $G = \{x \in \mathrm{GL}(n, \mathsf{K}) | \rho_x J = J\}$, $\mathfrak{g} = \{\mathsf{x} \in \mathfrak{gl}(n, \mathsf{K}) | J * \mathsf{x} \subset J\}$. [Compare J with the ideal I of $\mathsf{K}[\mathsf{T}_{ij}, 1/\det]$ vanishing on G, and adapt Lemmas 8.5, 9.4.]

2. Let $G = \mathrm{SL}(2, \mathsf{K})$, with $\mathfrak{g} = \mathfrak{sl}(2, \mathsf{K})$. Relative to the basis $\mathsf{h} = \begin{pmatrix} 1 & 0 \\ 0 & 1 \end{pmatrix}$, $\mathsf{x} = \begin{pmatrix} 0 & 1 \\ 0 & 0 \end{pmatrix}$, $\mathsf{y} = \begin{pmatrix} 0 & 0 \\ 1 & 0 \end{pmatrix}$ of $\mathfrak{g}$, compute the matrix of Ad x $(x \in G)$.

3. Let char $\mathsf{K} = p > 0$. A **Lie p-algebra** (or **restricted** Lie algebra) is a Lie algebra closed under the associative p^{th} power operation, e.g., $\mathfrak{gl}(n, \mathsf{K})$ or $\mathfrak{sl}(n, \mathsf{K})$.
   (a) If $\mathfrak{A}$ is any K-algebra, Der $\mathfrak{A}$ is a Lie p-algebra.
   (b) If G is an algebraic group, $\mathscr{L}(G)$ is a Lie p-algebra.
   (c) Let $G = \mathbf{G}_a$ (resp. $\mathbf{G}_m$), δ as in (9.3). Then $\delta^p = 0$ (resp. $\delta^p = \delta$).
   (d) If $\varphi : G \to G'$ is a morphism of algebraic groups, then $d\varphi$ is a homomorphism of Lie p-algebras, i.e., $d\varphi(x)^p = d\varphi(x^p)$.

Notes

This chapter is based on Borel [4, §3]; however, he places greater emphasis on dual numbers.

10. Differentiation

Passage from G to $\mathfrak{g}$ helps to "linearize" a number of questions. This technique is especially useful in dealing with morphisms. In the present section we shall compute the differentials of several morphisms and draw a few immediate conclusions, e.g., that the Lie algebra of a closed normal subgroup is an *ideal*.

10.1. Some Elementary Formulas

Let G be an algebraic group, with product $\mu: G \times G \to G$ and inverse $\iota: G \to G$. To calculate $d\mu_{(e,\,e)}$, recall first that $\mathcal{T}(G \times G)_{(e,\,e)}$ is canonically isomorphic to $\mathcal{T}(G)_e \oplus \mathcal{T}(G)_e = \mathfrak{g} \oplus \mathfrak{g}$ (5.1). If $f \in \mathsf{K}[G]$, write $\mu^* f = \sum f_i \otimes g_i$, so $f(xy) = \sum f_i(x) g_i(y)$. In particular: (*) $f = \sum g_i(e) f_i = \sum f_i(e) g_i$. Now if $(\mathsf{x}, \mathsf{y}) \in \mathcal{T}(G \times G)_{(e,\,e)}$ ($\mathsf{x}, \mathsf{y} \in \mathfrak{g}$), we have by definition $\mathsf{z} = d\mu_{(e,\,e)}(\mathsf{x}, \mathsf{y})$, where $\mathsf{z}(f) = (\mathsf{x}, \mathsf{y})(\sum f_i \otimes g_i) = \sum \mathsf{x}(f_i) g_i(e) + \sum f_i(e) \mathsf{y}(g_i)$. But in view of formula (*) this is the same as applying $\mathsf{x} + \mathsf{y}$ to f, so $d\mu_{(e,\,e)}(\mathsf{x}, \mathsf{y}) = \mathsf{x} + \mathsf{y}$. (This indicates that addition in $\mathfrak{g}$ is the infinitesimal analogue of multiplication in G.)

Next we compute $d\iota_e: \mathfrak{g} \to \mathfrak{g}$. Consider $G \xrightarrow{(1,\,\iota)} G \times G \xrightarrow{\mu} G$. Since the composite sends every x to e, the differential sends all $\mathsf{x} \in \mathfrak{g}$ to 0. Using the fact that $d(1) = 1$ along with the fact that $d(1, \iota) = (d(1), d(\iota))$, we get: $0 = d(\mu \circ (1,\,\iota))_e(\mathsf{x}) = d\mu_{(e,\,e)}(d(1,\,\iota)_e(\mathsf{x})) = d\mu_{(e,\,e)}(\mathsf{x}, d\iota_e(\mathsf{x})) = \mathsf{x} + d\iota_e(\mathsf{x})$, thanks to the above formula for $d\mu_{(e,\,e)}$. Therefore, $d\iota_e(\mathsf{x}) = -\mathsf{x}$ (i.e., negation in $\mathfrak{g}$ is the analogue of inversion in G).

Another interesting morphism is conjugation of a fixed $x \in G$: $\varphi_x(y) = yxy^{-1}$ ($y \in G$). This does *not* send e to e (unless $x = e$), so we consider first the morphism $\psi(y) = xy^{-1}x^{-1}$. Evidently $\psi = \mathrm{Int}\, x \circ \iota$, so $d\psi_e(\mathsf{x}) = d(\mathrm{Int}\, x)(d\iota_e(\mathsf{x})) = \mathrm{Ad}\, x(-\mathsf{x}) = -\mathrm{Ad}\, x(\mathsf{x})$, where $\mathrm{Ad}\, x$ is by definition $d(\mathrm{Int}\, x)$ (9.1). In turn, consider $\gamma_x(y) = yxy^{-1}x^{-1}$ (the **commutator morphism** relative to x), which may be realized as the composite $G \xrightarrow{(1,\,\psi)} G \times G \xrightarrow{\mu} G$. If $\mathsf{x} \in \mathfrak{g}$, $(d\gamma_x)_e(\mathsf{x}) = d\mu_{(e,\,e)}(\mathsf{x}, d\psi_e(\mathsf{x})) = \mathsf{x} - \mathrm{Ad}\, x(\mathsf{x}) = (1 - \mathrm{Ad}\, x)(\mathsf{x})$. Finally, the original conjugacy class morphism φ_x is just $G \xrightarrow{(\gamma_x,\,\varepsilon)} G \times G \xrightarrow{\mu} G$, where $\varepsilon(y) = x$ for all $y \in G$. (Later on we shall use this description in terms of γ_x to study the *separability* of φ_x.)

It may be well to summarize what we have obtained:

Proposition. *Let G be an algebraic group. Then for all* $\mathsf{x}, \mathsf{y} \in \mathfrak{g}$:
(a) $d\mu_{(e,\,e)}(\mathsf{x}, \mathsf{y}) = \mathsf{x} + \mathsf{y}$.
(b) $d\iota_e(\mathsf{x}) = -\mathsf{x}$.
(c) $(d\gamma_x)_e(\mathsf{x}) = (1 - \mathrm{Ad}\, x)(\mathsf{x})$, where $\gamma_x(y) = yxy^{-1}x^{-1}$ ($x, y \in G$). $\quad\square$

One further remark will be useful in adapting results about $\mathrm{GL}(n, \mathsf{K})$ to its closed subgroups: If H is a closed subgroup of an algebraic group G, then the differential of the restriction to H of a morphism φ of G (sending e to e) is the restriction to $\mathfrak{h}$ of the differential $d\varphi_e$ (i.e., restriction commutes with differentiation). This is obvious from the definitions.

10.2. Differential of Right Translation

Recall from (8.5) that G acts on $\mathsf{K}[G]$ by left and right translations: $(\lambda_x f)(y) = f(x^{-1}y)$, $(\rho_x f)(y) = f(yx)$. The resulting group homomorphisms

$\lambda, \rho: G \to \mathrm{GL}(\mathsf{K}[G])$ cannot themselves be regarded as morphisms of algebraic groups, since $\mathsf{K}[G]$ is infinite dimensional. However, $\mathsf{K}[G]$ is the union of finite dimensional subspaces stable under all ρ_x (or all λ_x). If E is such a subspace (stable under all ρ_x), with basis $(f_1, . , f_n)$, then we saw that $\rho_x f_i = \sum_j m_{ij}(x) f_j$, where the comorphism φ^* of the given action $\varphi: G \times E \to E$ sends f_i to $\sum_j m_{ij} \otimes f_j$. (Cf. Proposition 8.6 and the proof of Theorem 8.6.) It follows that $\psi(x) = (m_{ij}(x))$ defines a morphism $\psi: G \to \mathrm{GL}(n, \mathsf{K})$, the matrix representation associated with the given basis of E. Notice too that the subspace of $\mathsf{K}[G]$ spanned by the m_{ij} includes E and is stable under *left* translations; this follows from the calculation: $(\lambda_{x^{-1}} f_i)(y) = f_i(xy) = \sum_j m_{ij}(y) f_j(x)$ (remember that the action here of G on E involves *right* multiplication in G).

It is interesting to compute the differential $d\psi: \mathfrak{g} \to \mathfrak{gl}(n, \mathsf{K})$. If $\mathsf{x} \in \mathfrak{g}$, then by definition the matrix $d\psi(\mathsf{x})$ has as (i, j) entry the result of applying x to the pullback of the (i, j) coordinate function. But $\psi^* \mathsf{T}_{ij} = m_{ij}$, as remarked above, so $d\psi(\mathsf{x}) = (\mathsf{x}(m_{ij}))$. On the other hand, consider the action of x on the given basis of E by right convolution: $(f_i * \mathsf{x})(x) = \mathsf{x}(\lambda_{x^{-1}} f_i) = \mathsf{x}(\sum_j f_j(x) m_{ij}) = \sum_j f_j(x) \mathsf{x}(m_{ij})$, using the expression above for $\lambda_{x^{-1}} f_i$. In other words, x leaves E stable and has matrix $(\mathsf{x}(m_{ij}))$ relative to the basis $(f_1, . , f_n)$. To summarize:

Proposition. *Acting by right convolution, $\mathfrak{g}$ leaves stable every subspace of $\mathsf{K}[G]$ stable under right translation by G. On a finite dimensional, G-stable subspace, the differential of right translation is right convolution.* □

We could, of course, replace "right" by "left". Notice that the proposition sheds some light on the analogy between Lemma 8.5 and Lemma 9.4.

10.3. The Adjoint Representation

The action of G on $\mathsf{K}[G]$ by right translation is especially useful in constructing (finite dimensional) linear representations of G with good properties; this will become apparent in Chapter IV. However, the adjoint action of G on $\mathfrak{g}$ is more readily computable, and gives rather decisive information about G in case char $\mathsf{K} = 0$. It is in any case a valuable tool. In order to study this action, we begin with $\mathrm{GL}(n, \mathsf{K})$ and then adapt our findings to its closed subgroups.

Recall (9.1) that for a fixed $x \in G$, Ad x was defined to be the differential of the automorphism Int x of G; Ad x is therefore an automorphism of $\mathfrak{g}$ (as Lie algebra). If we think of $\mathfrak{g}$ as $\mathscr{L}(G)$, the left invariant derivations of $\mathsf{K}[G]$, then Ad x acts as $\delta \mapsto \rho_x \delta \rho_x^{-1}$ (9.1), which is suggestive of the concrete situation for $\mathrm{GL}(n, \mathsf{K})$.

Consider then the *special case* $G = \mathrm{GL}(n, \mathsf{K})$, with $\mathfrak{g} = \mathfrak{gl}(n, \mathsf{K})$. We begin by describing explicitly the action of G (resp. $\mathfrak{g}$) on $\mathsf{K}[G]$ by translation (resp. convolution). For this it is sufficient to look at the effect on the coordinate functions T_{ij}, since this determines the effect on $1/\det \mathsf{T}_{ij}$. Denote by T the matrix whose (i, j) entry is T_{ij}.

Lemma A. *Let $x \in \mathrm{GL}(n, \mathsf{K})$, $\mathsf{x} \in \mathfrak{gl}(n, \mathsf{K})$. Then $\rho_x \mathsf{T}_{ij}$ (resp. $\lambda_x \mathsf{T}_{ij}$, resp. $\mathsf{T}_{ij} * \mathsf{x}$) is the (i, j) entry of the matrix $\mathsf{T}x$ (resp. $x^{-1}\mathsf{T}$, resp. $\mathsf{T}x$).*

Proof. If $y \in \mathrm{GL}(n, \mathsf{K})$, $(\rho_x \mathsf{T}_{ij})(y) = \mathsf{T}_{ij}(yx) = \sum_h y_{ih} x_{hj} = \sum_h \mathsf{T}_{ih}(y) x_{hj} = (\mathsf{T}x)_{ij}(y)$, and similarly for left translation. Using this, we get

$$(\mathsf{T}_{ij} * \mathsf{x})(y) = \mathsf{x}(\lambda_{y^{-1}} \mathsf{T}_{ij}) = \mathsf{x}\left(\sum_h y_{ih} \mathsf{T}_{hj}\right) = \sum_h y_{ih} \mathsf{x}(\mathsf{T}_{hj})$$
$$= \sum_h y_{ih} \mathsf{x}_{hj} = \sum_h \mathsf{T}_{ih}(y) \mathsf{x}_{hj} = (\mathsf{T}\mathsf{x})_{ij}(y). \quad \square$$

With these descriptions in hand, it is easy to compute Ad x.

Lemma B. *Let $x \in \mathrm{GL}(n, \mathsf{K})$, $\mathsf{x} \in \mathfrak{gl}(n, \mathsf{K})$. Then $\mathrm{Ad}\, x(\mathsf{x}) = x\, \mathsf{x}\, x^{-1}$ (matrix product).*

Proof. It suffices to check that each side has the same effect on T_{ij}.

$$\begin{aligned}
\mathrm{Ad}\, x(\mathsf{x})(\mathsf{T}_{ij}) &= \rho_x(*\mathsf{x})\rho_x^{-1}(\mathsf{T}_{ij}) \\
&= \rho_x(*\mathsf{x})(\mathsf{T}x^{-1})_{ij} \\
&= \rho_x\left(\sum_h (\mathsf{T}\mathsf{x})_{ih} x_{hj}^{-1}\right) \\
&= \rho_x\left(\sum_h \sum_l \mathsf{T}_{il}\, \mathsf{x}_{lh}\, x_{hj}^{-1}\right) \\
&= \sum_h \sum_l \sum_m \left(\mathsf{T}_{im}\, x_{ml}\, \mathsf{x}_{lh}\, x_{hj}^{-1}\right) \\
&= (\mathsf{T}x\, \mathsf{x}\, x^{-1})_{ij} \\
&= x\, \mathsf{x}\, x^{-1}(\mathsf{T}_{ij}).
\end{aligned}$$

Here we have used Lemma A repeatedly. $\quad \square$

Corollary. $\mathrm{Ad} : \mathrm{GL}(n, \mathsf{K}) \to \mathrm{GL}(n^2, \mathsf{K})$ *is a morphism of algebraic groups* (*relative to any choice of basis for* $\mathfrak{gl}(n, \mathsf{K})$).

Proof. This is clear from the explicit formula given in Lemma B. $\quad \square$

Consider now the *general case*. Thanks to Theorem 8.6, any algebraic group G may be regarded as a closed subgroup of some $\mathrm{GL}(n, \mathsf{K})$; in this way $\mathfrak{g}$ becomes identified with a Lie subalgebra of $\mathfrak{gl}(n, \mathsf{K})$. Moreover, if $x \in G$, then $\mathrm{Int}_G(x)$ is the restriction to G of $\mathrm{Int}_{\mathrm{GL}(n, \mathsf{K})}(x)$. Since restriction commutes with differentiation (10.1), we conclude that $\mathrm{Ad}_G(x)$ is the restriction to $\mathfrak{g}$ of $\mathrm{Ad}_{\mathrm{GL}(n, \mathsf{K})}(x)$. Therefore Lemma B and its Corollary carry over intact to the general case:

Proposition. *Let G be an algebraic group. Then $\mathrm{Ad} : G \to \mathrm{GL}(\mathfrak{g})$ is a morphism of algebraic groups. In case G is a closed subgroup of $\mathrm{GL}(n, \mathsf{K})$, $\mathrm{Ad}\, x$ is just conjugation by x ($x \in G$).* $\quad \square$

10.4. Differential of Ad

Let G be an algebraic group. For each individual x in G, we obtained $\mathrm{Ad}\, x$ by differentiating the morphism $\mathrm{Int}\, x$ (at e). But in turn, $\mathrm{Ad} : G \to \mathrm{GL}(\mathfrak{g})$ turned out to be a morphism of algebraic groups (10.3), whose differential is therefore a Lie algebra homomorphism $\mathfrak{g} \to \mathfrak{gl}(\mathfrak{g})$ (also called a **representation**, since the image lies in some $\mathfrak{gl}(m, \mathsf{K})$).

Theorem. *The differential of* Ad *is* ad, *where* ad $\mathsf{x}(\mathsf{y}) = [\mathsf{x}, \mathsf{y}]$, *for* x, $\mathsf{y} \in \mathfrak{g}$.

Proof. As in (10.3), we treat first the special case $G = GL(n, \mathsf{K})$, $\mathfrak{g} = \mathfrak{gl}(n, \mathsf{K})$. If $x \in G$, then Ad x is the image of x under the composite map $G \overset{(1, \iota)}{\to} G \times G \overset{\sigma \times \tau}{\to} GL(\mathfrak{g}) \times GL(\mathfrak{g}) \overset{\mu}{\to} GL(\mathfrak{g})$, where $\iota(x) = x^{-1}$, μ is the product morphism, and $\sigma(x)$ (resp. $\tau(x)$) is the operation of left (resp. right) multiplication by x in $\mathfrak{g}$. It has to be observed that σ, τ are *morphisms* (of varieties), which send e to the identity matrix. We can make this obvious, and at the same time compute the differentials, if we work with the standard basis of $\mathfrak{g}$ consisting of the matrices e_{ij} (having 1 in the (i, j) position and 0 elsewhere). The (i, j) entry of the matrix xe_{lm} is just x_{il} (if $m = j$), 0 otherwise. Therefore the matrix of $\sigma(x)$ in $GL(n^2, \mathsf{K})$ may be divided into $n \times n$ blocks, a typical block containing in one column (say the mth) the entries from one column (say the lth) of x, and zeros elsewhere. This shows not only that σ (or τ) is a morphism, but also that the entries of $\sigma(x)$ (or $\tau(x)$) are *linear* polynomials of the simplest kind in the coefficients of x. It follows that, for example, $d\sigma_e(\mathsf{x}) =$ left multiplication by x in $\mathfrak{g}$ $(\mathsf{x} \in \mathfrak{g})$.

We already computed in (10.1) the differentials of ι and μ. Therefore, $d(\mathrm{Ad})$ is given by the recipe: $\mathsf{x} \mapsto (\mathsf{x}, -\mathsf{x}) \mapsto$ left multiplication by x minus right multiplication by x, i.e., $d(\mathrm{Ad})\mathsf{x}(\mathsf{y}) = \mathsf{x}\mathsf{y} - \mathsf{y}\mathsf{x} = [\mathsf{x}, \mathsf{y}]$.

In the general case, we use Theorem 8.6 to identify G with a closed subgroup of some $GL(n, \mathsf{K})$ and then use the fact (10.1) that restriction commutes with differentiation to adapt the preceding results to G. $\square$

It is worth remarking that, just as Ad x is a Lie algebra automorphism (not just an invertible linear transformation), so too ad x has a further property beyond being a linear transformation of $\mathfrak{g}$. It follows from (indeed, is equivalent to) the **Jacobi identity** for Lie algebras that ad x is a *derivation* of $\mathfrak{g}$.

It is also worth remarking that Ker ad is precisely the **center** $\mathfrak{z}(\mathfrak{g})$ of $\mathfrak{g}$ (those elements "commuting" with all elements of $\mathfrak{g}$). On the other hand, if $x \in Z(G)$, then Int $x = 1$ and hence Ad $x = 1$, i.e., $Z(G) \subset$ Ker Ad. It turns out that equality holds in characteristic 0 (13.4); otherwise there are exceptions (cf. Exercise 3).

The preceding theorem allows us to see more clearly the nature of the correspondence $G \mapsto \mathfrak{g}$ from the category of algebraic groups to the category of Lie algebras.

Corollary A. *Let H be a closed normal subgroup of the algebraic group G. Then $\mathfrak{h}$ is an* **ideal** *in $\mathfrak{g}$ (i.e., $[\mathsf{x}, \mathsf{y}] \in \mathfrak{h}$ whenever $\mathsf{x} \in \mathfrak{g}$, $\mathsf{y} \in \mathfrak{h}$).*

Proof. To say that H is normal is to say that all Int x $(x \in G)$ stabilize H; hence all Ad x $(x \in G)$ stabilize $\mathfrak{h}$. If we extend a basis of $\mathfrak{h}$ to a basis of $\mathfrak{g}$, the matrix of Ad x therefore has the form $\begin{pmatrix} * & * \\ \hline 0 & * \end{pmatrix}$, and clearly the matrix of $d(\text{Ad})x$ $(x \in \mathfrak{g})$ inherits this form. But $d(\text{Ad}) = \text{ad}$, so it follows that $[x, \mathfrak{h}] \subset \mathfrak{h}$ $(x \in \mathfrak{g})$. $\square$

As a variant of this corollary, we deduce:

Corollary B. *Let H be a closed subgroup of the algebraic group G, $N = N_G(H)$ its normalizer. Then $\mathfrak{n} \subset \mathfrak{n}_{\mathfrak{g}}(\mathfrak{h}) = \{x \in \mathfrak{g} | [x, \mathfrak{h}] \subset \mathfrak{h}\}$.*

Proof. We know that N is closed (Corollary 8.2), so $\mathfrak{n}$ is defined. Applying Corollary A to the normal subgroup H of N, we see that $\mathfrak{h}$ is an ideal in $\mathfrak{n}$, hence that $\mathfrak{n}$ normalizes $\mathfrak{h}$. $\square$

When char $K = 0$, the inclusion in the corollary is actually equality (Exercise 13.1). But this (like numerous other Lie algebra properties) has exceptions in prime characteristic, cf. Exercise 4.

10.5. Commutators

The commutator morphism $\gamma_x : G \to G$, defined by $\gamma_x(y) = y\,x\,y^{-1}x^{-1}$, has been seen to have differential (at e) $1 - \text{Ad } x : \mathfrak{g} \to \mathfrak{g}$. (Proposition 10.1 (c)).

Proposition. *Let A, B be closed subgroups of the algebraic group G, and let C be the closure of the group (A, B) generated by commutators. Then $\mathfrak{c}$ contains all elements $\mathsf{y} - \text{Ad } x(\mathsf{y})$, $\mathsf{x} - \text{Ad } y(\mathsf{x})$, $[\mathsf{x}, \mathsf{y}]$ $(x \in A, \mathsf{x} \in \mathfrak{a}, y \in B, \mathsf{y} \in \mathfrak{b})$.*

Proof. For $x \in A$, γ_x maps $B \to C$, so the differential $1 - \text{Ad } x$ maps $\mathfrak{b} \to \mathfrak{c}$. This yields all elements of the first type listed, and similarly for the second type. Next, for $\mathsf{x} \in \mathfrak{a}$, consider the morphism $\varphi : B \to \mathfrak{c}$ ($\mathfrak{c}$ viewed as an affine space) defined by $\varphi(y) = \mathsf{x} - \text{Ad } y(\mathsf{x})$. (We just showed that the image lies in $\mathfrak{c}$.) Since φ maps e to 0, we can compute $d\varphi_e : \mathfrak{b} \to \mathfrak{c}$ ($\mathfrak{c}$ being viewed as its own tangent space at 0). But $d(\text{Ad}) = \text{ad}$ (Theorem 10.4), so $d\varphi_e(\mathsf{y}) = -[\mathsf{y}, \mathsf{x}] = [\mathsf{x}, \mathsf{y}]$ lies in $\mathfrak{c}$. $\square$

Just as the sum and the negative in $\mathfrak{g}$ are infinitesimal analogues of the product and the inverse in G (10.1), the bracket $[\mathsf{x}, \mathsf{y}]$ is now seen to be analogous to the commutator in G (each measures the extent of departure from commutativity).

Corollary. *Let G be an algebraic group, H the closure of (G, G). Then $\mathfrak{h}$ includes $[\mathfrak{g}, \mathfrak{g}]$ ($=$ set of all linear combinations of elements $[\mathsf{x}, \mathsf{y}]$, $\mathsf{x}, \mathsf{y} \in \mathfrak{g}$).* $\square$

It will be seen later on (17.2) that (G, G) is always closed. The inclusion $\mathfrak{h} \supset [\mathfrak{g}, \mathfrak{g}]$ is actually an equality in case char $K = 0$, but may be strict otherwise (Exercise 3).

10.6. Centralizers

Consider further the commutator map $\gamma_x : G \to G$ for fixed $x \in G$. The fibre $\gamma_x^{-1}(e)$ is just $C_G(x)$. From the fact (10.1) that $(d\gamma_x)_e = 1 - \text{Ad } x$, we conclude that $\mathscr{L}(C_G(x))$ is contained in $\mathfrak{c}_\mathfrak{g}(x) = \{ \mathsf{x} \in \mathfrak{g} | \text{Ad } x(\mathsf{x}) = \mathsf{x} \}$ (the **infinitesimal centralizer** of x). As usual, this inclusion need not be an equality in prime characteristic (cf. Exercise 3). But in case $G = \text{GL}(n, \mathsf{K})$, the situation is always favorable. Indeed, the fixed points of $\text{Ad } x$ in $\mathfrak{g}$ are just the matrices commuting with x (Lemma B of (10.3)), so $C_G(x)$ consists of the invertible matrices in $\mathfrak{c}_\mathfrak{g}(x)$; in other words, $C_G(x)$ is a principal open subset of the affine space $\mathfrak{c}_\mathfrak{g}(x)$, so the dimensions are equal. Coupled with the inclusion just noted, we get:

Proposition. *Let G be an algebraic group, $x \in G$. Then $\mathscr{L}(C_G(x)) \subset \mathfrak{c}_\mathfrak{g}(x)$. If $G = \text{GL}(n, \mathsf{K})$, then equality holds.* $\quad\Box$

10.7. Automorphisms and Derivations

The purpose of this subsection is to make precise the following assertion: When an algebraic group G acts as a group of automorphisms, $\mathfrak{g}$ acts as a Lie algebra of derivations. This has already been observed in connection with the adjoint representation (10.4), for example.

We begin by examining more closely the actions of G on dual spaces and on tensor product spaces (8.2). For brevity, we shall denote by $x \cdot v$ the vector $\varphi(x)(v)$, when $\varphi : G \to \text{GL}(V)$ is a rational representation. Similarly, $\mathsf{x} \cdot v = d\varphi(\mathsf{x})(v)$, for $\mathsf{x} \in \mathfrak{g}$.

Proposition. *Let G be an algebraic group, and let $\varphi : G \to \text{GL}(V)$, $\psi : G \to \text{GL}(W)$ be rational representations.*

(a) G acts on V^* by the rule: $(x \cdot f)(v) = f(x^{-1} \cdot v)$, while $\mathfrak{g}$ acts by the rule: $(\mathsf{x} \cdot f)(v) = -f(\mathsf{x} \cdot v)$, where $x \in G$, $\mathsf{x} \in \mathfrak{g}$, $f \in V^*$, $v \in V$.

(b) G acts on $V \otimes W$ by the rule: $x \cdot (v \otimes w) = x \cdot v \otimes x \cdot w$, while $\mathfrak{g}$ acts by the rule: $\mathsf{x} \cdot (v \otimes w) = \mathsf{x} \cdot v \otimes w + v \otimes \mathsf{x} \cdot w$, where $x \in G$, $\mathsf{x} \in \mathfrak{g}$, and $v \otimes w$ is a typical generator of $V \otimes W$.

Proof. (a) The action of G is prescribed as in (8.2). If we fix a basis of V and write $\varphi(x)$ as a matrix, then it is clear that the matrix of x acting on V^* (relative to the dual basis) is just the transpose inverse matrix. Thanks to (10.1), the differential of $x \mapsto x^{-1}$ is $\mathsf{x} \mapsto -\mathsf{x}$, while the transposition map $\text{GL}(n, \mathsf{K}) \to \text{GL}(n, \mathsf{K})$ obviously has as differential the transposition map $\mathfrak{gl}(n, \mathsf{K}) \to \mathfrak{gl}(n, \mathsf{K})$ (the morphism in question being given by linear polynomials). So x acts on V^* as indicated.

(b) As remarked in (8.2), the action of G is defined (either via explicit choice of bases or via the universal property of tensor products). Fix bases $(v_1, \ldots, v_n)$ of V and $(w_1, \ldots, w_m)$ of W. Relative to these bases, let the action of a typical $x \in G$ be given by an $n \times n$ matrix (a_{ij}) and an $m \times m$ matrix

(b_{ij}). For $V \otimes W$ we take the basis elements $v_r \otimes w_s$ (in some order), so that G acts via a matrix representation $\tau: G \to \mathrm{GL}(mn, \mathsf{K})$. The matrix entry in the (i, j) row and (r, s) column is clearly $a_{ir}b_{js}$. τ factors as the composite of two morphisms $G \overset{\varphi \times \psi}{\longrightarrow} \mathrm{GL}(n, \mathsf{K}) \times \mathrm{GL}(m, \mathsf{K}) \to \mathrm{GL}(mn, \mathsf{K})$, the second being given by the coordinate recipe: $\mathsf{Z}_{ijrs} = \mathsf{X}_{ir}\mathsf{Y}_{js}$. Here we are just dealing with a morphism of affine $n^2 + m^2$-space into affine $(nm)^2$-space, so the differential is easy to compute using the method of (5.4). If (c_{ij}) and (d_{ij}) are typical elements of $\mathfrak{gl}(n, \mathsf{K})$ and $\mathfrak{gl}(m, \mathsf{K})$, the differential takes the pair of them to the matrix whose entry in the (i, j) row and (r, s) column is $\delta_{js}c_{ir} + \delta_{ir}d_{js}$. But this is precisely the rule asserted in the statement of the proposition. $\quad\square$

Part (b) of the proposition extends at once to tensor products involving three or more factors. In particular, the action of $\mathrm{GL}(V)$ on the *tensor algebra* $\mathfrak{T}(V)$ as a group of algebra automorphisms has as infinitesimal analogue an action of $\mathfrak{gl}(V)$ on $\mathfrak{T}(V)$ as a Lie algebra of derivations. Since $\mathrm{GL}(V)$ leaves stable the ideal of $\mathfrak{T}(V)$ which is factored out to obtain the *exterior algebra* $\wedge V$ (1.8), we obtain analogous actions of $\mathrm{GL}(V)$ and $\mathfrak{gl}(V)$ on the latter. (This will be important in (11.1) below.)

In order to reach the general assertion made at the outset, we have to note that whenever a closed subgroup of $\mathrm{GL}(V)$ fixes a nonzero vector $v \in V$, its Lie algebra kills that vector. This is easy to verify (cf. Exercise 2). In either case, we refer to the vector as an **invariant**.

Corollary. *Let $\mathfrak{A}$ be a finite dimensional K-algebra (not necessary associative), and let G be a closed subgroup of $\mathrm{GL}(\mathfrak{A})$ consisting of algebra automorphisms. Then $\mathfrak{g}$ consists of derivations of $\mathfrak{A}$.*

Proof. To give $\mathfrak{A}$ the structure of K-algebra is just to specify a bilinear map $\mathfrak{A} \times \mathfrak{A} \to \mathfrak{A}$, or equivalently, a K-linear map $\mathfrak{A} \otimes \mathfrak{A} \to \mathfrak{A}$. But there is a standard identification (cf. (8.2)) of $\mathrm{Hom}(V,W)$ with $V^* \otimes W$, so the K-algebra structure on $\mathfrak{A}$ is given by an element $t \in (\mathfrak{A} \otimes \mathfrak{A})^* \otimes \mathfrak{A} \cong \mathfrak{A}^* \otimes \mathfrak{A}^* \otimes \mathfrak{A}$. For $x \in \mathrm{GL}(\mathfrak{A})$ to be an automorphism of A, it is necessary and sufficient that t be an invariant of x (as is easy to check). Therefore, the preceding remark shows that t must be an invariant of $\mathfrak{g}$. We just have to unpack this a bit to conclude that $\mathfrak{g}$ consists of derivations.

From the proposition and the identifications just made, we deduce that the action of $\mathsf{x} \in \mathfrak{gl}(\mathfrak{A})$ on $\mathrm{Hom}(\mathfrak{A} \otimes \mathfrak{A}, \mathfrak{A})$ is given by the explicit rule: $(\mathsf{x} \,.\, f)(v \otimes w) = \mathsf{x} \,.\, (f(v \otimes w)) - f(\mathsf{x} \,.\, v \otimes w + v \otimes \mathsf{x} \,.\, w)$, where $f \in \mathrm{Hom}(\mathfrak{A} \otimes \mathfrak{A}, \mathfrak{A})$, $v, w \in \mathfrak{A}$. In particular, to say that $\mathsf{x} \,.\, t = 0$ is just to say that $\mathsf{x} \,.\, (vw) = (\mathsf{x} \,.\, v)w + v(\mathsf{x} \,.\, w)$, where the algebra product is indicated by juxtaposition. $\quad\square$

Exercises

1. Let G be a closed subgroup of $\mathrm{GL}(n, \mathsf{K})$. If G leaves stable a subspace W of K^n, prove that $\mathfrak{g} \subset \mathfrak{gl}(n, \mathsf{K})$ does likewise. Converse?

2. If all elements of a closed subgroup G of GL(n, K) fix a vector in K^n, all elements of $\mathfrak{g}$ send that vector to 0.

3. Let char K $= p > 0$. The following example shows that Ker Ad may be larger than $Z(G)$: $G \subset$ GL(3, K) consists of all matrices $\begin{pmatrix} a & 0 & 0 \\ 0 & a^p & b \\ 0 & 0 & 1 \end{pmatrix}$, $a \neq 0$.

   It is a closed connected, noncommutative group having dimension 2. The Lie algebra $\mathfrak{g}$ consists of all matrices $\begin{pmatrix} a & 0 & 0 \\ 0 & 0 & b \\ 0 & 0 & 0 \end{pmatrix}$ $(a, b \in$ K), and is commutative. Morever, Ker Ad $= G$.

4. Let char K $= 2$, $G =$ SL(2, K), $B =$ T(2, K) $\cap$ G. Verify that $N_G(B) = B$, whereas $\mathfrak{n}_\mathfrak{g}(\mathfrak{b}) = \mathfrak{g}$ (cf. Corollary B of Theorem 10.4).

5. Let H be a closed subgroup of the algebraic group G, $C = C_G(H)$. Prove that $\mathfrak{c} \subset \mathfrak{c}_\mathfrak{g}(\mathfrak{h}) = \{x \in \mathfrak{g} | [x, \mathfrak{h}] = 0\}$.

6. Let H be a closed subgroup of the algebraic group G, $x \in G$. Prove that Ad $x(\mathfrak{h}) = \mathscr{L}$ (Int $x(H)$).

7. Describe explicitly the matrices in GL(n, K) (resp. $\mathfrak{gl}(n$, K)) which commute with a given diagonal matrix (cf. Proposition 10.6).

Notes

The example in Exercise 3 is taken from Chevalley [4, pp. 145–146].

Chapter IV

Homogeneous Spaces

11. Construction of Certain Representations

Given an algebraic group G and a closed subgroup H, how can the homogeneous space G/H (or $H\backslash G$) be endowed with a "reasonable" structure of variety? This is a subtle question, in part because one is forced to accept non-affine varieties (for certain H). In this section we show how to construct a G-orbit in projective space having H as isotropy group, and then refine the construction in case H is a normal subgroup. In the next section we shall verify that the variety so constructed has the properties demanded of a "quotient".

11.1. Action on Exterior Powers

We shall need some simple facts from linear algebra.

Recall (10.7) how $\mathrm{GL}(n, \mathrm{K})$ and $\mathfrak{gl}(n, \mathrm{K})$ act on exterior powers of the vector space $W = \mathrm{K}^n$. If M is a d-dimensional subspace of W, it is especially useful to look at the action on $L = \wedge^d M$, which is a 1-dimensional subspace of $V = \wedge^d W$.

Lemma. *With the notation just introduced, let* $x \in \mathrm{GL}(W)$, $\mathsf{x} \in \mathfrak{gl}(W)$. *Then*

    (a) $(\wedge^d x)L = L$ *if and only if* $x(M) = M$,

    (b) $(d\wedge^d)(\mathsf{x})L \subset L$ *if and only if* $\mathsf{x}(M) \subset M$.

Proof. In each case, the "if" part is clear. For (a), choose a basis $(w_1, ., w_n)$ of W such that M is spanned by $w_1, ., w_d$ and $x(M)$ by $w_{l+1}, ., w_{l+d}$ for some $l \geqslant 0$. We just have to show that $l = 0$. Now $w_1 \wedge \cdots \wedge w_d$ spans L, and is sent by $\wedge^d x$ to a (nonzero) multiple of itself (by assumption). But clearly $\wedge^d x$ sends it to a multiple of $w_{l+1} \wedge \cdots \wedge w_{l+d}$, forcing $l = 0$.

For (b), choose a basis of M so that x maps the span of the first part of the basis back into M and sends all other vectors in M outside M. Then construct some $\mathsf{y} \in \mathfrak{gl}(W)$ agreeing with x on the first part of this basis, but acting as 0 on the remainder of the basis for M. Evidently y leaves M stable, so $(d\wedge^d)(\mathsf{y})$ leaves L stable. Therefore $\mathsf{x} - \mathsf{y}$ still verifies the hypothesis of (b). If we can show that $\mathsf{x} - \mathsf{y}$ stabilizes M, it will follow that x does. So we may replace x by $\mathsf{x} - \mathsf{y}$. This allows us to assume that M intersects $\mathsf{x}(M)$ trivially. Now let $w_1, ., w_d$ be a basis of M, chosen so that $\mathsf{x}(w_1), \ldots, \mathsf{x}(w_c)$ form a basis of $\mathsf{x}(M)$, while $\mathsf{x}(w_{c+1}) = \cdots = \mathsf{x}(w_d) = 0$. By definition, $(d\wedge^d)(\mathsf{x})(w_1 \wedge \cdots \wedge w_d) = \sum_i w_1 \wedge \cdots \wedge \mathsf{x}(w_i) \wedge \cdots \wedge w_d$. This is by assumption a multiple of $w_1 \wedge \cdots \wedge w_d$.

But the vectors $w_1 \wedge \cdots \wedge \mathsf{x}(w_i) \wedge \cdots \wedge w_d$ are linearly independent, or else 0, and none is equal to $w_1 \wedge \cdots \wedge w_d$. Conclusion: $\mathsf{x}(w_i) = 0$ $(1 \leqslant i \leqslant d)$, so x stabilizes M (by sending M to 0). $\square$

11.2. A Theorem of Chevalley

Theorem. *Let G be an algebraic group, H a closed subgroup. Then there is a rational representation $\varphi: G \to \mathrm{GL}(V)$ and a one dimensional subspace L of V such that:*

$$H = \{x \in G \,|\, \varphi(x)L = L\},$$
$$\mathfrak{h} = \{\mathsf{x} \in \mathfrak{g} \,|\, d\varphi(\mathsf{x})L \subset L\}.$$

Proof. Let I be the ideal in $\mathsf{K}[G]$ vanishing on H. Then I is finitely generated, and the span of a finite generating set lies in a *finite dimensional* subspace W of $\mathsf{K}[G]$ stable under all ρ_x $(x \in G)$ (Proposition 8.6). Set $M = W \cap I$ (so M generates I). Notice that M is stable under all ρ_x $(x \in H)$, since $H = \{x \in G \,|\, \rho_x I = I\}$ (Lemma 8.5). In turn, M is stable under all $*\mathsf{x}(\mathsf{x} \in \mathfrak{h})$ (Proposition 10.2). We claim, in fact, that $H = \{x \in G \,|\, \rho_x M = M\}$, $\mathfrak{h} = \{\mathsf{x} \in \mathfrak{g} \,|\, M*\mathsf{x} \subset M\}$. Say $\rho_x M = M$ $(x \in G)$. Since I is the ideal of $A = \mathsf{K}[G]$ generated by M, we have $\rho_x I = \rho_x(MA) = \rho_x(M)\rho_x(A) = MA = I$, forcing $x \in H$ (Lemma 8.5). Say $M*\mathsf{x} \subset M$ $(\mathsf{x} \in \mathfrak{g})$. Then the product rule for derivations yields: $I*\mathsf{x} = (MA)*\mathsf{x} \subset (M*\mathsf{x})A + M(A*\mathsf{x}) \subset MA = I$, and therefore $\mathsf{x} \in \mathfrak{h}$ (Lemma 9.4).

What we have just obtained is a good first approximation to the theorem. It remains only to compress M to a line, which can be achieved by passing to $V = \wedge^d W$, $L = \wedge^d M$ $(d = \dim M)$. Let $\varphi: G \to \mathrm{GL}(V)$ be the dth exterior power of the representation constructed above. The desired characterizations of H, $\mathfrak{h}$ then follow at once from Lemma 11.1. $\square$

This theorem of Chevalley is useful not only in the present context but also in the study of representations of G, where it helps to prove the existence of certain key representations, and in several other situations.

11.3. Passage to Projective Space

As in (11.2), let H be a closed subgroup of the algebraic group G. The choice of the rational representation $\varphi: G \to \mathrm{GL}(V)$ and the line L in V serves to identify H with the isotropy group in G of the point $[L]$ in $\mathbf{P}(V)$ corresponding to L. In this way the homogeneous space G/H is made to correspond (set-theoretically) to the orbit in $\mathbf{P}(V)$ of the point $[L]$. According to (8.3), such an orbit is itself a variety (*quasiprojective*, being open in its closure).

We have to make more explicit the infinitesimal behavior of the morphism $\pi: G \to \mathbf{P}(V)$ defined by $x \mapsto \varphi(x)[L] = [\varphi(x)L]$. The canonical map $\sigma: V - \{0\} \to \mathbf{P}(V)$ has as differential at v $(=$ any vector spanning $L)$ a linear map whose kernel is precisely L (5.4). Fix v, and consider the orbit map

$\omega : \mathrm{GL}(V) \to V$ defined by $\omega(x) = x(v)$. Relative to a basis $v = v_1, v_2, \ldots, v_n$ of V, ω just assigns to a matrix its first column. In particular, ω is given by *linear* polynomials, so $d\omega$ is given in the same way: $d\omega_e(\mathsf{x}) = \mathsf{x}(v)$ ($\mathsf{x} \in \mathfrak{gl}(V)$).

Combining these observations, we conclude that $d\pi_e$ sends $\mathsf{x} \in \mathfrak{g}$ to the image of $d\varphi(\mathsf{x})(v)$ in the tangent space to $\mathbf{P}(V)$ at $[L]$. In particular, thanks to Theorem 11.2, $\mathfrak{h}$ is precisely Ker $d\pi_e$. By dimension comparison, $d\pi_e$ is surjective, and therefore (5.5) the morphism π is *separable*. Notice too that π is G-equivariant (G acting on itself by left multiplication), cf. (8.2) and Exercise 8.6.

11.4. Characters and Semi-Invariants

In order to refine the preceding construction when H is a normal subgroup of G, we have to introduce the notion of **character** of an algebraic group G. This is by definition any morphism of algebraic groups $\chi : G \to \mathbf{G}_m$, e.g., $\det : \mathrm{GL}(n, \mathsf{K}) \to \mathbf{G}_m$. If χ_1, χ_2 are two characters of G, so is their "product" $\chi_1 \chi_2$, defined by $(\chi_1 \chi_2)(x) = \chi_1(x)\chi_2(x)$. Therefore, the set $\mathrm{X}(G)$ of all characters has the structure of commutative group. It should be noted that some groups have no nontrivial characters, e.g., $\mathbf{G}_a$ and $\mathrm{SL}(n, \mathsf{K})$ (Exercise 2), while other groups have many, e.g., $\mathrm{D}(n, \mathsf{K})$. This will be explored in §16.

Characters arise in connection with linear representations, as follows. Let G be a closed subgroup of $\mathrm{GL}(V)$. For each $\chi \in \mathrm{X}(G)$, define $V_\chi = \{v \in V \,|\, x \,.\, v = \chi(x)v$ for all $x \in G\}$. Evidently V_χ is a G-stable subspace of V (possibly 0). Any nonzero element of V_χ is called a **semi-invariant** of G, of **weight** χ. Conversely, if v is any nonzero vector which spans a G-stable line in V, then it is clear that $x \,.\, v = \chi(x)v$ defines a character χ of G.

More generally, if $\varphi : G \to \mathrm{GL}(V)$ is a rational representation, then the semi-invariants of G in V are by definition those of $\varphi(G)$. Notice that composition with φ induces an injective group homomorphism $\mathrm{X}(\varphi(G)) \to \mathrm{X}(G)$, so that V_χ may be defined in the obvious way for any $\chi \in \mathrm{X}(G)$ coming from a character of $\varphi(G)$.

Lemma. *Let $\varphi : G \to \mathrm{GL}(V)$ be a rational representation. Then the subspaces V_χ, $\chi \in \mathrm{X}(G)$, are linearly independent; in particular, only finitely many of them are nonzero.*

Proof. Otherwise we can choose minimal $n \geqslant 2$ and nonzero vectors $v_i \in V_{\chi_i}$ (for distinct χ_i, $1 \leqslant i \leqslant n$) such that $v_1 + \cdots + v_n = 0$. Since the χ_i are distinct, $\chi_1(x) \neq \chi_2(x)$ for some $x \in G$. But $0 = \varphi(x)(\sum v_i) = \sum \chi_i(x)v_i$, so $\sum \chi_1(x)^{-1}\chi_i(x)v_i = 0$. The coefficient of v_2 is different from 1; so we can subtract this equation from the equation $\sum v_i = 0$ to obtain a nontrivial dependence involving $\leqslant n - 1$ characters, contradicting the choice of n. $\quad\square$

Let us vary the situation slightly by assuming that N is a closed normal subgroup of G and by limiting our attention to the spaces V_χ for $\chi \in \mathrm{X}(N)$. We claim that each element of $\varphi(G)$ maps V_χ into some $V_{\chi'}$. For this we may as well assume that $G \subset \mathrm{GL}(V)$. If $x \in G$, $y \in N$, then $yx \cdot v = x(x^{-1}yx) \cdot v = x \cdot \chi(x^{-1}yx)v = \chi(x^{-1}yx)x \cdot v$ $(v \in V_\chi)$, and the function $y \mapsto \chi(x^{-1}yx)$ is evidently a character χ' of N. So x maps V_χ into $V_{\chi'}$. Indeed, this calculation shows that G acts naturally on $X(N)$, so we could write $\chi' = x \cdot \chi$.

11.5. Normal Subgroups

Theorem. *Let G be an algebraic group, N a closed normal subgroup of G. Then there is a rational representation $\psi : G \to \mathrm{GL}(W)$ such that $N = \mathrm{Ker}\ \psi$ and $\mathfrak{n} = \mathrm{Ker}\ d\psi$.*

Proof. Use Theorem 11.2 to construct a morphism $\varphi : G \to \mathrm{GL}(V)$ and a line L in V whose stabilizer in G (resp. $\mathfrak{g}$) is N (resp. $\mathfrak{n}$). Since N just acts on L by scalar multiplications, there is an associated character $\chi_0 \in X(N)$. Consider the sum in V of all nonzero subspaces V_χ $(\chi \in X(N))$: thanks to Lemma 11.4, this sum is direct (so only finitely many χ appear) and of course includes L. Moreover, we saw in (11.4) that $\varphi(G)$ permutes the various V_χ, since N is normal in G. So it is harmless to assume that V itself is the sum of the V_χ, N acting on each of these by scalar multiplications.

Now let W be the subspace of $\mathrm{End}\ V$ consisting of those endomorphisms which stabilize each V_χ, $\chi \in X(N)$. There is a natural isomorphism $W \cong \coprod \mathrm{End}\ V_\chi$. (Think of W as consisting of all block diagonal matrices with blocks of sizes $\dim V_\chi$.) If we view $\mathrm{End}\ V$ as $\mathfrak{gl}(V)$, then $\mathrm{GL}(V)$ acts via the adjoint representation, $\mathrm{Ad}\ x$ being conjugation by x (Proposition 10.3). Notice that the subgroup $\varphi(G)$ stabilizes W, since $\varphi(G)$ permutes the V_χ and W stabilizes each. Accordingly, we obtain a group homomorphism $\psi : G \to \mathrm{GL}(W)$. ψ is just φ, followed by Ad, followed by restriction to the subspace W of $\mathfrak{gl}(V)$; so ψ is a rational representation.

It remains to check $\mathrm{Ker}\ \psi$ and $\mathrm{Ker}\ d\psi$. If $x \in N$, then $\varphi(x)$ acts as a scalar on each V_χ, so conjugating by $\varphi(x)$ has no effect on W, i.e., $x \in \mathrm{Ker}\ \psi$. Conversely, let $x \in G$, $\psi(x) = e$. This means that $\varphi(x)$ stabilizes each V_χ and commutes with $\mathrm{End}\ V_\chi$. But the center of $\mathrm{End}\ V_\chi$ is the set of scalars, so $\varphi(x)$ acts on each V_χ as a scalar. In particular, $\varphi(x)$ stabilizes $L \subset V_{\chi_0}$, forcing $x \in N$ (Theorem 11.2). Therefore, $N = \mathrm{Ker}\ \psi$.

For $d\psi$ the argument is similar, using the fact that $d(\mathrm{Ad}) = \mathrm{ad}$ (Theorem 10.4). If $\mathrm{x} \in \mathfrak{n}$, then $d\varphi(\mathrm{x})$ acts (via $d\chi$) as a scalar on each V_χ, so that $\mathrm{ad}(d\varphi(\mathrm{x}))$ is 0 on W, and $\mathrm{x} \in \mathrm{Ker}\ d\psi$. Conversely, if $\mathrm{x} \in \mathrm{Ker}\ d\psi$, then $\mathrm{ad}(d\varphi(\mathrm{x}))$ is 0 on W, so $d\varphi(\mathrm{x})$ stabilizes each V_χ and commutes with $\mathrm{End}\ V_\chi$, hence acts on V_χ as a scalar. In particular, $d\varphi(\mathrm{x})L \subset L$, forcing $\mathrm{x} \in \mathfrak{n}$. $\square$

With the aid of this theorem, we can give the abstract group G/N the structure of (affine) algebraic group by identifying it with $\psi(G)$. However, some further work is needed in order to guarantee that this procedure is independent of the choices made and leads to good universal properties. In any case, the construction implies that the Lie algebra of $\psi(G)$ will be the quotient Lie algebra $\mathfrak{g}/\mathfrak{n}$.

Exercises

1. Prove that $X(\mathbf{G}_m)$ is an infinite cyclic group. What about the character group of $D(n, \mathsf{K})$?
2. Prove that $\mathbf{G}_a$ and $SL(n, \mathsf{K})$ have no nontrivial characters.

Notes

The material in this section (and the next) is drawn from Borel [4, Chapter II].

12. Quotients

12.1. Universal Mapping Property

Given a closed subgroup H of an algebraic group G, there is a separable, G-equivariant morphism $\pi: G \to Y$ (Y a G-orbit in some $\mathbf{P}(V)$) whose fibres are precisely the cosets xH (11.3). In case H is normal in G, Y may be taken to be an (affine) algebraic group and π a group homomorphism (11.5). The problem now is to show that the pair (π, Y) satisfies the obvious requirement:
(*). Given a variety X and a morphism $\varphi: G \to X$ whose nonempty fibres are unions of cosets xH, there exists a unique morphism $\psi: Y \to X$ such that $\psi \circ \pi = \varphi$.

This condition, applied to any other $\pi': G \to Y'$ we might have constructed by the method of §11, would show quickly that Y' has to be isomorphic (as variety) to Y by a unique ψ for which $\psi \circ \pi = \pi'$. This in turn would entitle us to call Y the **homogeneous space** G/H and to call π the **canonical morphism**. (An entirely analogous argument leads, of course, to the space $H\backslash G$ of cosets Hx.) It just remains to prove:

Theorem. *The pair (π, Y) satisfies the condition* (*).
For the proof, we make use of the facts about morphisms developed in (4.5) and (4.6), as well as a result from (5.3). The details will be given in (12.2), (12.3). But first we reformulate (*). The uniqueness of the required morphism $\psi: Y \to X$ (if it exists) will be obvious, since $\pi(x)$ has to be sent to $\varphi(x)$ for all $x \in G$ (and π is surjective). In order to construct ψ, three aspects of Y have to be dealt with (cf. (2.3)).

(1) Y as a *set*. By construction, Y corresponds to the set of cosets xH and π to the map $x \mapsto xH$. Thanks to the hypothesis on φ, this insures the existence of a set-theoretic map $\psi: Y \to X$ such that $\psi \circ \pi = \varphi$.

(2) Y as a *topological space*. The mapping $\psi: Y \to X$ in (1) must be shown to be continuous. For this, it is enough to show that Y has the quotient topology, i.e., that U is open in Y if (and only if) $\pi^{-1}(U)$ is open in G. Because we are dealing with a group, this is equivalent to the apparently stronger requirement that π be an *open* mapping. Indeed, let U be open in G. Then

$\pi^{-1}(\pi(U))$ is the union of those cosets which meet U; but this is the same as the union of all H-translates of U, which is a union of open sets and therefore open. So $\pi(U)$ must be open (if Y has the quotient topology). It will be verified in (12.2) that π is in fact open.

(3) Y as a *space with sheaf of functions*. Besides being continuous, ψ must induce homomorphisms $\mathcal{O}_X(U) \to \mathcal{O}_Y(\psi^{-1}(U))$ for all open sets $U \subset X$. The hypothesis on φ implies that for all such U, φ^* maps $\mathcal{O}_X(U)$ into the regular functions on $\varphi^{-1}(U)$ which take constant value on each coset xH. So it would be enough to know $\pi^*\mathcal{O}_Y(\psi^{-1}U)$ consists of *all* such functions: then $\psi^*(\mathcal{O}_X(U))$ would automatically be included in $\mathcal{O}_Y(\psi^{-1}U)$, thanks to the way ψ is defined. This will be verified in (12.3).

The preceding remarks reduce the proof of the theorem to checking two properties of (π, Y). We claim that it is enough to do this under the further assumption that G is *connected*. Indeed, suppose the connected case is done. Let $H' = G^{\circ} \cap H$, $Y' = $ orbit under G° of some point $y \in Y$, $\pi' : G^{\circ} \to Y'$ the orbit map. Since H' has finite index in H, its Lie algebra is also $\mathfrak{h}$. By the original construction of Y, H' is the full isotropy group of y in G°, while its Lie algebra $\mathfrak{h}$ is the full isotropy algebra of y in $\mathfrak{g} = \mathscr{L}(G^{\circ})$. The connected case then shows that the pair (π', Y') has the two properties discussed above. Now the G-equivariance of π allows us to transfer these properties to the pair (π, Y).

12.2. Topology of Y

Theorem 4.5 provides a criterion for a dominant morphism of irreducible varieties to be open. So we just have to verify the hypothesis there for the case $\pi : G \to Y$. First select an open subset $U \subset Y$ as in Theorem 4.3. According to that theorem, for each closed irreducible $W \subset Y$ which meets U, the components of $\pi^{-1}(W)$ meeting $\pi^{-1}(U)$ all have the "correct" dimension. Since G acts transitively on Y, the G-translates of U cover Y (and the G-translates of $\pi^{-1}(U)$ cover G). This makes it clear that *all* components of *all* $\pi^{-1}(W)$ have the correct dimension, so (4.5) applies.

12.3. Functions on Y

First we want to show that a polynomial function $f \in \mathsf{K}[G]$ which is constant on the cosets xH is the pullback under π^* of a rational function on Y. (Recall that G is being assumed to be connected, so that $\mathsf{K}(Y)$ is available.) For this, consider the set-up: $G \xrightarrow{\varphi} Y \times \mathbf{A}^1 \xrightarrow{\mathrm{pr}_1} Y$, where $\varphi(x) = (\pi(x), f(x))$. The composite is just π. If X is the closure in $Y \times \mathbf{A}^1$ of $\varphi(G)$, then X is irreducible and pr_1 induces a surjective morphism $\psi : X \to Y$. By construction, π is *separable*, so the same is true of ψ.

Now $\varphi(G)$ contains a (dense) open subset of X (4.3). Since f is constant on fibres of π, the restriction of ψ to this open set is clearly injective, as well as dominant and separable. So Theorem 4.6 may be applied. The conclusion is that ψ^* maps $\mathsf{K}(Y)$ isomorphically *onto* $\mathsf{K}(X)$. But in the above set-up,

$\mathrm{pr}_2: Y \times \mathbf{A}^1 \to \mathbf{A}^1$ induces on X a morphism $g: X \to \mathbf{A}^1$ (i.e., a regular function). Since $g \in \mathsf{K}(X)$, there exists a rational function $h \in \mathsf{K}(Y)$ for which $g = \psi^* h$. Finally, notice that $\varphi^* g = \varphi^* \psi^* h = \pi^* h$ agrees everywhere on G with the function f. So $f = \pi^* h$, as desired.

Next we want to show that the rational function $h \in \mathsf{K}(Y)$ just constructed is a regular function on Y. Since G acts transitively on Y, and Y has some simple points (5.2), all points of Y are simple. So Theorem 5.3B shows that unless h is everywhere defined on Y, $1/h$ is defined somewhere and then takes the value 0. But then $\pi^*(1/h) = 1/f$ must also take the value 0, which is absurd (since $f \in \mathsf{K}[G]$).

Notice that the preceding argument applies almost verbatim when we replace Y by an open subset U (whose points are again all simple!) and G by $\pi^{-1}(U)$. So in fact we have shown that each $f \in \mathcal{O}_G(\pi^{-1}(U))$ constant on cosets of H is the pullback $\pi^* h$ for some $h \in \mathcal{O}_Y(U)$. This completes the proof of Theorem 12.1. $\quad\square$

12.4. Complements

(1) The proof of Theorem 12.1 involved in effect the verification of the following assertion:

If G is an algebraic group, Y a variety on which G acts, $\pi: G \to Y$ a surjective, separable, G-equivariant morphism whose fibres are the cosets xH (H a closed subgroup of G), then the variety Y is determined uniquely up to isomorphism.

This may be applied in particular when $H = e$, to complete the discussion of $G = \mathrm{PGL}(2, \mathsf{K})$ in (6.4). Recall the orbit map $\varphi: G \to Y \subset \mathbf{P}^1 \times \mathbf{P}^1 \times \mathbf{P}^1$, which was shown to be bijective and separable. The conclusion now is that φ is an isomorphism.

(2) Given G and a closed subgroup H, it is natural to ask when the variety G/H is affine or projective or neither. When H is normal, G/H has been seen to be affine. On the other hand, important examples in which G/H is projective will emerge in Chapter VIII (cf. Exercise 3 below). Quite a bit is known about the nature of G/H in general (see Notes).

12.5. Characteristic 0

In case char $\mathsf{K} = 0$, questions of separability do not arise. This has important implications for the algebraic group—Lie algebra correspondence, to be explored in the following chapter.

Theorem. *Let G be an algebraic group, with closed subgroups A, B. If* char $\mathsf{K} = 0$, *then* $\mathfrak{a} \cap \mathfrak{b} = \mathcal{L}(A \cap B)$.

Proof. Let $\pi: G \to G/B$ be the canonical morphism. So Ker $d\pi_e = \mathfrak{b}$. Consider the restriction $\pi': A \to \pi(A)$. This morphism has as fibres the cosets

$x(A \cap B)$, and is automatically separable (char $\mathsf{K} = 0$). The remark (1) in (12.4) allows us to identify π' with the canonical morphism $A \to A/(A \cap B)$. Therefore Ker $d\pi'_e = \mathscr{L}(A \cap B)$. But evidently Ker $d\pi'_e = \mathfrak{a} \cap$ Ker $d\pi_e = \mathfrak{a} \cap \mathfrak{b}$. $\square$

Exercises

1. Let G be an algebraic group with closed subgroups A, B. Prove that $\mathfrak{a} \cap \mathfrak{b} = \mathscr{L}(A \cap B)$ if and only if the restriction to A of the canonical morphism $\pi: G \to G/B$ is again separable.
2. Let G be a connected algebraic group, H a closed subgroup. Then H acts naturally on $\mathsf{K}(G)$ as a group of automorphisms, and $\mathsf{K}(G/H)$ is isomorphic to the subfield $\mathsf{K}(G)^H$ fixed by H.
3. Prove that $GL(n, \mathsf{K})/T(n, \mathsf{K})$ is isomorphic to the flag variety $\mathfrak{F}(\mathsf{K}^n)$ (1.8). Use this to compute the dimension of the latter (cf. Exercise 3.2).
4. Let X, Y be varieties on which an algebraic group G acts transitively, and let $\varphi: X \to Y$ be a surjective, G-equivariant morphism. Prove that φ is an open map.

Notes

Concerning the nature of G/H, see Bialynicki-Birula [1], Borel [3, pp. 45–49], Matsushima [1], Rosenlicht [4], as well as the references for Hilbert's 14th Problem listed in the Notes to §14 below.

Chapter V

Characteristic 0 Theory

This chapter constitutes a brief digression into the case when char $K = 0$. None of the results proved here will be essential elsewhere in the text.

In §13 only a mild acquaintance with Lie algebras is presupposed, but in §14 the reader is expected to know some of the main theorems about semi-simple Lie algebras. A few notions will be met here which make their formal appearance somewhat later in the text.

13. Correspondence Between Groups and Lie Algebras

13.1. The Lattice Correspondence

Some examples in the previous chapters (e.g., Exercises 10.3, 10.4) have indicated that algebraic groups and their Lie algebras can behave in divergent ways in prime characteristic. More evidence of this will surface in Chapter VI. Inseparability of certain field extensions is to blame; this being absent in characteristic 0, the Lie algebra reflects quite faithfully what goes on in the group.

Separability of morphisms implies on the one hand that whenever $\varphi : G \to G'$ is a morphism of algebraic groups, Ker $d\varphi = \mathscr{L}(\mathrm{Ker}\ \varphi)$. On the other hand, the separability of certain quotient morphisms yields Theorem 12.5, affirming that the Lie algebra of an intersection of closed subgroups is the intersection of the Lie algebras. These facts are the key to everything we prove in the present section.

Theorem. *Let G be a connected algebraic group. The correspondence $H \mapsto \mathfrak{h}$ is 1–1 and inclusion preserving between the collection of closed connected subgroups H of G and the collection of their Lie algebras, regarded as subalgebras of $\mathfrak{g}$.*

Proof. Use Theorem 12.5. $\square$

A subalgebra $\mathscr{L}(H) = \mathfrak{h}$ of $\mathfrak{g}$ is referred to as **algebraic**. It is not an easy matter to determine precisely which subalgebras of $\mathfrak{g}$ come under this heading, e.g., when $G = \mathrm{GL}(n, K)$, $\mathfrak{g} = \mathfrak{gl}(n, K)$. However, quite a bit is known (see Notes below).

The theorem is deficient in another respect: it does not spell out the behavior of normal subgroups. In any characteristic, it is true that the Lie algebra of a closed normal subgroup of G is an ideal in $\mathfrak{g}$ (Corollary 10.4A). We shall prove shortly that the converse is true in characteristic 0, provided we limit our attention to *connected* groups (why is this necessary?).

87

13.2. Invariants and Invariant Subspaces

Suppose H is a closed subgroup of some $GL(V)$. It is true in any characteristic (cf. Exercises 10.1, 10.2) that $\mathfrak{h}$ leaves stable the subspaces of V stabilized by H and kills the vectors which are fixed by H. Both of these statements can be strengthened in the following way. Consider a subspace W of V, and let $G_W = \{x \in GL(V)| x(W) = W\}$, $\mathfrak{g}_W = \{x \in \mathfrak{gl}(V)| x(W) \subset W\}$. Similarly, for $v \in V$, define $G_v = \{x \in GL(V)| x(v) = v\}$ and $\mathfrak{g}_v = \{x \in \mathfrak{gl}(V)| x(v) = 0\}$. The groups G_W, G_v are closed in $GL(V)$ (8.2), while $\mathfrak{g}_W$ and $\mathfrak{g}_v$ are Lie subalgebras of $\mathfrak{gl}(V)$. We claim that (in any characteristic): $\mathscr{L}(G_W) = \mathfrak{g}_W$ and $\mathscr{L}(G_v) = \mathfrak{g}_v$. The proofs are based on dimension comparisons. Take a basis $v_1, \ldots, v_n$ of V such that $v_1, \ldots, v_m$ span W (resp. such that $v_1 = v$). Then $\mathfrak{g}_W$ (resp. $\mathfrak{g}_v$) may be identified with a linear variety of dimension $n(n - m) + m^2$ (resp. $n^2 - n$) in $\mathbf{A}^{n^2}$, and G_W (resp. G_v) with an affine open subset of this linear variety. (Strictly speaking, in the case of G_v one has to translate by the identity matrix.) Since $\mathscr{L}(G_W) \subset \mathfrak{g}_W$, $\mathscr{L}(G_v) \subset \mathfrak{g}_v$ (by the exercises cited), equality must hold.

Theorem. *Let H be any closed subgroup of $GL(V)$. If W is a subspace of V (resp. $v \in V$), define closed subgroups G_W, G_v of $GL(V)$ as above, and let $H_W = H \cap G_W$, $H_v = H \cap G_v$. Then $\mathscr{L}(H_W) = \mathfrak{h} \cap \mathfrak{g}_W$, $\mathscr{L}(H_v) = \mathfrak{h} \cap \mathfrak{g}_v$.*

Proof. Apply the previous remarks along with Theorem 12.5. $\square$

Notice that if $\varphi : H \to GL(V)$ is a rational representation, the fact that $d\varphi(\mathfrak{h}) = \mathscr{L}(\varphi(H))$ allows us to apply the theorem in the form: H and $\mathfrak{h}$ leave stable the same subspaces of V (resp. have the same invariants in V).

Invariants were used in (10.7) to show that the Lie algebra of the automorphism group of a finite dimensional K-algebra consists of derivations. This can be strengthened in characteristic 0.

Corollary. *Let $\mathfrak{A}$ be a finite dimensional K-algebra, $H = $ Aut $\mathfrak{A}$. Then $\mathfrak{h} = $ Der $\mathfrak{A}$, the full algebra of derivations.*

Proof. The proof of Corollary 10.7 shows that $x \in GL(A)$ is an automorphism if and only if it fixes a certain tensor $t \in \mathfrak{A}^* \otimes \mathfrak{A}^* \otimes \mathfrak{A}$, while $x \in \mathfrak{gl}(\mathfrak{A})$ is a derivation if and only if it kills t. Now apply the theorem above (and the comment following it). $\square$

For example, when $\mathfrak{A}$ is the Cayley algebra (a certain 8-dimensional nonassociative algebra over K), Aut $\mathfrak{A}$ has Der $\mathfrak{A}$ as its Lie algebra. The latter is known to be a simple Lie algebra of type G_2, which implies that Aut $\mathfrak{U}$ is "almost simple"; cf. §14 for further applications in this direction.

13.3. Normal Subgroups and Ideals

Theorem. *Let G be a connected algebraic group, H a closed connected subgroup. Then $\mathfrak{h}$ is an ideal in $\mathfrak{g}$ if and only if H is normal in G.*

Proof. The "if" part is already known. Suppose $\mathfrak{h}$ is an ideal in $\mathfrak{g}$, and consider $N = \{x \in G | \text{Ad } x(\mathfrak{h}) = \mathfrak{h}\}$. According to Theorem 13.2 (and the remark following it), $\mathfrak{n} = \{x \in \mathfrak{g} | d(\text{Ad})x(\mathfrak{h}) \subset \mathfrak{h}\} = \{x \in \mathfrak{g} | \text{ad } x(\mathfrak{h}) \subset \mathfrak{h}\} = \mathfrak{g}$ (since ad is the differential of Ad (10.4)). Because G is connected, this forces $N = G$. Now if $x \in G$, $\mathscr{L}(\text{Int } x(H)) = \text{Ad } x(\mathfrak{h}) = \mathfrak{h}$. The two connected groups H and xHx^{-1} have the same Lie algebra, so they must coincide (Theorem 13.1). $\square$

A variation on this proof shows that the Lie algebra of $N_G(H)$ is the normalizer of $\mathfrak{h}$ in $\mathfrak{g}$ (Exercise 1).

13.4. Centers and Centralizers

Theorem. *Let G be a connected algebraic group.*
(a) *If $x \in G$, $\mathscr{L}(C_G(x)) = \mathfrak{c}_\mathfrak{g}(x) (= \{x \in \mathfrak{g} | \text{Ad } x(x) = x\})$.*
(b) $\text{Ker Ad} = Z(G)$, *and its Lie algebra is* $\text{Ker ad} = \mathfrak{z}(\mathfrak{g})$.

Proof. (a) Proposition 10.6 shows that this is true when $G = \text{GL}(n, \mathsf{K})$, $\mathfrak{g} = \mathfrak{gl}(n, \mathsf{K})$. In general, embed G as a closed subgroup of some $\text{GL}(n, \mathsf{K})$ (Theorem 8.6). Then $C_G(x) = G \cap C_{\text{GL}(n, \mathsf{K})}(x)$, while $\mathfrak{c}_\mathfrak{g}(x) = \mathfrak{g} \cap \mathfrak{c}_{\mathfrak{gl}(n, \mathsf{K})}(x)$. So Theorem 12.5 applies.

(b) Separability of morphisms implies that $N = \text{Ker Ad}$ has as Lie algebra $\text{Ker } d(\text{Ad}) = \text{Ker ad}$ (10.4), but this is by definition $\mathfrak{z}(\mathfrak{g})$. It remains to show that $N = Z(G)$. We already know that $\text{Ad } x = 1$ for $x \in Z(G)$ (since $\text{Ad } x$ is the differential of Int x). Conversely, if $\text{Ad } x = 1$, then $\mathfrak{c}_\mathfrak{g}(x) = \mathfrak{g}$; but this coincides (by (a)) with $\mathscr{L}(C_G(x))$, forcing $C_G(x) = G$, or $x \in Z(G)$. $\square$

13.5. Semisimple Groups and Lie Algebras

Theorem 13.4(b) shows that a connected algebraic group G is commutative if and only if its Lie algebra is **commutative**, i.e., $[x, y] = 0$ for all $x, y \in \mathfrak{g}$. By definition, a Lie algebra of positive dimension is **semisimple** if it has no nonzero commutative ideal. This suggests that we call a connected algebraic group of positive dimension **semisimple** if it has no closed connected commutative normal subgroup except e. (A different, but equivalent, definition will be given in (19.5).) Now we can prove:

Theorem. *A connected algebraic group G is semisimple if and only if $\mathfrak{g}$ is semisimple.*

Proof. Let $\mathfrak{g}$ be semisimple. If N is a closed connected commutative normal subgroup of G, then $\mathfrak{n}$ is a commutative ideal of $\mathfrak{g}$ (use the easy half of Theorem 13.3 and the remarks above). So $\mathfrak{n} = 0$, forcing $N = e$.

Conversely, let G be semisimple and let $\mathfrak{n}$ be a commutative ideal of $\mathfrak{g}$. Define $H = C_G(\mathfrak{n})^\circ$. Then $\mathfrak{h} = \mathfrak{c}_\mathfrak{g}(\mathfrak{n})$ (this follows from Theorem 13.2, cf. Exercise 2). Since $\mathfrak{n}$ is an ideal, the Jacobi identity shows at once that $\mathfrak{h}$ is also an ideal (including $\mathfrak{n}$, since $\mathfrak{n}$ is commutative, as part of its center). Thanks to Theorem 13.3, H must be normal in G; as a result $Z = Z(H)^\circ$ is also normal in G. By Theorem 13.4(b), $\mathfrak{z}$ is the center of $\mathfrak{h}$, and therefore includes $\mathfrak{n}$. But G is semisimple, so $Z = e$, $\mathfrak{z} = 0$. This forces $\mathfrak{n} = 0$. $\square$

Quite a lot is known about semisimple Lie algebras, so this theorem is especially significant, as we shall see in the following section.

Exercises

1. Let G be a connected algebraic group, H a closed connected subgroup. Prove that $\mathscr{L}(N_G(H)) = \mathfrak{n}_\mathfrak{g}(\mathfrak{h})$ (cf. Corollary 10.4B) and that $\mathscr{L}(C_G(H)) = \mathfrak{c}_\mathfrak{g}(\mathfrak{h})$ (cf. Exercise 10.5).
2. Let G be a connected algebraic group, $\mathfrak{n}$ a subalgebra of $\mathfrak{g}$. Prove that $\mathscr{L}(C_G(\mathfrak{n})) = \mathfrak{c}_\mathfrak{g}(\mathfrak{n})$ $(= \{x \in \mathfrak{g} | [x, y] = 0 \text{ for all } y \in \mathfrak{n}\}$.
3. Prove that SL(2, K) is semisimple.

Notes

The correspondence between linear algebraic groups and Lie algebras was first developed in depth by Chevalley [4] [5], relying on formal exponential power series; for another approach, cf. Hochschild [8]. Further results may be found in Borel [4, §7] and in Demazure, Gabriel [1, II §6]. Some questions about algebraic Lie algebras are treated in Hochschild [2] [7], Humphreys [1], Seligman [1] [2] [3].

14. Semisimple Groups

As in §13, we assume that char $K = 0$. We also assume that the reader is familiar with the theory of semisimple Lie algebras.

14.1. The Adjoint Representation

In general, Ker Ad $= Z(G)$ and its Lie algebra is Ker ad $= \mathfrak{z}(\mathfrak{g})$ (13.4). When G is semisimple, $\mathfrak{g}$ is also semisimple (13.5), so $\mathfrak{z}(\mathfrak{g}) = 0$ and $Z(G)$ is finite. More interesting in this case are the images of Ad and ad.

Theorem.　*Let G be semisimple. Then* Ad $G = (\text{Aut } \mathfrak{g})^\circ$ *and* ad $\mathfrak{g} = $ Der $\mathfrak{g}$.

Proof.　It is known that all derivations of a semisimple Lie algebra are "inner" (i.e., of the form ad x), since the Killing form is nondegenerate. Thus ad $\mathfrak{g} = $ Der $\mathfrak{g}$. In turn, Corollary 13.2 says that Der $\mathfrak{g}$ is the Lie algebra of Aut $\mathfrak{g}$. Since Ker Ad $= Z(G)$ is finite, while Ker ad $= 0$, the dimensions of G, $\mathfrak{g}$, Ad G, Aut $\mathfrak{g}$, Der $\mathfrak{g}$ all coincide. In particular, Ad $G \subset (\text{Aut } \mathfrak{g})^\circ$ forces equality to hold. $\square$

This theorem goes a long way toward a classification of semisimple algebraic groups (in characteristic 0). It shows already that if G, G' are semisimple groups whose Lie algebras $\mathfrak{g}$, $\mathfrak{g}'$ are isomorphic, then $G/Z(G)$ is isomorphic to $G'/Z(G')$. (Thus G and G' are "locally isomorphic", to borrow the language of Lie group theory.) Now the classification of semisimple Lie algebras is completely known, thanks to the classical work of W. Killing and E. Cartan: $\mathfrak{g}$ is determined (up to isomorphism) by its "root system" (see Appendix), which is a disjoint union of irreducible root systems belonging to the (unique) simple ideals of which the Lie algebra is a direct sum. The irreducible root systems comprise four infinite families A_ℓ, B_ℓ, C_ℓ, D_ℓ and five exceptional types E_6, E_7, E_8, F_4, G_2.

To complete the picture, one of course has to pin down $Z(G)$. It turns out (via representation theory, for example) that $Z(G)$ must be isomorphic to a subgroup of the "fundamental group" of the root system (cf. Appendix), defined to be the weight lattice modulo the root lattice. For example, in type A_1 this group has order 2; so SL(2, K) and PGL(2, K) will be the only semi-simple groups of type A_1.

Remarkably enough, the classification of semisimple groups in prime characteristic (Chapter XI) leads to virtually the same results. The Lie algebra no longer plays so much of a role, but the root system emerges in due course as a crucial invariant of the group. All of this, however, requires a considerable amount of further work. This further work does lead to much more detailed information about the groups than can be gotten from the Lie algebra alone in characteristic 0, so the extra effort is worthwhile even for the reader interested just in this case.

14.2. Subgroups of a Semisimple Group

Let G again be a semisimple group. It is interesting to see what the known structure of $\mathfrak{g}$ can tell us about the closed subgroups of G. As stated in (13.1), it is generally difficult to decide whether a given subalgebra $\mathfrak{h}$ of $\mathfrak{g}$ plays the role of $\mathscr{L}(H)$. But the nice correspondence between normalizers (or centralizers) can sometimes be exploited.

Consider, for example, the canonical decomposition of $\mathfrak{g}$ as a direct sum of simple ideals $\mathfrak{g}_1, \ldots, \mathfrak{g}_t$, to which we alluded in (14.1). (A nonzero Lie algebra is **simple** if it is noncommutative and has no proper ideals.) Here $\mathfrak{c}_{\mathfrak{g}}(\mathfrak{g}_2 \oplus \cdots \oplus \mathfrak{g}_t) = \mathfrak{g}_1$, and this is the Lie algebra of $C_G(\mathfrak{g}_2 \oplus \cdots \oplus \mathfrak{g}_t)$ (cf. Exercise 13.2). Call the identity component of the latter group G_1. Then G_1 is normal in G (13.3). If G_1 had a proper closed connected normal subgroup H, then $\mathfrak{h}$ would be a proper ideal of $\mathfrak{g}_1$, contradicting the simplicity. So G_1 is **almost simple** (i.e., noncommutative and having no proper closed connected normal subgroups). Construct $G_2, \ldots, G_t$ in similar fashion. Since $[\mathfrak{g}_i\, \mathfrak{g}_j] = 0$ for $i \neq j$, we see quickly that G_i centralizes G_j. Moreover, $G = G_1 \cdots G_t$ (since the right side is a closed subgroup whose Lie algebra includes $\mathfrak{g}$), and each G_i intersects the remaining product in a central (hence finite) subgroup. So G is decomposed as an "almost direct" product.

In another direction, the reader familiar with Cartan and Borel subalgebras of $\mathfrak{g}$, each of which is its own normalizer, will have no difficulty in constructing corresponding subgroups of G. Such subgroups will be introduced in arbitrary characteristic later in the text, as (respectively) "maximal tori" and "Borel subgroups". It will then be shown, using some algebraic geometry, that all subgroups of G of either species are conjugate in G. As a result, in characteristic 0 the Cartan (resp. Borel) subalgebras of $\mathfrak{g}$ must all be conjugate under Ad G. This latter result is of course well known (but not easy to prove!), if Ad G is defined directly as the subgroup of Aut $\mathfrak{g}$ generated by "inner" automorphisms exp ad $\mathfrak{n}$ (ad $\mathfrak{n}$ nilpotent).

14.3. Complete Reducibility of Representations

A basic theorem of Weyl states that every (finite dimensional) representation $\rho:\mathfrak{g} \to \mathfrak{gl}(V)$ of a semisimple Lie algebra is **completely reducible**. This means that each subspace of V stable under $\rho(\mathfrak{g})$ has a complement with the same property. Thus V can be written as a direct sum of subspaces stable under $\rho(\mathfrak{g})$, each having no proper subspace stable under $\rho(\mathfrak{g})$. In view of Theorem 13.2, rational representations of a semisimple algebraic group are also completely reducible. This fact has an important application to **Hilbert's Fourteenth Problem**, which we now sketch.

$GL(n, \mathsf{K})$ acts naturally on K^n and hence on the symmetric algebra of this vector space, which may be identified with the polynomial algebra $R = \mathsf{K}[\mathsf{T}_1, \ldots, \mathsf{T}_n]$. This is a "rational" action in the sense that R is the union of finite dimensional subspaces stable under $GL(n, \mathsf{K})$, on each of which $GL(n, \mathsf{K})$ acts rationally: To see this, notice that R has a natural grading (based on degree) which the group action respects. If G is an arbitrary subgroup of $GL(n, \mathsf{K})$, denote by R^G the set of its fixed points in R. Evidently R^G is a K-algebra. Hilbert's question was: *Must R^G be finitely generated* (as a K-algebra)? The answer is affirmative in many cases of interest, as we shall see; but in 1958 M. Nagata produced an example for which the answer is no (K of any characteristic).

Since G and its Zariski closure have the same invariants, G might as well be assumed to be closed in $GL(n, \mathsf{K})$.

Theorem. *Let G be a closed subgroup of* $GL(n, \mathsf{K})$, *acting naturally on* $R = \mathsf{K}[\mathsf{T}_1, \ldots, \mathsf{T}_n]$. *Assume that all rational representations of G are completely reducible. Then R^G is a finitely generated K-algebra.*

Proof.   First observe that if G fixes a polynomial, it also fixes the homogeneous parts of this polynomial, so R^G inherits the grading from R. Complete reducibility of the representation of G on the individual graded pieces of R

then allows us to write $R = R^G \oplus S$, where S is the sum of all subspaces of R on which G acts irreducibly but nontrivially. Evidently $R^G \cap R^G S = 0$. Since $R^G S$ is stable under the action of G, complete reducibility further implies that $R^G S \subset S$.

Now we can show that the ring R^G is *noetherian*. If $f_1, \ldots, f_t \in R^G$, the preceding argument implies that $(\sum f_i R) \cap R^G = \sum f_i R^G$. So extending and then restricting a finitely generated ideal of R^G gets us back to the given ideal. This implies that R^G satisfies ACC on ideals, since R does.

In particular, the homogeneous ideal in R^G consisting of polynomials of positive degree has a finite set of generators, which may be assumed to be homogeneous. Call them $f_1, \ldots, f_t$. We claim that every homogeneous $f \in R^G$ of positive degree is a polynomial in the f_i, from which it will follow that $R^G = \mathsf{K}[f_1, \ldots, f_t]$. Start with $f = \sum a_i f_i$, where $a_i \in R^G$ is homogeneous and can be assumed to be of degree $= \deg f - \deg f_i \geqslant 0$ (or else $a_i = 0$ if $\deg f < \deg f_i$). Then either $a_i \in \mathsf{K}$ or $0 < \deg a_i < \deg f$, in which case induction can be used. $\square$

One reason for being interested in the finite generation of rings of invariants is this: When an algebraic group G acts on an affine variety X, the set Y of its orbits in X can often be given the structure of a variety in such a way that the canonical map $G \to Y$ is a morphism with good universal properties. For example, X could be an algebraic group containing G, with G acting on X by left or right translations; then Y is the set of right or left cosets (cf. Chapter IV). Once such a "quotient" is constructed, it is natural to ask whether the underlying variety is *affine* or not. If it is to be affine, the only plausible candidate for its affine algebra is the ring of invariants $\mathsf{K}[X]^G$ (the polynomial functions on X taking constant value on each G-orbit). So one is led to ask whether this ring is finitely generated. Conversely, if G acts on X and if $\mathsf{K}[X]^G$ is known to be finitely generated, one can attempt to impose a corresponding structure of affine variety on the set of G-orbits, in such a way that the orbit map has a suitable universal property. For this one may have to add supplementary conditions, e.g., that the orbits all be closed in X.

The preceding theorem deals only with the case in which the variety X has $\mathsf{K}[\mathsf{T}_1, \ldots, \mathsf{T}_n]$ as its ring of polynomial functions. But the general case can be reduced to this one (Exercise 2), since $\mathsf{K}[X]$ is a homomorphic image of such a ring. The conclusion is that $\mathsf{K}[X]^G$ is finitely generated, provided all rational representations of G are completely reducible.

Exercises

1. Let G be an algebraic group, R a finitely generated (commutative) K-algebra on which G acts as a group of algebra automorphisms. We say that G acts "rationally" on R if each element of R lies in a finite dimensional G-stable subspace which affords a rational representation of G. Let G act rationally on two such algebras R, R', and let $\varphi : R \to R'$ be

a G-equivariant epimorphism. Assuming that all rational representations of G are completely reducible, prove that φ maps R^G onto R'^G.

2. Let the algebraic group G act on an affine variety X. Assuming that all rational representations of G are completely reducible, prove that $\mathsf{K}[X]^G$ is a finitely generated K-algebra. [Use Exercise 1 to reduce this to Theorem 14.3.]

Notes

For the theory of semisimple Lie algebras consult Jacobson [1], Serre [2], Humphreys [6]. The literature on invariant theory and Hilbert's 14th Problem is extensive, cf. Dieudonné, Carrell [1], Fogarty [1], Grosshans [4], Hochschild, Mostow [4], Mumford [1], Nagata [1] [2] [3], Seshadri [1] [2] [3], Takeuchi [1].

Chapter VI

Semisimple and Unipotent Elements

15. Jordan-Chevalley Decomposition

The Jordan normal form (diagonal plus nilpotent) for a single linear transformation has a multiplicative counterpart (diagonal times unipotent) in $GL(n, K)$. The main result of this section (Theorem 15.3) asserts that a closed subgroup of $GL(n, K)$ contains these components of each of its elements. From this we can deduce to some extent the structure of a *commutative* algebraic group (15.5).

15.1. Decomposition of a Single Endomorphism

Let $x \in \mathrm{End}\ V$, V a finite dimensional vector space over K. Then x is **nilpotent** if $x^n = 0$ for some n (equivalently, if 0 is the sole eigenvalue of x). At the other extreme, x is called **semisimple** if the minimal polynomial of x has distinct roots (if and only if x is diagonalizable over K). Evidently, 0 is the only endomorphism of V which is both nilpotent and semisimple. Indeed, the following (**additive**) **Jordan decomposition** is well known:

Lemma A. *Let $x \in \mathrm{End}\ V$.*
(a) *There exist unique x_s, $x_n \in \mathrm{End}\ V$ satisfying the conditions: $x = x_s + x_n$, x_s is semisimple, x_n is nilpotent, $x_s x_n = x_n x_s$.*
(b) *There exist polynomials $p(T)$, $q(T)$ in one variable, without constant term, such that $x_s = p(x)$, $x_n = q(x)$. In particular, x_s and x_n commute with any endomorphism of V which commutes with x.*
(c) *If $A \subset B \subset V$ are subspaces, and x maps B into A, then so do x_s and x_n.*
(d) *If $xy = yx$ $(y \in \mathrm{End}\ V)$, then $(x + y)_s = x_s + y_s$ and $(x + y)_n = x_n + y_n$.* $\square$

In case $x \in GL(V)$, its eigenvalues are nonzero and thus $x_s \in GL(V)$. Therefore, we can write $x_u = 1 + x_s^{-1} x_n$, thereby obtaining a (**multiplicative**) **Jordan decomposition** $x = x_s x_u$. Notice that $x_s^{-1} x_n$ is nilpotent (since x_n is, and x_s^{-1} commutes with it). We call an invertible endomorphism **unipotent** if it is the sum of the identity and a nilpotent endomorphism, or equivalently, if its sole eigenvalue is 1. We just showed how to write x as product of commuting semisimple and unipotent endomorphisms. If $x = su$ is any such decomposition, then $x = s(1 + n)$ (n nilpotent, $ns = sn$), or $x = s + sn$ (s semisimple, sn nilpotent, s commuting with sn). By part (a) of Lemma A,

95

$s = x_s$ and $sn = x_n$, whence also $u = 1 + n = x_u$. This proves (in view of Lemma A):

Lemma B.   *Let $x \in \mathrm{GL}(V)$.*
 (a) *There exist unique $x_s, x_u \in \mathrm{GL}(V)$ satisfying the conditions: $x = x_s x_u$, x_s is semisimple, x_u is unipotent, $x_s x_u = x_u x_s$.*
 (b) *x_s, x_u commute with any endomorphism of V which commutes with x.*
 (c) *If A is a subspace of V stable under x, then A is stable under x_s, x_u.*
 (d) *If $xy = yx$ ($y \in \mathrm{GL}(V)$), then $(xy)_s = x_s y_s$, $(xy)_u = x_u y_u$.* $\square$

We call x_s the **semisimple part** and x_u the **unipotent part** of x. Note that if x is both semisimple and unipotent, then $x = 1$.

It is sometimes useful to allow V to be *infinite dimensional*, even though the notions "semisimple" and "unipotent" do not carry over directly to this case. If $x \in \mathrm{GL}(V)$, and if V is the union of finite dimensional subspaces stable under x, then the decompositions $x|W = (x|W)_s (x|W)_u$ exist for all such subspaces W. Moreover, the restriction of a semisimple (resp. unipotent) endomorphism to an intersection $W \cap W'$ is of the same type, so it is clear that we can piece together the various $(x|W)_s$ (resp. $(x|W)_u$) to obtain invertible endomorphisms of V, of which x is the product. These may again be denoted x_s, x_u, and called the Jordan parts of x. It is important to observe (using part (c) of Lemma B) that x_s and x_u leave stable every subspace of V, finite dimensional or not, which is stable under x. Similar remarks of course apply to the additive Jordan decomposition.

The behavior of semisimple endomorphisms is much the same in both zero and prime characteristics. But not so for unipotents. When $0 < p = \mathrm{char}\ \mathsf{K}$, we claim that $x \in \mathrm{GL}(V)$ is unipotent if and only if $x^{p^t} = 1$ for some $t \geqslant 0$. Indeed, if $x = 1 + n$ (n nilpotent), then for large enough t, $n^{p^t} = 0$, whence $x^{p^t} = (1 + n)^{p^t} = 1 + n^{p^t} = 1$. (The argument is reversible.) When $\mathrm{char}\ \mathsf{K} = 0$, the picture is totally different, since a unipotent element $\neq 1$ of $\mathrm{GL}(V)$ must have infinite order (Exercise 5). In this case the nature of unipotent elements can be made clear by introducing exponential and logarithmic power series. If $n \in \mathrm{End}\ V$ is nilpotent, the series $\exp n = \sum_{i=0}^{\infty} n^i/i!$ is a polynomial and therefore defines an element of $\mathrm{GL}(V)$, which is unipotent (being the sum of 1 and a nilpotent endomorphism). If $x \in \mathrm{GL}(V)$ is unipotent, then $x - 1$ is nilpotent and the series $\log x = \sum_{i=1}^{\infty} (-1)^{i+1}(x - 1)^i$ is a polynomial which defines a nilpotent endomorphism of V. Thanks to the usual formal identities for power series, $\exp(\log x) = x$ and $\log(\exp n) = n$.

Lemma C.   *Let $\mathrm{char}\ \mathsf{K} = 0$. If $x \in \mathrm{GL}(V)$ is unipotent, $x \neq 1$, the map $\varphi(a) = \exp(an)$ ($n = \log x$) defines an isomorphism of $\mathbf{G}_a$ onto a closed subgroup of $\mathrm{GL}(V)$, which is the unique smallest closed subgroup of $\mathrm{GL}(V)$ containing x. In particular, every connected closed 1-dimensional subgroup of $\mathrm{GL}(V)$ which contains a unipotent element $\neq 1$ must be isomorphic to $\mathbf{G}_a$.*

Proof. The assignment $\varphi(a) = \exp(an)$ defines a morphism of varieties $\mathbf{G}_a \to \mathrm{GL}(V)$, which is a homomorphism by virtue of the formal identity $\exp((T + T')n) = (\exp Tn)(\exp T'n)$. (This holds in the formal power series ring in two variables T, T' over the commutative subring $K[n]$ of End V.) Moreover, $\varphi(1) = x$. To show that φ is an isomorphism, we just note that

$$an = \log \exp(an) = \log \varphi(a) = \sum_{i=1}^{\infty} (-1)^{i+1}(\varphi(a)-1)^i \text{ is a polynomial in}$$

$\varphi(a)$. Finally, since x has infinite order, the closure of the subgroup it generates in $\mathrm{GL}(V)$ has dimension at least 1, hence must coincide with the closed subgroup $\varphi(\mathbf{G}_a)$ of dimension 1. $\square$

15.2. GL(*n*, K) and gl(*n*, K)

If G is an arbitrary subgroup of $\mathrm{GL}(n, K)$, there is no reason to expect G to contain the semisimple and the unipotent part of each of its elements. If G is *closed*, however, this does turn out to be the case. (And, for good measure, g also turns out to contain the semisimple and the nilpotent part in $\mathrm{gl}(n, K)$ of each of its elements.) The reason for this becomes transparent if we recall the criterion for membership in G (resp. g) given by Lemma 8.5 (resp. Lemma 9.4). For example, given $x \in G$, one wants to know that ρ_{x_s} leaves stable the ideal $I \subset K[\mathrm{GL}(n, K)]$ vanishing on G. In view of the discussion in (15.1), it would be enough to know that ρ_{x_s} is the "semisimple part" of ρ_x. Accordingly, we look first at this question (which has nothing to do with G).

Proposition. *Let* $G = \mathrm{GL}(n, K)$, $\mathrm{g} = \mathrm{gl}(n, K)$. *If* $x \in G$ *(resp.* x $\in$ g*), then* ρ_x *(resp.* *x*) *has Jordan decomposition* $\rho_{x_s}\rho_{x_u}$ *(resp.* *x*$_s$ + *x*$_n$*).*

Proof. Since $K[G]$ is the union of finite dimensional subspaces stable under all ρ_x (hence under all *x, cf. Proposition 10.2), Jordan decompositions do exist. Moreover, $\rho_x = \rho_{x_s}\rho_{x_u}$ (resp. *x = *x$_s$ + *x$_n$), and the operators commute, so it will be enough to show that ρ_{x_s} and *x$_s$ are semisimple, that ρ_{x_u} is unipotent, and that *x$_n$ is nilpotent.

$K[G]$ is the ring of quotients of the polynomial ring $K[T]$ in n^2 indeterminates T_{ij} obtained by allowing powers of $d = \det(T_{ij})$ as denominators. It is clear from Lemma 10.3A that $K[T]$ is stable under both right translation and right convolution. Moreover, it is easy to describe how G and g act on d: First, $(\rho_x d)(y) = \det(yx) = \det y \cdot \det x$, so $\rho_x d = \det x \cdot d$. This shows that the space spanned by d is G-stable, hence (Proposition 10.2) g-stable, with $d*$x $= \mathrm{Tr}(x)d$, in view of the fact (5.4) that the differential of det is trace. From this we see how to describe the action of ρ_x or *x on $K[G]$ once the action on $K[T]$ is known.

In particular, if $\rho_x|K[T]$ is semisimple, then ρ_x is semisimple (d being an eigenvector of ρ_x in any case). If $\rho_x|K[T]$ is unipotent, then its eigenvalue det x on the space spanned by d must be 1, so it is clear that ρ_x is unipotent.

Similarly, if $*x|K[T]$ is semisimple (resp. nilpotent), then so is $*x$. (Recall that $*x$ acts on $f(T)/d^t$ by the quotient rule for derivations.) Therefore it will be enough to consider the actions of ρ_x, $*x$ on $K[T]$ in place of $K[G]$.

Set $E = \text{End } V$, where $V = K^n$, and regard G as the subset $GL(V)$ of E, while $\mathfrak{g} = \mathfrak{gl}(V) = E$. $K[T]$ may be identified with the symmetric algebra $\mathfrak{S}(E^*)$ on the dual space of E. If $x \in E$, define an endomorphism $r_x : E \to E$ by $r_x(y) = yx$, and let $r_x^* : E^* \to E^*$ be its transpose. Then we see that ρ_x (resp. $*x$) is just the canonical extension of r_x^* (resp. r_x^*) to an automorphism (resp. derivation) of $\mathfrak{S}(E^*)$, cf. Lemma 10.3A. So it just has to be verified that the property of being semisimple or unipotent (resp. semisimple or nilpotent) is preserved at each step when we pass from x to r_x to r_x^* to ρ_x (resp. from x to r_x to r_x^* to $*x$). The passage from r_x to r_x^* poses no problem. To go from the action on E^* to the action on $\mathfrak{S}(E^*)$ we first act on tensor powers of E^* (cf. Exercise 1), then factor out from the tensor algebra the appropriate ideal (which is stable under the given action) to reach $\mathfrak{S}(E^*)$; so again there is no problem. It remains to treat the passage from x to r_x $(x \in E)$.

Consider the separate cases. If $x \in E$ is *semisimple*, choose a basis $(v_1, \ldots, v_n)$ of V consisting of eigenvectors, so that $x \cdot v_i = a_i v_i$ for some $a_i \in K$. Take the corresponding basis e_{ij} for E, where $e_{ij}(v_k) = \delta_{jk} v_i$, so $r_x e_{ij} = a_j e_{ij}$. This shows that r_x is semisimple. If $x \in E$ is *nilpotent*, say $x^t = 0$, then $(r_x)^t(y) = yx^t = 0$ for all $y \in E$, so r_x is nilpotent. If $x = 1 + n$ is *unipotent*, say $n^t = 0$, then $r_x = 1 + r_n$, so $(r_x - 1)^t = 0$ and r_x is unipotent. (As an alternative to treating these cases, notice that when E is identified canonically with $V^* \otimes V$, r_x corresponds to the endomorphism $x^* \otimes 1$.) $\square$

15.3. Jordan Decomposition in Algebraic Groups

In order to treat the general case, we look more closely at right translation and convolution, for arbitrary algebraic groups. These are, in the first place, *faithful* representations of G, $\mathfrak{g}$. In the second place, translation and convolution behave well when we pass from G to a closed subgroup H. Indeed, if $x \in H$, then the following diagram commutes:

$$\begin{array}{ccc} K[G] & \xrightarrow{\rho_x} & K[G] \\ \downarrow & & \downarrow \\ K[H] & \xrightarrow[\rho_x]{} & K[H] \end{array}$$

(where ρ_x on top is right translation for x viewed as an element of G, and where the vertical maps are canonical). For right convolution one has the same sort of diagram. Similarly, let $\varphi : G \to G'$ be an epimorphism of algebraic groups, so φ^* identifies $K[G']$ with a subring of $K[G]$. If $x \in G$, then clearly $\rho_{\varphi(x)}$ may be viewed as the restriction of ρ_x to this subring (and $*d\varphi(x)$ as the restriction of $*x$).

Theorem.   *Let G be an algebraic group.*

(a) *If $x \in G$, there exist unique elements s, $u \in G$ such that: $x = su$, s and u commute, ρ_s is semisimple, ρ_u is nilpotent. (Then we call s the* **semisimple part** *of x, written x_s, and u the* **unipotent part**, *written x_u.)*

(b) *If $\mathsf{x} \in \mathfrak{g}$, there exist unique elements s, $\mathsf{n} \in \mathfrak{g}$ such that: $\mathsf{x} = \mathsf{s} + \mathsf{n}$, $[\mathsf{s}, \mathsf{n}] = 0$, $*\mathsf{s}$ is semisimple, $*\mathsf{n}$ is nilpotent. (We write $\mathsf{s} = \mathsf{x}_s$, $\mathsf{n} = \mathsf{x}_n$, and call these the* **semisimple** *and* **nilpotent parts** *of x.)*

(c) *If $\varphi : G \to G'$ is a morphism of algebraic groups, then $\varphi(x)_s = \varphi(x_s)$, $\varphi(x)_u = \varphi(x_u)$, $d\varphi(\mathsf{x})_s = d\varphi(\mathsf{x}_s)$, and $d\varphi(\mathsf{x})_n = d\varphi(\mathsf{x}_n)$ for all $x \in G$, $\mathsf{x} \in \mathfrak{g}$.*

Proof.   We may embed G as a closed subgroup of some $GL(n, \mathsf{K})$ (Theorem 8.6). If I is the ideal in $\mathsf{K}[GL(n, \mathsf{K})]$ defining G, then the criterion for $x \in GL(n, \mathsf{K})$ (resp. $\mathsf{x} \in \mathfrak{gl}(n, \mathsf{K})$) to be in G (resp. $\mathfrak{g}$) is that ρ_x (resp. $*\mathsf{x}$) stabilize I (Lemmas 8.5, 9.4). Now let $x \in G$, and let $x = su$ be its Jordan decomposition in $GL(n, \mathsf{K})$. Thanks to Proposition 15.2, $(\rho_x)_s = \rho_s$ and $(\rho_x)_u = \rho_u$. The remarks in (15.1) then show that both of these operators stabilize I, so that s, $u \in G$. This yields the desired decomposition of x in G, in view of the preceding remarks (the uniqueness being a consequence of the faithfulness of ρ). For $\mathsf{x} \in \mathfrak{g}$, the argument is parallel. So (a) and (b) are proved.

Next consider $\varphi : G \to G'$, which factors into two morphisms: the epimorphism $G \to \varphi(G)$, followed by the inclusion $\varphi(G) \to G'$. It suffices to treat each of these cases separately. In case φ is an epimorphism, we noted above that right translation by $\varphi(x)$ is essentially the restriction of ρ_x to $\mathsf{K}[G']$ viewed as a subring of $\mathsf{K}[G]$ (and similarly for right convolution). But the restriction of a semisimple (resp. unipotent) operator to a subspace is again of the same type. It follows that $\rho_{\varphi(x)} = \rho_{\varphi(x_s)}\rho_{\varphi(x_u)}$ is the Jordan decomposition of $\rho_{\varphi(x)}$, hence that $\varphi(x_s) = \varphi(x)_s$, $\varphi(x_u) = \varphi(x)_u$ (and similarly for $d\varphi(\mathsf{x})$). In case φ is an inclusion, the situation is just like the one considered above (G viewed as a closed subgroup of $GL(n, \mathsf{K})$). $\square$

The theorem shows that in any algebraic group G, the subsets $G_s = \{x \in G \mid x = x_s\}$ and $G_u = \{x \in G \mid x = x_u\}$ are intrinsically defined and intersect in e. (Similarly, we can define $\mathfrak{g}_s$ and $\mathfrak{g}_n$; these intersect in 0.) Part (c) of the theorem insures that morphisms of algebraic groups (and their differentials) preserve these sets: $\varphi(G_s) = \varphi(G)_s$, etc. Moreover, G_u and $\mathfrak{g}_n$ are *closed* sets ($\mathfrak{g}$ being given the topology of an affine space). To see this, just observe that the set of all unipotent (resp. nilpotent) matrices in $GL(n, \mathsf{K})$ (resp. $M(n, \mathsf{K})$) is closed, being the zero set of the polynomials implied by $(x - 1)^n = 0$ (resp. $x^n = 0$). By contrast, G_s is rarely a closed subset of G.

15.4. Commuting Sets of Endomorphisms

A subset M of $M(n, \mathsf{K})$ is said to be **diagonalizable** (resp. **trigonalizable**) if there exists $x \in GL(n, \mathsf{K})$ such that $xMx^{-1} \subset \mathfrak{d}(n, \mathsf{K})$ (resp. $\mathfrak{t}(n, \mathsf{K})$).

Proposition. *If $M \subset \mathrm{M}(n, \mathsf{K})$ is a commuting set of matrices, then M is trigonalizable. In case M consists of semisimple matrices, M is even diagonalizable.*

Proof. Set $V = \mathsf{K}^n$ and proceed by induction on n.

If $x \in M$, $a \in \mathsf{K}$, the subspace $W = \mathrm{Ker}\,(x - a \cdot 1)$ is evidently stable under the endomorphisms of V which commute with x (hence is stable under M). Unless M consists of scalar matrices (then we are done), it is possible to choose x and a so that $0 \neq W \neq V$. By induction, there exists $v_1 \in W$ such that $\mathsf{K}v_1$ is stable under M. Applying the induction hypothesis next to the induced action of M on $V/\mathsf{K}v_1$, we obtain $v_2, \ldots, v_n \in V$ completing the basis for V, such that M stabilizes each subspace $\mathsf{K}v_1 + \cdots + \mathsf{K}v_i$ ($1 \leqslant i \leqslant n$). The transition from the canonical basis of V to $(v_1, \ldots, v_n)$ therefore trigonalizes M.

In case M consists of semisimple matrices, for each $x \in M$ we can write $V = V_1 \oplus \cdots \oplus V_r$, $V_i = \mathrm{Ker}\,(x - a_i \cdot 1)$, where $a_1, \ldots, a_r$ are the distinct eigenvalues of x. As before, each V_i is M-stable. Unless M consists of scalar matrices (hence is diagonal already), it is possible to choose x so that $r > 1$. Then the induction hypothesis allows us to diagonalize the action of M on each V_i. $\quad\square$

15.5. Structure of Commutative Algebraic Groups

Theorem. *Let G be a commutative algebraic group. Then G_s, G_u are closed subgroups, connected if G is, and the product map $\varphi : G_s \times G_u \to G$ is an isomorphism of algebraic groups. Moreover, $\mathscr{L}(G_s) = \mathfrak{g}_s$ and $\mathscr{L}(G_u) = \mathfrak{g}_n$.*

Proof. That G_s, G_u are *subgroups* of G follows from Lemma B (d) of (15.1). We already observed (15.3) that G_u is *closed*. Moreover, Theorem 15.3 makes it clear that φ is a group isomorphism. Now embed G in some $\mathrm{GL}(n, \mathsf{K})$. Proposition 15.4 allows us to assume without loss of generality that $G \subset \mathrm{T}(n, \mathsf{K})$ and then that $G_s \subset \mathrm{D}(n, \mathsf{K})$. This forces $G_s = G \cap \mathrm{D}(n, \mathsf{K})$, so G_s is also closed. Moreover, φ is a morphism of algebraic groups.

It has to be shown that the reverse map is a *morphism*. For this it suffices to show that $x \mapsto x_s$ and $x \mapsto x_u$ are morphisms. But the second is if the first is, since $x_u = x_s^{-1}x$. Now if $x \in G$, the eigenvalues of x (with multiplicity) are those of its diagonal part. But x_s is also diagonal and has precisely these same eigenvalues; moreover, x_s is a polynomial in x (15.1), so x_s must simply be the diagonal part of x. It follows that $x \mapsto x_s$ is a morphism. Furthermore, when G is *connected*, so are G_s, G_u (these being the images of G under the morphisms $x \mapsto x_s$, $x \mapsto x_u$).

The chosen embedding of G in $\mathrm{T}(n, \mathsf{K})$ shows also that $\mathscr{L}(G_s) \subset \mathfrak{d}(n, \mathsf{K})$, $\mathscr{L}(G_u) \subset \mathfrak{n}(n, \mathsf{K})$. Therefore, $\mathscr{L}(G_s) \subset \mathfrak{g}_s$ and $\mathscr{L}(G_u) \subset \mathfrak{g}_n$. But φ is an isomorphism, so $\mathscr{L}(G_s) \oplus \mathscr{L}(G_u) = \mathfrak{g}$. Since also $\mathfrak{g} = \mathfrak{g}_s + \mathfrak{g}_n$ (with uniqueness of expression), this forces each inclusion to be equality. $\quad\square$

Exercises

1. Let $x \in GL(V)$, $y \in GL(W)$. Prove that $(x \otimes y)_s = x_s \otimes y_s$ and that $(x \otimes y)_u = x_u \otimes y_u$.
2. Let $\mathfrak{A}$ be a finite dimensional K-algebra. If $x \in \operatorname{Aut} \mathfrak{A}$ (resp. $x \in \operatorname{Der} \mathfrak{A}$), then $x_s, x_u \in \operatorname{Aut} \mathfrak{A}$ (resp. $x_s, x_n \in \operatorname{Der} \mathfrak{A}$).
3. Find a subgroup of $GL(2, K)$ which does not contain the semisimple and unipotent parts of all its elements.
4. Show that the semisimple (resp. unipotent) elements of $SL(n, K)$ $(n \geqslant 2)$ do not form a subgroup.

In the remaining exercises, char $K = 0$.

5. An element of $GL(n, K)$ having finite order must be semisimple.
6. Prove that any closed subgroup of $GL(n, K)$ consisting of unipotent elements is connected.
7. Prove that every nontrivial rational representation $\mathbf{G}_a \to GL(V)$ has the form indicated in Lemma 15.1C.
8. Prove that exp and log set up a bijection between the set of nilpotent matrices in $M(n, K)$ and the set of unipotent matrices in $GL(n, K)$.
9. An element $x \in GL(V)$ is unipotent if and only if there exists a rational representation $\mathbf{G}_a \to GL(V)$ whose image contains x.
10. If U is a closed subgroup of $GL(V)$ consisting of unipotent elements, show that $\log x$ belongs to $\mathfrak{u}$.
11. Prove that any 1-dimensional algebraic group consisting of unipotent elements is isomorphic to $\mathbf{G}_a$.

Notes

The treatment here follows Borel [4, §4]; cf. the predecessor of Borel, Springer [1], which appears in AGDS. The proof of Theorem 15.3 in Chevalley [8, 4–10] is more complicated and less transparent. For the origins of this theorem, see Kolchin [2], Chevalley [4], Borel [1].

16. Diagonalizable Groups

16.1. Characters and d-Groups

It is natural to call an algebraic group G **diagonalizable** if it is isomorphic to a closed subgroup of the diagonal group $D(n, K)$ for some n. In that case G is evidently commutative and consists of semisimple elements. Conversely, if G has these two properties and if we embed G in some $GL(n, K)$, then it follows from (15.4) that G is conjugate in $GL(n, K)$ to a subgroup of $D(n, K)$. Notice that *closed subgroups and homomorphic images of a diagonalizable group are again diagonalizable* (since such groups are commutative and consist of semisimple elements).

The precise structure of a diagonalizable group G is not immediately obvious from the definition. For example, we would like to know that if G is n-dimensional and connected, then G is isomorphic to $D(n, K)$. We would also like to know how (in principle) to describe all the closed subgroups of $D(n, K)$, for a given n. Recall (11.4) the notion of (rational) **character** of an algebraic group G. This is by definition any morphism of algebraic groups $\chi: G \to \mathbf{G}_m$. If χ, ψ are two characters of G, we obtain another character by multiplying their values: $(\chi\psi)(x) = \chi(x)\psi(x)$. In this way we obtain an abelian group $X(G)$, the **character group** of G. Notice that $X(G)$ can be viewed as a subset of $K[G]$.

$D(n, K)$ has plenty of characters, e.g., the various coordinate functions $\chi_i: \mathrm{diag}\,(a_1, \ldots, a_n) \mapsto a_i$. (Indeed, $X(D(n, K))$ is free abelian of rank n, with basis $\chi_1, \ldots, \chi_n$.) By contrast, a group such as $SL(n, K)$ which is perfect (equal to its own derived group) has no nontrivial character (Exercise 3). The idea now is to characterize diagonalizable groups as those algebraic groups possessing enough characters to separate points. We shall need the following familiar lemma.

Lemma. *Let G be an (abstract) group, X the set of all homomorphisms $G \to K^*$. Then X is a linearly independent subset of the space of all K-valued functions on G.*

Proof. Suppose not. Let $\chi_1, \ldots, \chi_n \in X$ be linearly dependent, with $n > 1$ as small as possible; say $\sum_{i=1}^{n-1} a_i\chi_i + \chi_n = 0$ $(a_i \in K)$. Since $\chi_1 \neq \chi_n$, we can find $y \in G$ such that $\chi_1(y) \neq \chi_n(y)$. For arbitrary $x \in G$ we get two equations:

$$\sum_{i=1}^{n-1} a_i\chi_i(x)\chi_i(y) + \chi_n(x)\chi_n(y) = 0,$$

$$\sum_{i=1}^{n-1} a_i\chi_i(x)\chi_n(y) + \chi_n(x)\chi_n(y) = 0.$$

Subtracting, we are left with $\sum_{i=1}^{n-1} a_i(\chi_i(y) - \chi_n(y))\chi_i = 0$, not all coefficients 0, which contradicts the minimality of n. $\quad\square$

Now define an algebraic group G to be a *d*-**group** if $K[G]$ has a basis consisting of characters. (The lemma shows that this is equivalent to requiring that $X(G)$ be a basis of $K[G]$.) For example, $D(n, K)$ is a *d*-group, since any polynomial function is a combination of monomials constructed from the coordinate functions and their inverses. When G, G' are *d*-groups, a morphism $\varphi: G \to G'$ of algebraic groups yields a group homomorphism $\varphi^\circ: X(G') \to X(G)$, the restriction of $\varphi^*: K[G'] \to K[G]$. Conversely, a group homomorphism $X(G') \to X(G)$ extends (since these are bases) to a homomorphism of K-algebras $K[G'] \to K[G]$ and in turn defines a morphism $G \to G'$.

Proposition. (a) *If H is a closed subgroup of a d-group G, then H is also a d-group and is the intersection of kernels of some characters of G. In particular, diagonalizable groups are d-groups.*

(b) *Any d-group is diagonalizable.*

Proof. (a) The canonical homomorphism $\varphi : \mathsf{K}[G] \to \mathsf{K}[H]$ induced by restriction is surjective. Evidently the restriction to H of a character of G is a character of H, so $\mathsf{K}[H]$ is spanned by characters and H is a d-group. Moreover, if a linear combination of characters of G vanishes on H, then so does each individual character (thanks to the lemma above, applied to the group H). We conclude that the kernel $\mathscr{I}(H)$ of φ is spanned by the kernel of the restriction map $\mathsf{X}(G) \to \mathsf{X}(H)$. Since $\mathsf{D}(n, \mathsf{K})$ is a d-group, we also conclude that diagonalizable groups are d-groups.

(b) Let G be a d-group. Since $\mathsf{K}[G]$ is a finitely generated K-algebra and spanned by $\mathsf{X}(G)$, it is generated as K-algebra by some finite set of characters $\chi_1, \ldots, \chi_n$. Define $\varphi : G \to \underbrace{\mathbf{G}_m \times \cdots \times \mathbf{G}_m}_{n \text{ copies}} \cong \mathsf{D}(n, \mathsf{K})$ by $\varphi(x) = (\chi_1(x), \ldots, \chi_n(x))$. Evidently φ is a morphism of algebraic groups, with trivial kernel (because the χ_i generate $\mathsf{K}[G]$). In particular, G must be commutative and must consist of semisimple elements, i.e., G is diagonalizable. (We do not claim, however, that φ is itself an isomorphism of G onto a subgroup of $\mathsf{D}(n, \mathsf{K})$ unless char $\mathsf{K} = 0$; cf. Exercise 8). $\quad\square$

There is a notion dual to "character" which will be handy later on. If G is a d-group, any morphism of algebraic groups $\lambda : \mathbf{G}_m \to G$ is called a **one parameter multiplicative subgroup** of G (by abuse of language), abbreviated **1-psg**. The set of these is denoted $\mathsf{Y}(G)$. It becomes an abelian group if we define a product $(\lambda\mu)(a) = \lambda(a)\mu(a)$. Notice that the composite of a 1-psg λ with a character χ of G yields a morphism of algebraic groups $\mathbf{G}_m \to \mathbf{G}_m$, i.e., an element of $\mathsf{X}(\mathbf{G}_m) \cong \mathbb{Z}$. This allows us to define a natural pairing $\mathsf{X}(D) \times \mathsf{Y}(D) \to \mathbb{Z}$, denoted $\langle \chi, \lambda \rangle$, under which $\mathsf{X}(D)$ and $\mathsf{Y}(D)$ become dual $\mathbb{Z}$-modules if D is *connected* (Exercise 4). As a matter of notation, we shall write the groups $\mathsf{X}(D)$, $\mathsf{Y}(D)$ additively from now on.

16.2. Tori

The characters of a d-group ($=$ diagonalizable group) may be used to good advantage.

Lemma A. *Let G be a d-group. Then $\mathsf{X}(G)$ is a finitely generated abelian group.*

Proof. If $G \subset \mathsf{D}(n, \mathsf{K})$, $\mathsf{X}(G)$ is a homomorphic image of $\mathsf{X}(\mathsf{D}(n, \mathsf{K})) \cong \mathbb{Z}^n$. (The lemma actually holds for arbitrary G: Exercise 12.) $\quad\square$

The structure of a finitely generated abelian group A is well known: $A = \mathbb{Z}^r \times B$, where B is the torsion subgroup of A ($=$ set of all elements of finite order) and $r = \operatorname{rank} A$ is a well determined nonnegative integer. Observe that when char $\mathsf{K} = p \neq 0$, the character group $A = \mathrm{X}(G)$ of an algebraic group has no p-torsion (since K has no nontrivial p^{th} roots of unity). This shows that not every group A can appear as an $\mathrm{X}(G)$; but we shall see below that p-torsion creates the only exceptions.

Lemma B. *If G is a connected algebraic group, $\mathrm{X}(G)$ is torsion-free.*

Proof. If $\chi \in \mathrm{X}(G)$, the image $\chi(G)$ in $\mathbf{G}_m$ is connected. The only connected subgroups of $\mathbf{G}_m$ being itself and 1, we conclude that $\chi^n \neq 1$ $(n > 0)$ unless $\chi = 1$. $\square$

The reader may be familiar with the duality theory of locally compact abelian groups, in which (compact) tori correspond to (discrete) free abelian groups. This analogy and other considerations lead us to call an algebraic group a **torus** if it is isomorphic to some $\mathrm{D}(n, \mathsf{K})$. The following lemma will be useful in (27.4).

Lemma C. *Let T be a torus. If $\chi_1, \ldots, \chi_r \in \mathrm{X}(T)$ are linearly independent, and $c_1, \ldots, c_r \in \mathbf{G}_m$, there exists $t \in T$ for which $\chi_i(t) = c_i$ $(1 \leqslant i \leqslant r)$.*

Proof. It makes sense to describe characters as "linearly independent", since $\mathrm{X}(T)$ is free abelian. Define $\varphi : T \to \mathbf{G}_m \times \cdots \times \mathbf{G}_m$ (r copies) by $\varphi(t) = (\chi_1(t), \ldots, \chi_r(t))$. The characters $\sum m_i \chi_i$ $(m_i \in \mathbb{Z})$ are all distinct, so the functions $\prod \chi_i^{m_i}$ (in multiplicative notation) are linearly independent over K in $\mathsf{K}[T]$ (Lemma 16.1). It follows that $\chi_1, \ldots, \chi_r$ are algebraically independent in $\mathsf{K}(T)$, so φ^* is injective and φ is dominant. Being a morphism of algebraic groups, φ is therefore surjective. $\square$

It is time to lay bare the structure of an arbitrary d-group.

Theorem. *Let G be a d-group. Then $G = G^\circ \times H$ (direct product of algebraic groups), where G° is a torus and H is a finite group (not uniquely determined in general) of order prime to p ($=$ char exp K). In particular, a connected d-group is a torus.*

Proof. We must find a suitable complement H for the torus G°. To begin with we may assume that G is a closed subgroup of $D = \mathrm{D}(n, \mathsf{K})$ for some n. Let $\mathrm{X}(G)$ have rank r ($= \dim G$). If $r = 0$, there is nothing to prove. In general, the inclusion $G^\circ \hookrightarrow D$ induces a surjective restriction homomorphism $\mathbb{Z}^n = \mathrm{X}(D) \overset{\varphi}{\to} \mathrm{X}(G^\circ) = \mathbb{Z}^r$. This allows us to use the characteristic property of free ($=$ projective) $\mathbb{Z}$-modules to obtain a splitting $\mathrm{X}(D) = \operatorname{Ker} \varphi \oplus \mathbb{Z}^r$. As a subgroup of the free abelian group $\mathrm{X}(D)$, $\operatorname{Ker} \varphi$ has a basis $\chi_1, \ldots, \chi_{n-r}$, to which we may add a basis $\chi_{n-r+1}, \ldots, \chi_n$ of $\mathbb{Z}^r$ (mapped by φ onto a basis of $\mathrm{X}(G^\circ)$). It is clear that $x \mapsto \operatorname{diag}(\chi_1(x), \ldots, \chi_n(x))$ defines an automorphism of D, sending G° onto the subgroup of diagonal matrices whose first $n - r$ coordinates are equal to 1.

If $D' = \{\text{diag } (a_1, \ldots, a_n) | a_{n-r+1} = \cdots = a_n = 1\}$, then $D = D' \times G°$ (direct product of algebraic groups). Therefore, $G = H \times G°$, where $H = D' \cap G \cong G/G°$. Finally, H has order prime to p, since 1 is the only p^{th} root of unity. $\square$

It may be well at this point to sum up in categorical language the connection between d-groups and finitely generated abelian groups having no p-torsion. Call $\mathscr{D}$ and $\mathscr{A}$ the respective categories having these groups as their objects and having the obvious morphisms. To a d-group D we of course associate $X(D)$, while to a morphism $\varphi : D \to D'$ of d-groups we associate the group homomorphism $\varphi° : X(D') \to X(D)$ gotten by composing φ with characters of D'. This yields a (contravariant) functor $F : \mathscr{D} \to \mathscr{A}$.

How can we go in the reverse direction? Given a finitely generated abelian group A, we can always form its **group algebra** $K[A]$, which is described informally as consisting of the finite formal K-linear combinations of elements of A, the product of two such elements being determined by the distributive law and the product in A. Rigorously, $K[A]$ is defined to be the vector space of K-valued functions on A having finite support, with convolution product. Since A is finitely generated, $K[A]$ is obviously a finitely generated commutative K-algebra. It is easy to see that the absence of p-torsion in A is equivalent to the absence of nilpotent elements in $K[A]$. In particular, we obtain an affine algebra for each object A of $\mathscr{A}$, thereby defining an affine variety. But $K[A]$ also has a Hopf algebra structure (coming from the group law in A), which gives us an affine algebraic group (cf. (7.6)), call it $D(A)$. What are its characters? To $\chi : D(A) \to \mathbf{G}_m$ corresponds $\chi^* : K[\mathbf{G}_m] \to K[D(A)] = K[A]$, which in turn induces a homomorphism of character groups $\mathbb{Z} \to A$. But Hom $(\mathbb{Z}, A) \cong A$, so we conclude that $X(D(A)) \cong A$. It follows readily that $D(A)$ is a d-group and that $A \mapsto D(A)$ (along with the induced action on morphisms) is a contravariant functor $\mathscr{A} \to \mathscr{D}$ whose composite in either order with F is equivalent to the identity. Therefore $\mathscr{A}$ *and* $\mathscr{D}$ *are canonically equivalent under* F.

16.3. Rigidity of Diagonalizable Groups

We have shown that a d-group is isomorphic to a direct product of algebraic groups $\mathbf{G}_m \times \cdots \times \mathbf{G}_m \times H$, with H finite (of order relatively prime to the characteristic exponent p of K). This description is quite satisfactory if the d-group is taken in isolation. But more often we shall be interested in d-groups of more complicated algebraic groups, e.g., $D(n, K)$ in $GL(n, K)$. The main result to be established in this subsection says that a d-group is "rigid", i.e., admits "few" nontrivial automorphisms; in particular, the normalizer of a d-group in an ambient algebraic group is not much bigger than the centralizer.

Rigidity will follow from two basic facts about elements of finite order in a d-group: (a) *they are a dense subset,* and (b) *there are only finitely many of them of given order.* To prove these facts, notice that it suffices to treat the case $G = \mathbf{G}_m$. The elements of K^* of finite order are just the mth roots of unity ($m \in \mathbb{Z}^+$), of which there are at most m for each fixed m. Since K is algebraically closed, there exist infinitely many distinct roots of unity in K^*; since $\mathbf{G}_m$ is connected and one dimensional, the closure of this torsion subgroup is therefore $\mathbf{G}_m$. (One can be a bit more precise about counting roots of unity: if $(p, m) = 1$, there are exactly $\varphi(m)$ elements of order m in K^*, φ the Euler function.)

The following argument is axiomatized to emphasize the precise assumptions needed.

Proposition. *Let $\varphi: V \times G \to H$ be a morphism of varieties, where:*

(a) G is an algebraic group whose elements of finite order form a dense subset;

(b) H is an algebraic group containing only finitely many elements of order m (for each $m > 0$);

(c) V is a connected variety;

(d) for $x \in V$, $\varphi_x: G \to H$ (sending $y \mapsto \varphi(x, y)$) is a group homomorphism. Then $x \mapsto \varphi_x$ is constant ($x \in V$).

Proof. Define for $y \in G$, $\psi_y(x) = \varphi(x, y)$, so $\psi_y: V \to H$ is a morphism. When y has finite order, the image of ψ_y is finite, thanks to (b) (d). But V is connected, by (c), so this image must reduce to a point. In other words, for y of finite order in G, $\varphi(x, y) = \varphi(x', y)$ for all $x, x' \in V$, i.e., $\varphi_x(y) = \varphi_{x'}(y)$. Therefore $\varphi_{x'}\varphi_x^{-1}$ sends a dense subset of G (thanks to (a)) onto e. Conclusion: $\varphi_x = \varphi_{x'}$ for all $x, x' \in V$. $\square$

Corollary. *Let G be a diagonalizable subgroup of an algebraic group G'. Then $N_{G'}(G)^\circ = C_{G'}(G)^\circ$.*

Proof. Set $G = H$ in the proposition, so (a) (b) hold by the preceding remarks. Then let $V = N_{G'}(G)^\circ$, and define $\varphi: V \times G \to G$ by $\varphi(x, y) = xyx^{-1}$. Evidently (c) and (d) are satisfied. That φ_x is constant implies that $\varphi_x = \varphi_e$, i.e., that all $x \in N_{G'}(G)^\circ$ actually centralize G. Conversely, of course, $C_{G'}(G)^\circ \subset N_{G'}(G)^\circ$. $\square$

As an example, the normalizer of $\mathsf{D}(n, \mathsf{K})$ in $\mathsf{GL}(n, \mathsf{K})$ is the group of monomial matrices; the centralizer of $\mathsf{D}(n, \mathsf{K})$ being itself, the quotient is finite, isomorphic to S_n. On the other hand, $\mathsf{D}(n, \mathsf{K})$ is a self-normalizing subgroup of the group $\mathsf{T}(n, \mathsf{K})$.

16.4. Weights and Roots

Let D be a diagonalizable subgroup of an algebraic group G. We have just seen that D is rather "rigid" under the action of G by inner automorphisms. It is also important to turn the question around and to see how D acts on G. For example, when D is the diagonal subgroup of $G = \mathsf{SL}(n, \mathsf{K})$, it can be seen that D stabilizes certain one dimensional unipotent subgroups of G corresponding to the root system in the Lie algebra case (char $\mathsf{K} = 0$).

In general, Int x $(x \in D)$ is an automorphism of G, so its differential Ad x is an automorphism of $\mathfrak{g}$. Ad being a morphism of algebraic groups (10.3), Ad D is a diagonalizable subgroup of Aut $\mathfrak{g} \subset \mathrm{GL}(\mathfrak{g})$. Whenever we are given a d-group H in some $\mathrm{GL}(V)$, it is convenient to write V as a direct sum of subspaces V_α, $\alpha \in \mathrm{X}(H)$, where

$$V_\alpha = \{v \in V \,|\, x \,.\, v = \alpha(x)v \text{ for all } x \in H\}.$$

Those α for which $V_\alpha \neq 0$ are the **weights** of H in V (11.4). In particular, returning to the situation $D \subset G$, Ad $D \subset \mathrm{GL}(\mathfrak{g})$, we obtain weights of D (i.e., of Ad D) in $\mathfrak{g}$. The *nonzero* ones are called the **roots of G relative to D**, the set of these being denoted $\Phi(G, D)$ (or just Φ when no confusion can exist). Corresponding to the weight 0 we have the fixed point space $\mathfrak{g}^D$, which of course includes $\mathfrak{d}$ (but may be bigger). With this notation we have a decomposition: $\mathfrak{g} = \mathfrak{g}^D \oplus \coprod_{\alpha \in \Phi} \mathfrak{g}_\alpha$, $\Phi = \Phi(G, D)$. When G is "semisimple" and D is chosen to be a torus of maximal dimension, we shall eventually be able to show that Φ is an **abstract root system** in the sense of the Appendix and that G is (almost) characterized by Φ. Indeed, the decomposition of $\mathfrak{g}$ just obtained will turn out to be the classical Cartan decomposition when char $\mathsf{K} = 0$. But all of this requires a substantial amount of groundwork.

To conclude this section we prove a modest, but useful, proposition.

Proposition. *Let D be a connected d-subgroup of G (i.e., a torus), H a closed subgroup of G stabilized by* Int D. *Then there exists $x \in D$ such that $C_H(x) = C_H(D)$ and $\mathfrak{c}_\mathfrak{h}(x) = \mathfrak{c}_\mathfrak{h}(D)$. In particular, $C_G(D)$ coincides with the centralizer of some element of D.*

Proof. We can assume that $G = \mathrm{GL}(V)$ for some V. Write $V = V_1 \oplus \cdots \oplus V_r$, where V_i belongs to the weight $\alpha_i \in \mathrm{X}(D)$. Since D is connected, Ker $(\alpha_i \alpha_j^{-1})$ $(i \neq j)$ has codimension one in D; it is therefore possible to find $x \in D$ not in any of these (finitely many) subgroups. View $M = \mathrm{GL}(V_1) \times \cdots \times \mathrm{GL}(V_r)$ as a closed subgroup of $\mathrm{GL}(V)$ in the obvious way. Then it follows readily that $C_H(x) = M \cap H = C_H(D)$, $\mathfrak{c}_\mathfrak{h}(x) = \mathfrak{m} \cap \mathfrak{h} = \mathfrak{c}_\mathfrak{h}(D)$ (by direct calculation). $\square$

If D is a d-group which acts as a group of automorphisms of G (not necessarily inner), then D also acts as a group of automorphisms of $\mathfrak{g}$. This leads again to a set $\Phi(G, D)$ consisting of the nonzero weights of D in $\mathfrak{g}$.

Exercises

1. Give an example of an algebraic group which consists of semisimple elements but is not diagonalizable.
2. If G is a d-group, $\mathfrak{g}$ consists of semisimple elements.
3. If $G = (G, G)$, then $\mathrm{X}(G) = 0$.

4. Let T be a torus. Show that $X(T)$ and $Y(T)$ are dual $\mathbb{Z}$-modules under the natural pairing $\langle \chi, \lambda \rangle \in \mathbb{Z}$. How does this fail if T is replaced by a nonconnected d-group?

5. Let T be a torus, m a positive integer. Prove that the morphism $x \mapsto x^m$ maps T *onto* T and is bijective if m is a power of p, separable if $(p, m) = 1$ ($p = \text{char exp } \mathsf{K}$).

6. Prove that a connected subgroup of a d-group is a direct factor. [Cf. the proof of Theorem 16.2.]

7. Assume K is not the algebraic closure of a finite field. If T is a torus, show that T is the closure of the cyclic subgroup generated by one of its elements. [Note that K^* contains free abelian groups of arbitrarily large finite rank.]

8. The differential of a character $\chi : G \to \mathbf{G}_m$ is a linear function $d\chi : \mathfrak{g} \to \mathsf{K}$. If G is a torus and $X(G)$ has basis $\chi_1, \ldots, \chi_n$, prove that $d\chi = 0$ if and only if $\chi = p\sum a_i \chi_i$ ($a_i \in \mathbb{Z}$).

9. Let $T = \mathrm{SL}(n, \mathsf{K}) \cap \mathrm{D}(n, \mathsf{K})$, $\chi : T \to \mathbf{G}_m$ the character sending diag $(a_1, \ldots, a_n) \mapsto a_i a_{i+1}^{-1}$ for fixed $i < n$. Construct a 1-psg $\lambda : \mathbf{G}_m \to T$ such that $\langle \chi, \lambda \rangle = 2$.

10. For which $x \in \mathrm{D}(n, \mathsf{K})$ is it true that $C_{\mathrm{GL}(n, \mathsf{K})}(x) = C_{\mathrm{GL}(n, \mathsf{K})}(\mathrm{D}(n, \mathsf{K}))$?

11. Verify the details of the equivalence between $\mathscr{D}$ and $\mathscr{A}$ outlined in (16.2).

12. Let G be an algebraic group, $H = \bigcap_{\chi \in X(G)} \mathrm{Ker}\, \chi$. Prove that (a) H is a closed normal subgroup, (b) G/H is diagonalizable, (c) $X(G) \cong X(G/H)$. In particular, $X(G)$ is finitely generated. Describe $X(\mathrm{GL}(n, \mathsf{K}))$.

13. The elements $x \in D$ described in Proposition 16.4 form a dense subset of D.

Notes

The categorical framework sketched in (16.2) is developed extensively in SGAD and in Demazure, Gabriel [1].

Chapter VII

Solvable Groups

17. Nilpotent and Solvable Groups

The usual notions of "nilpotent" and "solvable" group adapt very well to the study of algebraic groups, because the various commutator groups involved turn out to be *closed* (17.2). To prove this we need a purely group-theoretic fact, which is developed in (17.1) to avoid a later interruption. In (17.5) and (17.6) we shall see to what extent a solvable linear group can be put in triangular form. The result of (17.5) is essential later on, while that of (17.6) will be proved again in §21 using some high powered machinery (the proof here is, by contrast, quite elementary).

17.1. A Group-Theoretic Lemma

Recall that (x, y) denotes the **commutator** $xyx^{-1}y^{-1}$, when x, y are elements of a group G. If A, B are two subgroups of G, the subgroup of G generated by all (x, y), $x \in A$, $y \in B$, will be denoted by (A, B). The identity

$$(*) \qquad z(xyx^{-1}y^{-1})z^{-1} = (zxz^{-1})(zyz^{-1})(zxz^{-1})^{-1}(zyz^{-1})^{-1}$$

shows that (A, B) is normal in G if both A and B are normal.

Our first result will be a corollary of the main lemma, but is needed to obtain the latter.

Lemma A. *Let* $[G:Z(G)] = n < \infty$. *Then* (G, G) *is finite.*

Proof. Let S be the set of all commutators in G, so S generates (G, G). It is clear that (x, y) depends only on the cosets of x, y modulo $Z(G)$; in particular, Card $S \leqslant n^2$. Given a product of commutators, any two can be made adjacent by suitable conjugation, e.g., $(x_1, y_1)(x_2, y_2)(x_3, y_3) = (x_1, y_1)(x_3, y_3)(z^{-1}x_2z, z^{-1}y_2z)$, where $z = (x_3, y_3)$. Therefore it is enough to show that the $(n + 1)^{st}$ power of an element of S is the product of n elements of S, in order to conclude that each element of (G, G) is the product of at most n^3 factors from S. This in turn will clearly force (G, G) to be finite.

Now $(x, y)^n \in Z(G)$, allowing us to rewrite $(x, y)^{n+1}$ as $y^{-1}(x, y)^n y(x, y) = y^{-1}((x, y)^{n-1}(x, y^2))y$, a product of n commutators. $\square$

It is only the *finiteness* of S which counts in this lemma, not the size of $G/Z(G)$, but this fact requires a more subtle proof. The following lemma of R. Baer goes through under an even weaker hypothesis (see Notes), but we need only the stated version.

109

Lemma B.   *Let A, B be normal subgroups of G, and suppose the set $S = \{(x, y) | x \in A, y \in B\}$ is finite. Then (A, B) is finite.*

Proof.   It does no harm to assume that $G = AB$. G permutes the set S (acting via inner automorphisms, as in (∗) above). If H is the kernel of the resulting homomorphism $G \to$ symmetric group on S, then clearly H is normal and of finite index in G. Moreover, H centralizes $C = (A, B)$. It follows that $H \cap C$ is central in C and of finite index. By Lemma A, (C, C) is *finite* (as well as normal in G, since $C \lhd G$). So we can replace G by $G/(C, C)$, i.e., we can assume that C is abelian.

Now the commutators (x, y), $x \in A$, $y \in C$, lie in S and *commute* with each other. C being abelian, $(x, y)^2 = (xyx^{-1})^2 y^{-2} = (x, y^2)$ is another such commutator. This clearly forces (A, C) to be finite (as well as normal in G). Replacing G by $G/(A, C)$, we may further assume that A centralizes C. This implies that the square of an arbitrary commutator is again a commutator. It follows that (A, B) is finite. $\square$

17.2. Commutator Groups

If A and B are arbitrary closed (e.g., finite) subgroups of an algebraic group G, the group (A, B) generated by commutators $xyx^{-1}y^{-1}$ ($x \in A, y \in B$) unfortunately need not be closed (Exercise 1). The situation improves when A or B is *connected*, or when both A and B are *normal* in G. (In fact, if A is only required to normalize B, part (b) of the following proposition remains true; but this requires a more elaborate version of Lemma 17.1B.)

Proposition.   *Let A, B be closed subgroups of an algebraic group G.*
(a) *If A is connected, then (A, B) is closed and connected.*
(b) *If A and B are normal in G, then (A, B) is closed (and normal in G).*
In particular, (G, G) is always closed.

Proof.   (a) Associate with each $y \in B$ the morphism $\varphi_y : A \to G$ defined by $\varphi_y(x) = xyx^{-1}y^{-1}$. Since A is connected, and $\varphi_y(e) = e$, Proposition 7.5 shows that the group generated by all $\varphi_y(A)$ ($y \in B$) is closed and connected; but this is by definition (A, B).

(b) It follows from part (a) that (A°, B) and (A, B°) are closed, connected (as well as normal) subgroups of G, so their product C has the same properties (cf. Corollary 7.4). To show that (A, B) is closed, it therefore suffices to show that C has finite index in (A, B), which is a purely group-theoretic question. In the abstract group G/C, the image of A° (resp. B°) centralizes the image of B (resp. A). Since $[A:A^\circ]$ and $[B:B^\circ]$ are finite, this implies that there are only finitely many commutators in G/C constructible from the images of A and B. Lemma 17.1B then guarantees that $(A, B)/C$ is finite. $\square$

17.3. Solvable Groups

Recall that an abstract group G is **solvable** if its **derived series** terminates in e, this series being defined inductively by $\mathscr{D}^0 G = G$, $\mathscr{D}^{i+1}G = (\mathscr{D}^i G, \mathscr{D}^i G)$

($i \geqslant 0$). In case G is an algebraic group, $\mathscr{D}^1 G = (G, G)$ is a *closed* normal subgroup of G, *connected* if G is (Proposition 17.2). By induction, the same holds true for all $\mathscr{D}^i G$. Therefore, the abstract notion of solvability is well suited for use here. For example, if G is a connected solvable algebraic group of positive dimension, then dim $(G, G) <$ dim G. (This fact is handy in inductive arguments.) It is easy to see that an algebraic group G is solvable if and only if there exists a chain $G = G_0 \supset G_1 \supset \cdots \supset G_n = e$ of closed subgroups for which $(G_i, G_i) \subset G_{i+1}$ ($0 \leqslant i < n$), cf. Exercise 4.

The following group-theoretic facts are well known.

Lemma. (a) *Subgroups and homomorphic images of a solvable group are solvable.*

(b) *If N is a normal solvable subgroup of G for which G/N is solvable, then G is itself solvable.*

(c) *If A, B are normal solvable subgroups of G, so is AB.* $\square$

Since the theory of finite solvable groups is rather elaborate in its own right, it is unrealistic to expect a detailed description of all solvable algebraic groups. But the *connected* ones turn out to be fairly manageable. A good example to keep in mind is the upper triangular group $T(n, \mathsf{K})$. Since the diagonal entries in the product of two upper triangular matrices are just the respective products of diagonal entries, it is clear that the derived group of $T = T(n, \mathsf{K})$ lies in $U = U(n, \mathsf{K})$. On the other hand, we have seen that U is generated by the matrices $u_{ij}(a) = 1 + ae_{ij}$ ($i < j$; $a \in \mathsf{K}$). If t is the diagonal matrix with i^{th} entry 2 and all other diagonal entries 1, then a quick calculation shows that $tu_{ij}(a)t^{-1} = u_{ij}(2a)$, hence that $(t, u_{ij}(a)) = u_{ij}(a)$. It follows that U is precisely (T, T).

Consider in turn the commutation in U. It is convenient to approach this a bit indirectly. Denote by $\mathfrak{T}$ the full set of upper triangular matrices, viewed as an associative subalgebra of $M(n, \mathsf{K})$ (rather than as the Lie algebra $\mathfrak{t}(n, \mathsf{K})$). Evidently the subset $\mathfrak{N}$ of matrices with 0 diagonal is a 2-sided ideal of $\mathfrak{T}$ (as a Lie algebra, $\mathfrak{N}$ is $\mathfrak{n}(n, \mathsf{K})$). So each ideal power $\mathfrak{N}^h$ ($h \geqslant 1$), consisting of sums of h-fold products from $\mathfrak{N}$, is again a 2-sided ideal. It is easy to see that $\mathfrak{N}^h$ is the linear span of all e_{ij} ($j - i \geqslant h$). Now $U = 1 + \mathfrak{N}$, so we conclude from this that $U_h = 1 + \mathfrak{N}^h$ is a closed normal subgroup of U, with $(U_h, U_l) \subset U_{h+l}$. In particular, U is *solvable*. This shows that T is solvable.

As a sort of converse, it will be seen in (17.6) that every *connected* solvable algebraic group is isomorphic to a closed subgroup of some $T(n, \mathsf{K})$.

17.4. Nilpotent Groups

The **descending central series** of a group G is defined inductively by $\mathscr{C}^0 G = G$, $\mathscr{C}^{i+1} G = (G, \mathscr{C}^i G)$. Each $\mathscr{C}^i G$ is normal in G. Thanks again to Proposition 17.2, these are *closed* subgroups when G is an algebraic group (connected if G is). G is called **nilpotent** provided $\mathscr{C}^n G = e$ for some n. Evidently a commutative group is nilpotent; on the other hand, a nilpotent group is solvable, because of the inclusion $\mathscr{D}^i G \subset \mathscr{C}^i G$. The calculations made in (17.3) actually show that U(n, K) is nilpotent, whereas T(n, K) ($n \geqslant 2$) is not. For later use, define $G^\infty = \cap \, \mathscr{C}^i(G)$.

We recall a few familiar facts from group theory.

Lemma. (a) *Subgroups and homomorphic images of a nilpotent group are nilpotent.*

(b) *If $G/Z(G)$ is nilpotent, then G is nilpotent.*

(c) *Let G be nilpotent. If n is the greatest index for which $\mathscr{C}^n G \neq e$, then $\mathscr{C}^n G \subset Z(G)$; in particular, $Z(G) \neq e$ (unless $G = e$).*

(d) *Let G be nilpotent, H a proper subgroup of G. Then H is properly included in $N_G(H)$.* $\square$

For algebraic groups, there are a couple of important refinements:

Proposition. *Let G be a connected nilpotent algebraic group of positive dimension. Then:*

(a) *$Z(G)$ has positive dimension.*

(b) *If H is a proper closed connected subgroup of G, dim $H <$ dim $N_G(H)$. In particular, if* codim $H = 1$, *then H is normal in G.*

Proof. (a) Since all $\mathscr{C}^i G$ are connected, part (c) of the lemma shows that $Z(G)^\circ \neq e$.

(b) Use induction on dim G. If $Z = Z(G)^\circ$ is included in H, pass to the connected nilpotent group G/Z (of lower dimension than G, thanks to (a)). By induction, the image of H is properly contained in its normalizer, being a proper subgroup of G/Z; so the same holds true for H. If Z is not included in H, then ZH is a subgroup of $N_G(H)$ having larger dimension than H. $\square$

17.5. Unipotent Groups

A subgroup of an algebraic group is called **unipotent** if all its elements are unipotent (15.1). Example: U(n, K). We intend to prove that a unipotent linear group (closed or not) is conjugate to a subgroup of U(n, K), hence is nilpotent. This may be viewed as a generalization of the fact that finite p-groups are nilpotent, since in the case char K $= p > 0$, an endomorphism is unipotent if and only if its order in GL(V) is a power of p.

Theorem. *Let G be a unipotent subgroup of* GL(V), $0 \neq V$ *finite dimensional. Then G has a common eigenvector in V, i.e., a nonzero vector fixed by all elements of G.*

Proof. Suppose V has a proper nonzero subspace W stable under G. Then the image of G in GL(W) is unipotent, and a vector in W fixed by this group is also a vector in V fixed by G. Therefore, we may assume from the outset that V is an *irreducible G-module*. This will allow us to invoke a standard (though nontrivial) theorem of Burnside: If R is a subalgebra of End V

which acts irreducibly on V, then $R = \text{End } V$. (For a proof, cf. S. Lang, Algebra, XVII, §3.)

Now the assumption that G is unipotent implies $\text{Tr}(x) = \text{Tr}(1) = (\dim V) \cdot 1$ for all $x \in G$. Writing a typical $x \in G$ as $1 + n$ (n nilpotent), we deduce that for arbitrary $y \in G$, $\text{Tr}(y) = \text{Tr}(xy) = \text{Tr}((1 + n)y) = \text{Tr}(y) + \text{Tr}(ny)$, or $\text{Tr}(ny) = 0$. The K-linear combinations y of elements of G therefore satisfy the same equation. But these form a subalgebra R of End V, acting irreducibly on V (since G does). Burnside's Theorem implies that for all $x = 1 + n \in G$, all $y \in \text{End } V$, $\text{Tr}(ny) = 0$. By taking n in matrix form and letting y vary over the standard unit matrices e_{ij}, we see that all entries of n are 0, i.e., G consists only of 1 (and dim $V = 1$). $\square$

Under the assumption that $G \subset \text{GL}(V)$ is unipotent, the theorem yields a common eigenvector v_1. If $V_1 = \text{K}v_1$, G acts on V/V_1, the image of G in $\text{GL}(V/V_1)$ being again unipotent. Induction on dim V therefore allows us to conclude that V has a full flag of subspaces stable under G. In matrix terms:

Corollary.   *If G is a unipotent subgroup of* $\text{GL}(n, \text{K})$, *then G is conjugate to a subgroup of* $\text{U}(n, \text{K})$. *In particular, G is nilpotent.* $\square$

There is no corresponding theorem for subgroups of $\text{GL}(V)$ consisting of *semisimple* elements; indeed, examples of such groups are quite varied (cf. Exercise 6). (So we do not want to waste the term "semisimple group" on this concept! Cf. (19.5) below.) After further study of solvable groups we shall, in partial compensation, be able to prove that a *closed connected* subgroup of $\text{GL}(V)$ consisting of semisimple elements must be a torus (Exercise 21.2).

17.6. Lie-Kolchin Theorem

Theorem 17.5 and its corollary may be viewed as the precise analogue of Engel's Theorem for Lie algebras. We can also prove a good analogue of Lie's Theorem (if we impose a connectedness assumption); this works in arbitrary characteristic, although the Lie algebra theorem does not.

Theorem.   *Let G be a connected solvable subgroup of* $\text{GL}(V)$, $0 \neq V$ *finite dimensional. Then G has a common eigenvector in V.*

Proof.   We may assume that G is closed (cf. Exercise 2). The theorem follows at once from Proposition 15.4, if G is commutative. To proceed in general, we use induction on $n = \dim V$ and on $d = $ derived length of G (the least d for which $\mathscr{D}^d G = e$). The cases $n = 1$, $d = 1$ have just been disposed of.

Suppose V has a nonzero proper subspace W stable under G. If a basis of W is extended to a basis of V, the matrices representing G have the form

$$\begin{pmatrix} \varphi(x) & * \\ \hline 0 & \psi(x) \end{pmatrix}$$

Evidently the group of all such matrices is closed in $GL(n, K)$, and the homomorphism $x \mapsto \varphi(x)$ is a morphism of algebraic groups. G being *connected*, $\varphi(G) \subset GL(W)$ is also connected, as well as solvable (17.3). Since $n > \dim W$, we can find a common eigenvector for $\varphi(G)$ in W, hence for G in V.

This leaves the case in which G acts *irreducibly* on V. As in the proof of Theorem 17.5, we have to show now that $n = 1$. Set $G' = (G, G)$, so G' is normal in G, connected (17.2), and solvable of derived length $d - 1$. By induction, G' has a common eigenvector in V, i.e., a *semi-invariant* (11.4). Let W be the span of all such vectors in V. Then Lemma 11.4 implies that G' acts diagonally on W. The discussion in (11.4) shows also that W is G-stable, so $W = V$ by irreducibility. Since G' acts diagonally on V, G' is commutative (i.e., $d \leqslant 2$).

Take $y \in G'$. What we have just said shows that for each $x \in G$, the matrix of xyx^{-1} (relative to a basis of V on which y acts diagonally) is again diagonal. Being conjugate to the matrix of y, it has the same eigenvalues (possibly permuted), which limits xyx^{-1} to a finite number of possibilities. Thus, the image of G under the morphism $x \mapsto xyx^{-1}$ is both finite and connected, hence consists only of y. In other words, $G' \subset Z(G)$.

Now we appeal to Schur's Lemma, which is standard (and also very easy to prove): cf. S. Lang, Algebra, XVII, §1. This guarantees (since K is algebraically closed and G acts irreducibly on V) that G' consists of scalars. But any commutator of matrices has determinant 1, so the scalars in G' are n^{th} roots of unity (of which K contains at most n). This forces G' to be finite, as well as connected, i.e., $G' = e$ and $d = 1, n = 1$. $\square$

This theorem, usually referred to as the **Lie-Kolchin Theorem**, will be proved again in §21, in a way which does not depend on the intervening application in §19. So the present proof, though elementary, could be omitted.

Exercises

1. Find an algebraic group G having two closed subgroups A, B for which (A, B) is not closed. [Cf. Exercise 7.10.]

2. Let G be an algebraic group, H a (not necessarily closed) subgroup. If H is commutative (resp. nilpotent, solvable, unipotent), prove that $\bar{H}$ has the same property.

3. Let G be an algebraic group. Use Lemma 17.1A to prove directly that (G, G) is closed, avoiding Lemma 17.1B.

4. An algebraic group G is solvable if and only if there is a chain $G = G_0 \supset G_1 \supset \cdots \supset G_n = e$ of closed subgroups such that $(G_i, G_i) \subset G_{i+1}$ $(0 \leqslant i < n)$. Give an analogous criterion for G to be nilpotent.

5. Let G be a nilpotent algebraic group (not necessarily connected). Prove that $Z(G)$ has positive dimension if G does.

6. Prove that $SO(3, \mathbb{R})$ ($=$ group of 3×3 real orthogonal matrices of det 1) is a connected subgroup of $SL(3, \mathbb{C})$ consisting of semisimple elements, but is not commutative.

7. Show as follows that $U = U(n, K)$ has a chain of closed connected subgroups, each normal in U and of codimension 1 in the preceding one. Order the pairs (i, j) with $i < j$ by the rule: $(i, j) < (k, l)$ if $j < l$ or if $j = l$ and $i > k$. So the ascending order is: $(1, 2)$; $(2, 3)$, $(1, 3)$; $(3, 4)$, $(2, 4)$, $(1, 4)$; $\ldots$, $(2, n)$, $(1, n)$. Set $U_{ij} = \{x \in U | x_{kl} = 0 \text{ for } (k, l) \leqslant (i, j)\}$, and verify that the chain $U \supset U_{12} \supset U_{23} \supset \cdots \supset U_{1n} \supset e$ has the asserted properties. Verify too that each U_{ij} is normalized by $D(n, K)$.

8. Let G be a unipotent algebraic group acting on an affine variety X. Prove that all G-orbits are closed in X. In particular, all conjugacy classes of G are closed. [Consider the action of G on $K[X]$ by translations (8.6). If Y is a nonclosed orbit, find $f \in K[X]$ vanishing on $\bar{Y} - Y$ but not on Y. The G-translates of f span a finite dimensional G-stable space consisting of functions which vanish on $\bar{Y} - Y$. Apply Theorem 17.5 to this representation of G.]

Notes

Lemma 17.1A goes back to I. Schur, while Lemma 17.1B (in a stronger version) is due to R. Baer, cf. Rosenlicht [5] or Borel [4, pp. 111–114]. Theorem 17.5 was proved by Kolchin [2, pp. 775–776] (for another approach, see Demazure, Gabriel [1, IV §2]). Theorem 17.6 appears in Kolchin [1, pp. 19–20].

18. Semisimple Elements

In §16 we studied the structure of a commutative algebraic group consisting of semisimple elements and looked at how such a group acts upon (or is acted on by) an ambient group. This action can be described to a certain extent in terms of a single semisimple element s. We have to analyze now some of the more subtle aspects of this situation, with a view toward proving a structure theorem for solvable groups in §19.

To motivate what follows, it is instructive to look at the way $\mathfrak{g}$ decomposes under Ad s (which is also semisimple): $\mathfrak{g} = \mathfrak{c}_{\mathfrak{g}}(s) \oplus \mathfrak{n}$, where $\mathfrak{n}$ is the sum of eigenspaces for eigenvalues of Ad s different from 1. Is there a corresponding decomposition in G? On the one hand, the group $C_G(s)$ ought to correspond to $\mathfrak{c}_{\mathfrak{g}}(s)$, i.e., the latter ought to be the Lie algebra of the former. This is true for semisimple s (18.1) (but fails sometimes for other elements of G when char $K \neq 0$). What should correspond to $\mathfrak{n}$ is less obvious. The conjugacy class $Cl_G(s)$ has the right dimension. It turns out to be closed (not just locally closed) when s is semisimple (18.2), and its translate $M = Cl_G(s)s^{-1}$ turns out to have $\mathfrak{n}$ as its tangent space at e. When we generalize this situation to the case where s acts (by conjugation) on a connected unipotent subgroup U of G we obtain finally a very precise decomposition in U of the above type (18.3).

18.1. Global and Infinitesimal Centralizers

Let $G = \mathrm{GL}(n, \mathsf{K})$, $\mathfrak{g} = \mathfrak{gl}(n, \mathsf{K})$. It was shown in (10.6) that for arbitrary $x \in G$, $C_G(x)$ has as Lie algebra precisely $\mathfrak{c}_\mathfrak{g}(x)$. When char $\mathsf{K} = 0$, this is true for any algebraic group (13.4), but unfortunately it can fail in characteristic p. The problem arises in connection with the bad behavior of unipotent elements, and disappears when we limit attention to *semisimple* elements.

Consider a closed subgroup H of $G = \mathrm{GL}(n, \mathsf{K})$ which is normalized by a semisimple element $s \in G$. Evidently $C_H(s) = H \cap C_G(s)$, forcing $\mathscr{L}(C_H(s)) \subset \mathfrak{h} \cap \mathscr{L}(C_G(s)) = \mathfrak{h} \cap \mathfrak{c}_\mathfrak{g}(s) = \mathfrak{c}_\mathfrak{h}(s)$. To prove the reverse inclusion we must in effect show that $\mathfrak{c}_\mathfrak{h}(s)$ is not "too big". But since $\mathrm{Ad}\ s$ is semisimple, we have a decomposition $\mathfrak{h} = \mathfrak{c}_\mathfrak{h}(s) \oplus \mathfrak{n}'$ (cf. beginning of this section), $\mathfrak{n}'$ being the sum of nontrivial eigenspaces, and it is therefore sufficient to show that $\mathfrak{n}'$ is not "too small".

The conjugacy class $\mathrm{Cl}_G(s)$ is a locally closed set (8.3) of dimension complementary to dim $C_G(s)$ (cf. Theorem 4.3). Its translate $M = \mathrm{Cl}_G(s)s^{-1}$ passes through e, so we can consider its tangent space $\mathscr{T}(M)_e = \mathfrak{m}$ to be a subspace of $\mathscr{T}(G)_e = \mathfrak{g}$. We shall see below that $\mathfrak{m}$ coincides with the space $\mathfrak{n}$ defined earlier: $\mathfrak{g} = \mathfrak{c}_\mathfrak{g}(s) \oplus \mathfrak{n}$. Now $M \cap H \supset M'$, the analogous subvariety of H (of dimension complementary to dim $C_H(s)$), and the tangent space $\mathfrak{m}'$ lies in $\mathfrak{m} \cap \mathfrak{h} = \mathfrak{n} \cap \mathfrak{h} = \mathfrak{n}'$. This time the inclusion works in our favor: dim $\mathfrak{m}' \leqslant$ dim $\mathfrak{n}'$ implies that codim $C_H(s)$ in H = dim $M' \leqslant$ dim $\mathfrak{n}'$ = codim $\mathfrak{c}_\mathfrak{h}(s)$ in $\mathfrak{h}$, whence dim $\mathfrak{c}_\mathfrak{h}(s) \leqslant$ dim $C_H(s)$, as desired.

It remains to be shown that $\mathfrak{m} = \mathfrak{n}$. (This concerns only G, not H.) Recall that the morphism $\gamma_s : G \to G$ defined by $\gamma_s(x) = xsx^{-1}s^{-1}$ has as differential the linear map $1 - \mathrm{Ad}\ s : \mathfrak{g} \to \mathfrak{g}$ (10.1). Evidently $M = \gamma_s(G)$, so $(1 - \mathrm{Ad}\ s)\mathfrak{g} \subset \mathfrak{m}$. On the other hand, it is plain that $\mathrm{Ker}\ (d\gamma_s) = \mathfrak{c}_\mathfrak{g}(s)$ and $\mathrm{Im}\ (d\gamma_s) = \mathfrak{n}$, so in particular $\mathfrak{m} \supset \mathfrak{n}$. But we already know all the dimensions involved, since $\mathfrak{c}_\mathfrak{g}(s) = \mathscr{L}(C_G(s))$: dim $\mathfrak{m}$ = dim M = codim $C_G(s)$ = codim $\mathfrak{c}_\mathfrak{g}(s)$ = dim $\mathfrak{n}$. Therefore, $\mathfrak{m} = \mathfrak{n}$. To summarize:

Proposition. *Let H be a closed subgroup of $\mathrm{GL}(n, \mathsf{K})$, normalized by a semisimple element $s \in \mathrm{GL}(n, \mathsf{K})$. Then $\mathfrak{c}_\mathfrak{h}(s) = \mathscr{L}(C_H(s))$, and the sum of all nontrivial eigenspaces of $\mathrm{Ad}\ s$ in $\mathfrak{h}$ is the tangent space of $\mathrm{Cl}_H(s)s^{-1}$.* $\square$

(Notice that the proposition remains true if we substitute for $\mathrm{GL}(n, \mathsf{K})$ any algebraic subgroup containing H and s.)

It is worthwhile to interpret this result in geometric terms. With G, H, s as before, consider the conjugacy class morphism

$$\varphi : H \to \mathrm{Cl}_H(s) \subset H,$$

where $\varphi(x) = xsx^{-1}$. This is just the composite of γ_s and translation by s; the latter being an isomorphism, we see that $\mathrm{Ker}\ (d\varphi)_e = \mathrm{Ker}\ (d\gamma_s)_e = \mathfrak{c}_\mathfrak{h}(s) = \mathscr{L}(C_H(s))$. Since the fibres of φ are the cosets of $C_H(s)$, this implies that φ is separable (5.5). From remark (1) in (12.4) we then conclude:

Corollary A. *G, H, s as before. The conjugacy class morphism $\varphi:H \to$ $Cl_H(s)$ may be identified with the canonical morphism $H \to H/C_H(s)$.* $\square$

At the outset we discussed the problem of imitating the decomposition $\mathfrak{h} = \mathfrak{c}_\mathfrak{h}(s) \oplus \mathfrak{n}'$ in H. Since $\mathfrak{n}'$ is precisely the tangent space of $M' = Cl_H(s)s^{-1}$, by the above proof, and since $\mathfrak{c}_\mathfrak{h}(s)$ is the Lie algebra of $C_H(s)$, we obtain the following partial solution:

Corollary B. *G, H, s as before. Let $\tau:C_H(s) \times M' \to H$ be the product morphism. Then $d\tau_{(e, e)}$ is an isomorphism $\mathfrak{c}_\mathfrak{h}(s) \oplus \mathfrak{m}' \overset{\sim}{\to} \mathfrak{h}$. Moreover, $d(\gamma_s|M') = (1 - \mathrm{Ad}\ s)_{|\mathfrak{m}'}:\mathfrak{m}' \to \mathfrak{m}'$ is an isomorphism.* $\square$

18.2. Closed Conjugacy Classes

Proposition. *Let H be a closed subgroup of $G = \mathrm{GL}(n, \mathsf{K})$ normalized by a semisimple element $s \in G$. Then $Cl_H(s)$ is closed.*

Proof. If x is an endomorphism of a finite dimensional vector space, T an indeterminate, denote by $m(x, \mathsf{T})$ (resp. $c(x, \mathsf{T})$) the minimal (resp. characteristic) polynomial of x. Recall that x is semisimple if and only if $m(x, \mathsf{T})$ has no repeated roots.

Now define $W = \{x \in N_G(H)|m(s, x) = 0$ and $c(\mathrm{Ad}\ x|\mathfrak{h}, \mathsf{T}) = c(\mathrm{Ad}\ s|\mathfrak{h}, \mathsf{T})\}$. Evidently W is a *closed* subset of $N_G(H)$ (hence of G), stable under conjugation by H. The first condition defining W requires that each $x \in W$ satisfy the minimal polynomial of s, which has no multiple roots (by assumption); therefore $m(x, \mathsf{T})$ has no multiple roots, and x is also semisimple. Then Proposition 18.1 applies to the action of x on H: dim $Cl_H(x) =$ codim $C_H(x) =$ codim $\mathfrak{c}_\mathfrak{h}(x) =$ dim $\mathfrak{h} - m_1(x)$, where $m_1(x) =$ multiplicity of the eigenvalue 1 of $\mathrm{Ad}\ x|\mathfrak{h}$. But the second condition defining W forces $m_1(x) = m_1(s)$, whence all H-orbits in W have equal dimension. By Proposition 8.3, all orbits $Cl_H(x)$ are closed in W, hence in G. In particular, $Cl_H(s)$ is closed. $\square$

There is an analogous result for the H-orbit of a semisimple element of $\mathfrak{g}$: Exercise 6. As in (18.1) we note that $\mathrm{GL}(n, \mathsf{K})$ may be replaced by a closed subgroup G containing H, s.

The reader may wonder to what extent conjugacy classes of non-semisimple elements fail to be closed. This information is not needed elsewhere in the book, so we shall only indicate briefly what is true. If H is a commutative algebraic group, then each class consists of a single element, hence is surely closed. The classes of a unipotent group are also closed (Exercise 17.8). But if H is reductive (19.5), then the only closed classes are the semisimple ones. It is easy to see why this should be so in some cases. For example, in $\mathrm{GL}(n, \mathsf{K})$ the matrices $1 + ae_{ij}$ for fixed $i \neq j$ form a one dimensional unipotent group normalized by $\mathrm{D}(n, \mathsf{K})$, and since K is closed under taking square roots, all such matrices with $a \neq 0$ are easily seen to be conjugate. But then the closure of the conjugacy class of $1 + e_{ij}$ in $\mathrm{GL}(n, \mathsf{K})$ will include 1 (Exercise 5), which is in a class by itself.

18.3. Action of a Semisimple Element on a Unipotent Group

When H is a closed subgroup of an algebraic group G, normalized by a semisimple element $s \in G$, Proposition 18.2 shows that $M = Cl_H(s)s^{-1}$ is *closed* in H (and irreducible whenever H is connected, being the image of H under the morphism γ_s). On the other hand, $C = C_H(s)$ is a closed subgroup of H. If we know that H is connected and that the product morphism $\tau : C \times M \to H$ is bijective, we can conclude that C is also *connected*: The image of $C^\circ \times M$ (call it D) is constructible (4.4) and its finitely many left translates are disjoint and cover H. Since $\bar{D}$ is irreducible, $\bar{D} = H$. But then the left translates of D contain disjoint open subsets of H, which is possible only if $D = H$. Unfortunately, τ need not be bijective and C need not be connected, unless we restrict the choice of H. We are going to require now that H be *unipotent*. This allows us to exploit fully the Jordan decomposition and also to use induction on dim H, based on the fact that H is a *nilpotent* group (Corollary 17.5).

Theorem.   *Let U be a connected unipotent subgroup of an algebraic group G, $s \in G$ a semisimple element normalizing U. Define $\gamma_s(x) = xsx^{-1}s^{-1}$ $(x \in G)$, $M = \gamma_s(U)$, $C = C_U(s)$. Then:*

(a) *C is a closed subgroup and M a closed connected subvariety of U.*

(b) *The product morphism $\tau : C \times M \to U$ is bijective and (hence) C is connected.*

(c) *$\gamma = \gamma_s | M$ is a bijection of M onto itself.*

Proof.   The preceding remarks imply (a), while for (b) it will suffice to show that τ is bijective. For (c), we must show γ bijective. We proceed in steps, the first of which is obvious.

(1) *If $x \in U$, $y \in C$, then $\gamma_s(xy) = \gamma_s(x) = \gamma_s(x)\gamma_s(y)$.*

(2) *If $x \in Z(U)$, $y \in U$, then $\gamma_s(xy) = \gamma_s(x)\gamma_s(y)$, whence $\gamma_s(x^{-1}) = \gamma_s(x)^{-1}$.* Write $\gamma_s(xy) = xysy^{-1}x^{-1}s^{-1} = x(ysy^{-1}s^{-1})x^{-1}(xsx^{-1}s^{-1})$. When x is central, this equals $\gamma_s(y)\gamma_s(x)$. But $\gamma_s(xy) = \gamma_s(yx)$, so the assertion follows.

(3) *In case U is abelian, τ is bijective.* Since $U = Z(U)$, part (2) shows that $\gamma_s : U \to U$ is a group homomorphism. Its image is M, by definition, and its kernel is evidently C. Since M is now a group, τ is a group homomorphism, therefore *surjective* (since dim U = dim C + dim M). If $x \in C$, $y \in M$ and $\tau(x, y) = xy = e$, then writing $y = \gamma_s(u)$ for some $u \in U$, we have $s^{-1}x = us^{-1}u^{-1}$; but s and x commute, so the left side must be the Jordan decomposition of the semisimple element $us^{-1}u^{-1}$, forcing $x = e = y$. This shows that τ is also *injective*.

(4) *τ is always bijective.* Here we use induction on dim U. U being nilpotent, there is a nontrivial connected subgroup V of $Z(U)$ normalized by s (e.g., the identity component of $Z(U)$, cf.(17.4)(a)). If $V = Z(U) = U$, we can appeal to (3). Otherwise we consider the smaller dimensional connected unipotent group $U' = U/V$. Notice that s, U both lie in $N_G(V)$, while V is a normal subgroup, so we can set $s' = \pi(s)$, $\pi : N_G(V) \to N_G(V)/V$

the canonical morphism. The triple (G', U', s'), where $G' = N_G(V)/V$, satisfies the original assumptions on (G, U, s), so by induction the product morphism $\tau': C_{U'}(s') \times M' \to U'$ is bijective, where $M' = \gamma_{s'}(U') = \pi(M)$. The induction hypothesis applies equally well to (G, V, s), so that $\tau_0: C_V(s) \times \gamma_s(V) \to V$ is bijective.

Now we can prove τ *injective*. Let $z_1 x = z_2 y$ $(z_i \in C, x, y \in M)$, or $zx = y$ $(z = z_2^{-1} z_1 \in C)$. Apply π to this equation; since $\pi(z) \in C_{U'}(s')$ and $\pi(x)$, $\pi(y) \in M'$, the injectivity of τ' yields $\pi(z) = e$, or $z \in V$. Write $x = \gamma_s(u)$, $y \in \gamma_s(v)$, so $z(usu^{-1}) = vsv^{-1}$. The left side is the Jordan decomposition of the semisimple element vsv^{-1} (since $V \subset Z(U)$), whence $z = e$ and $x = y$, as desired.

Finally we have to show τ *surjective*. Notice first that $\gamma_s(V) \subset M \cap V$. Conversely, any $v \in V$ has the form zy $(y \in \gamma_s(V), z \in C_V(s))$ because τ_0 is surjective; if in addition $v \in M$, then $v = y \in \gamma_s(V)$ since τ is injective. This equality $\gamma_s(V) = M \cap V$ can be used in turn to show that $C_{U'}(s') = \pi(C)$, as follows. Let $\pi(x) = x'$ $(x \in U)$ commute with $\pi(s) = s'$. Then $\gamma_{s'}(x') = e$, or $\gamma_s(x) \in \operatorname{Ker} \pi \cap M = V \cap M = \gamma_s(V)$, or $\gamma_s(x) = \gamma_s(v)$ $(v \in V)$. But $V \subset Z(U)$, so (2) implies that $v^{-1}x \in C$, $\pi(v^{-1}x) = \pi(x)$ as claimed.

To show that τ is surjective is now easy:

$$
\begin{aligned}
U &= CMV && (\tau' \text{ is surjective, } \pi(M) = M', \pi(C) = C_{U'}(s')) \\
&= CVM && (V \subset Z(U)) \\
&= CC_V(s)\gamma_s(V)M && (\tau_0 \text{ is surjective}) \\
&= CM && (\text{since } \gamma_s(V)M = M \text{ by (2)}).
\end{aligned}
$$

(5) γ *is bijective.* By part (1) and the surjectivity of τ in (4), $M = \gamma_s(U) = \gamma_s(CM) = \gamma_s(M)$, so γ is surjective. If $\gamma_s(x) = \gamma_s(y)$ for $x, y \in M$, then $xy^{-1} \in C$ and $\tau(e, x) = \tau(xy^{-1}, y)$, whence $xy^{-1} = e$ (τ being injective), $x = y$. $\square$

The *connectedness* of $C_U(s)$ is going to be crucial later on. Instead of obtaining it directly we resorted to an admittedly roundabout argument; this illustrates the fact that connectedness properties of centralizers (etc.) are often quite subtle. However, it should be pointed out that when $0 = \operatorname{char} K$, a unipotent group (such as $C_U(s)$) is always connected (Exercise 15.6). The bijectivity of τ and of γ would, in any case, still have to be proved more or less as above, for later use.

18.4. Action of a Diagonalizable Group

Let G be any algebraic group. Besides studying the fixed points of a single semisimple element of G on a closed subgroup H, we shall have to investigate the action of a d-group D. For example, do global and infinitesimal centralizers still correspond as in (18.1)?

Proposition A. *Let H be a closed subgroup of G normalized by a d-group D. Then $\mathscr{L}(C_H(D)) = \mathfrak{c}_{\mathfrak{h}}(D)$.*

Proof. If $H^\circ \subset C_H(D)$, there is nothing to prove. Otherwise use induction on dim H, starting at dimension 0. Find $s \in D$ for which $H' = C_H(s)$ has smaller dimension than H. Of course, $C_H(D) \subset H'$. Thanks to Proposition 18.1, $\mathfrak{c}_\mathfrak{h}(s) = \mathfrak{h}'$; therefore $\mathfrak{c}_\mathfrak{h}(D) \subset \mathfrak{h}'$. Using these facts and induction, we have $\mathscr{L}(C_H(D)) = \mathscr{L}(C_{H'}(D)) = \mathfrak{c}_{\mathfrak{h}'}(D) = \mathfrak{c}_\mathfrak{h}(D) \cap \mathfrak{h}' = \mathfrak{c}_\mathfrak{h}(D)$. $\square$

Whenever a d-group D acts on an algebraic group H, Proposition A implies that the Lie algebra of the closed subgroup H^D consists of the fixed points of D on $\mathfrak{h}$. Indeed, just take G to be the semidirect product of D with H.

Corollary. *Let the d-group D act on algebraic groups H, H', and let $\varphi : H \to H'$ be an epimorphism equivariant with respect to these actions of D. Then φ maps the identity component of H^D onto that of H'^D.*

Proof. φ may be factored as $H \to H/K \to H'$, each morphism D-equivariant ($K = \mathrm{Ker}\, \varphi$). Since the second arrow is bijective, we may assume that φ is separable, $H' = H/K$. This means that $\mathrm{Ker}\, d\varphi = \mathfrak{k}$. Since D acts diagonalizably on $\mathfrak{h}$, we can find a D-invariant subspace $\mathfrak{n}$ complementary to $\mathfrak{k}$; then $\mathfrak{n}$ is mapped isomorphically onto $\mathfrak{h}'$ by $d\varphi$, the fixed points of D on $\mathfrak{n}$ corresponding to the fixed points of D on $\mathfrak{h}'$. The proposition shows that dim $K^D = $ dim $\mathfrak{k}^D$, while dim $H'^D = $ dim $\mathfrak{h}'^D = $ dim $\mathfrak{h}^D$. Combining these, we get dim $H'^D = $ dim $\mathfrak{h}^D - $ dim $K^D = $ dim $H^D - $ dim $K^D = $ dim $\varphi(H^D)$. Since φ maps the identity component of H^D into that of H'^D, this yields the desired surjectivity. $\square$

Finally, we consider the analogue of (18.3).

Proposition B. *Let U be a connected closed unipotent subgroup of G, normalized by a d-group D. Then $C_U(D)$ is connected.*

Proof. If D centralizes U there is nothing to prove. Otherwise use induction on dim U. Find $s \in D$ not centralizing U. Then $U' = C_U(s)$ is *connected* (Theorem 18.3) and includes $C_U(D)$, so we may replace U by U' and appeal to induction. $\square$

It is worth remarking that Proposition B and the Corollary of Proposition A make no mention of Lie algebras. However, neither of these would be so easy to prove without the aid of infinitesimal techniques.

Exercises

1. Let $G = \mathrm{PGL}(2, \mathbf{C})$, $s = \mathrm{diag}(i, -i)$, where $i^2 = -1$. Show that $C_G(s)$ has two connected components.
2. Let $G = \mathrm{GL}(V)$, H a closed subgroup of G. Assume: (*) $\mathfrak{h}$ has a complementary subspace $\mathfrak{w}$ in $\mathfrak{g}$ invariant under Ad H. Prove that if $x \in H$, then $Cl_G(x) \cap H$ is the union of finitely many H-conjugacy classes. Prove

an analogous statement for $Cl_G(\mathsf{x}) \cap \mathfrak{h}$, $\mathsf{x} \in \mathfrak{h}$. [Compare dimensions as in (18.1).]

3. Use Exercise 2 to show that $SL(n, \mathsf{K})$ has only finitely many classes of unipotent elements, provided char K does not divide n. (Or appeal to the Jordan normal form for arbitrary n.) What can you say about other classical groups?

4. Under assumption (*) of Exercise 2, prove that global and infinitesimal centralizers in H, $\mathfrak{h}$ correspond for arbitrary $x \in G$.

5. Prove the assertion made in (18.2), that the class of $1 + e_{ij}$ in $GL(n, \mathsf{K})$ (or $SL(n, \mathsf{K})$) fails to be closed. Exhibit nonclosed unipotent classes in the other classical groups.

6. Let H be a closed subgroup of an algebraic group an algebraic group G. An element $\mathsf{x} \in \mathfrak{g}$ is said to *normalize* H if $\mathrm{Ad}\, y(\mathsf{x}) - \mathsf{x} \in \mathfrak{h}$ for all $y \in H$ (and in that case $[\mathsf{x}\mathfrak{h}] \subset \mathfrak{h}$). Prove that if x is semisimple and normalizes H, then $\{\mathrm{Ad}\, y(\mathsf{x}) \mid y \in H\}$ is closed in $\mathfrak{g}$. [Imitate the proof of Proposition 18.2.]

7. Let $G = SL(2, \mathsf{K})$, $s = \mathrm{diag}\,(a, a^{-1})$, $a \neq \pm 1$. Then $C_G(s) = D(2, \mathsf{K}) \cap G$. Show that $C_G(s)$ intersects $Cl_G(s)s^{-1}$ in $\{1, \mathrm{diag}(a^{-2}, a^2)\}$. Is the map τ of Theorem 18.3 surjective in this case?

8. Prove the analogue of Theorem 18.3 when s is replaced by a unipotent element u and U is replaced by a torus T, assuming that $\mathscr{L}(C_T(u)) = c_t(u)$.

9. In Theorem 18.3, prove that τ and γ are isomorphisms of varieties. [For separability, cf. Corollary B of Proposition 18.1.]

10. Let G be a connected 1-dimensional algebraic group. Prove first that each semisimple element of G lies in $Z(G)$, then deduce that $G = Z(G)$.

Notes

Theorem 18.3 appears in Borel [1], but the emphasis on the correspondence between global and infinitesimal centralizers (18.1) is more recent, cf. Borel [4, §9], Borel, Tits [1, §10], Borel, Springer [1] (and its predecessor in AGDS). Further discussion of closed conjugacy classes may be found in Borel, Harish-Chandra [1]. The techniques of (18.1) (cf. Exercises 2, 3, 4) are applied to the study of unipotent conjugacy classes in Richardson [1]; cf. Springer, Steinberg [1, I.5.2].

19. Connected Solvable Groups

In this section we obtain a description of an arbitrary connected solvable algebraic group as semidirect product of a torus and a (normal) unipotent subgroup. For this we make rather full use of the results of §16 and §18, along with the fundamental Lie-Kolchin Theorem (17.6).

19.1. An Exact Sequence

If G is a connected solvable algebraic group, the Lie-Kolchin Theorem allows us to regard G as a subgroup of some $T(n, K)$. For the latter there is a (split) exact sequence:

$$1 \to U(n, K) \to T(n, K) \xrightarrow{\pi} D(n, K) \to 1.$$

Restriction to G yields immediately:

$$(*) \qquad\qquad 1 \to G_u \to G \xrightarrow{\pi} T' \to 1.$$

Here $G_u = G \cap U(n, K)$ is the closed normal subgroup of G consisting of all unipotent elements of G, while $T' = \pi(G)$ is a closed connected subgroup of $D(n, K)$, hence a torus (Proposition 16.2). G/G_u being abelian, G_u includes (G, G).

We claim that G_u is *connected*. This is true in the commutative case (15.5), in particular for the group $G/(G, G) = G' = G'_s \times G'_u$. Let $\varphi: G \to G'$ be the canonical morphism. Evidently $G_u \subset \varphi^{-1}(G'_u)$. Conversely, the inclusion $(G, G) = \operatorname{Ker} \varphi \subset G_u$ forces $\varphi^{-1}(G'_u) \subset G_u$. Now (G, G) and G'_u are both connected (cf. (17.2)), so our claim follows at once.

By intersecting G_u with successive terms of a suitable normal series of $U(n, K)$ (Exercise 17.7), taking identity components, and eliminating repetitions, we see that G_u has a chain of connected subgroups, each normal in G and each of codimension one in the preceding.

It has to be shown that the exact sequence is *split*. For this it will suffice to prove that G actually includes a torus T of the same dimension as T', for in that case $T \cap G_u = e$ is obvious, $\mathscr{L}(T) \cap \mathscr{L}(G_u) = 0$ is almost as obvious (15.3), and $\pi(T) = T'$ for reasons of dimension. We shall also prove that any two maximal tori of G are conjugate. (This result eventually plays the role which is played in Lie algebra theory by the conjugacy of Cartan subalgebras.) The abelian case is already done (15.5), so we treat next the nilpotent case.

19.2. The Nilpotent Case

When G is solvable, the set G_s of all semisimple elements need not be closed and need not be a subgroup (cf. $G = T(n, K)$, $n > 1$).

Proposition. *Let G be a connected solvable algebraic group. Then G is nilpotent if and only if G_s is a subgroup, in which case G_s is closed and $G = G_s \times G_u$.*

Proof. Let G_s be a subgroup of G. In the exact sequence $(*)$ of (19.1), π injects G_s into T', so $\bar{G}_s$ is a d-group. But π is also surjective, so G_s is also closed (and connected). In particular, $(*)$ splits and G is a semidirect product $G_s \ltimes G_u$. But G_s is obviously *normal* in G, so in fact the product is direct. In particular, G_s and G_u commute elementwise, so $G_s \subset Z(G)$ (this follows also from the rigidity of d-groups, Corollary 16.3). Now $G/Z(G)$ is isomorphic to a subgroup of the nilpotent group G_u (17.5), so G is nilpotent (Lemma 17.4(b)).

In the other direction, suppose G is nilpotent. We have to show that G_s is a subgroup; for this it is clearly enough to show that any pair of semisimple elements x, y commute (the product of commuting semisimple elements being again semisimple). But $xyx^{-1}y^{-1} = z \in (G, G) \subset G_u$, $xyx^{-1} = zy$. If we can show that y commutes with all unipotents, it will follow that the right side is the Jordan decomposition of the semisimple element xyx^{-1}, whence $z = e$ and $xy = yx$ as desired. To show that $y \in G_s$ centralizes G_u, recall the set-up of Theorem 18.3: y acts on the connected unipotent group G_u, $M = \gamma_y(G_u)$, and $\gamma_y : M \to M$ is *bijective* (part (c)). But then $M \subset G^\infty$. Since G is nilpotent, $M = \{e\}$ and y centralizes G_u. $\square$

It is worth emphasizing that a connected nilpotent group has a *unique maximal torus* (namely, G_s), so the question of conjugacy does not arise here.

19.3. The General Case

Theorem.   *Let G be a connected solvable algebraic group. Then:*

(a) *G_u is a closed connected normal subgroup of G including (G, G), and G_u has a chain of closed connected subgroups, each normal in G and each of codimension one in the next.*

(b) *The maximal tori of G are conjugate under G^∞, and if T is one of these,* $G = T \ltimes G_u$.

Proof.   The assertions in (a) have been proved in (19.1). For (b), use induction on dim G (starting at dimension 0). The first task is to locate a torus in G which projects onto T' in the exact sequence (*) of (19.1), from which it will follow that $G = T \ltimes G_u$. If G is nilpotent, Proposition 19.2 does the job. Otherwise it is clear that we can find a *noncentral* semisimple element $s \in G$, so $H = C_G(s)^\circ$ has lower dimension than G. If D is the closure of $\langle s \rangle$ in G, the morphism $\pi : G \to G/G_u = T'$ is equivariant with respect to D, the action of $\pi(D)$ on T' being trivial since T' is commutative. Applying the Corollary of Proposition 18.4A, we see that $\pi(H) = T'$. This shows that H_u has codimension equal to dim T'. By induction, H includes a torus of this dimension.

Now fix a maximal torus T of G. As a first step in showing that all other maximal tori are conjugate to T, we claim that any semisimple element $s \in G$ has a conjugate (under G^∞) in T. If G is nilpotent, $T = G_s$ (19.2), and there is nothing to prove. Otherwise $G^\infty \subset G_u$ is nontrivial, as well as connected and unipotent (hence *nilpotent* thanks to Corollary 17.5). The identity component N of $Z(G^\infty)$ therefore has positive dimension, and is clearly normal in G. Under the canonical map $\varphi : G \to G/N = G'$, G^∞ maps *onto* G'^∞. Write $G' = S \ltimes G'_u$, $S = \varphi(T)$. By induction, $\varphi(s)$ has a conjugate $\varphi(x)\varphi(s)\varphi(x)^{-1}$ in S, $\varphi(x) \in G'^\infty$ (where we may assume $x \in G^\infty$). This means that s is G^∞-conjugate to $xsx^{-1} \in TN$. Changing notation, it will suffice to prove that any semisimple element $s \in TN$ is conjugate under N to an element of T. Write $s = tn$ $(t \in T, n \in N)$ and apply Theorem 18.3(b) to the action of t^{-1} on $N : n = z\gamma_{t^{-1}}(u)$, for some $z \in C_N(t)$, $u \in N$, whence $s = tzt^{-1}utu^{-1} = z(utu^{-1})$. But z is unipotent and (N being abelian) commutes with u as well as t, so the right side is the Jordan decomposition of s, forcing $z = e$. Conclusion: $u^{-1}su = t \in T$.

Finally, let S be any other maximal torus of G. By Proposition 16.4, there exists $s \in S$ having the same centralizer in G as S. The preceding paragraph allows us to assume, after replacing S by a conjugate, that $s \in T$. But then $T \subset C_G(s) = C_G(S)$. Therefore, $(ST)^\circ$ is a torus ($=$ connected d-group) including the maximal tori S, T. Conclusion: $S = T$. $\square$

In proving that all maximal tori of G are conjugate, we actually showed more:

Corollary (of proof). *Let G be a connected solvable group. Then each semisimple (resp. unipotent) element of G lies in a maximal torus (resp. a maximal connected unipotent subgroup).* $\square$

Later on we will be able to remove the word "solvable" from this statement, by proving that any connected algebraic group is the union of connected solvable subgroups (22.2).

The structure theorem we have proved gives a satisfactory picture of an arbitrary connected solvable group, with one important exception: It does not pin down the exact structure of a *one dimensional* group. In fact, $\mathbf{G}_a$ and $\mathbf{G}_m$ are the only possibilities (this will be shown in the following section), but the unipotent case (when char $\mathsf{K} > 0$) is a bit tricky to settle.

19.4. Normalizer and Centralizer

To round out our discussion of solvable groups, we record some useful information which is peculiar to the solvable case (cf. Exercise 1).

Proposition. *Let H be a subgroup (not necessarily closed) of a connected solvable group G, H consisting of semisimple elements. Then*
  (a) *H is included in a torus;*
  (b) *$C_G(H) = N_G(H)$ is connected.*

Proof. The canonical map $\pi : G \to G/G_u$, restricted to the subgroup H, is injective. Therefore H is commutative and its closure $\bar{H}$ is a d-group. Since $H, \bar{H}$ have the same normalizer (resp. centralizer) in G, we may as well assume now that H is closed. Rigidity of d-groups (Corollary 16.3) already implies that $N_G(H)^\circ = C_G(H)^\circ$, but we can argue more precisely (using π): If $x \in N_G(H)$, then since $\pi(x)$ commutes with $\pi(H)$, we have $xyx^{-1} = yz$ for $y \in H$ ($z \in \mathrm{Ker}\, \pi = G_u$), while $xyx^{-1} \in H$. Thus $z \in H \cap G_u = \{e\}$, $x \in C_G(H)$.

Now fix a maximal torus T of G. If $H \subset Z(G)$, Corollary 19.3 shows that each element of H already lies in T (and, of course, $C_G(H) = G$ is connected). Otherwise find $s \in H$, $s \notin Z(G)$. Corollary 19.3 allows us to replace T by a conjugate which contains s, so $T \subset C_G(s)$. This makes it clear that $C_G(s) = T \ltimes C_{G_u}(s)$. But $C_{G_u}(s)$ is *connected* (Theorem 18.3), so $C_G(s)$ is connected, includes H, and has smaller dimension than G. By induction on dim G, H is conjugate in $C_G(s)$ to a subgroup of T, and moreover, $C_G(H) = C_{C_{G(s)}}(H)$ is connected. $\square$

19.5. Solvable and Unipotent Radicals

Modulo §20, we are now prepared to tackle the general structure theory of algebraic groups. This is a good time to introduce the key notions of "semisimple" group and "reductive" group. We could have done so a little earlier, but that would not have had any practical value. In conjunction with Corollary 7.4, Lemma 17.3(c) shows that an arbitrary algebraic group G possesses a unique largest normal solvable subgroup, which is automatically closed. Its identity component is then the largest connected normal solvable subgroup of G; we call it the **radical** of G, denoted $R(G)$. The subgroup of $R(G)$ consisting of all its unipotent elements is normal in G; we call it the **unipotent radical** of G, denoted $R_u(G)$. It may be characterized as the largest connected normal unipotent subgroup of G.

If $R(G)$ is trivial and $G \neq e$ is connected, we call G **semisimple** (example: SL(n, K)). If $R_u(G)$ is trivial and $G \neq e$ is connected, we call G **reductive** (examples: GL(n, K), any torus, any semisimple group). Starting with an arbitrary connected algebraic group G, we obviously get a semisimple group $G/R(G)$ and a reductive group $G/R_u(G)$, unless $G = R(G)$ or $G = R_u(G)$. Therefore, the study of an arbitrary G reduces to some extent to the study of a finite group G/G°, a semisimple (or reductive) group, and a connected solvable (or unipotent) group. Of course, the way in which these pieces fit together (the *extension problem*) also has to be considered; it may in practice be rather intractable. The remainder of this book will be concerned largely with reductive groups. It is very difficult (as we shall see) to make use of the assumption that G has no normal connected unipotent subgroup. But some preliminary remarks can be made.

Lemma. *Let G be reductive. Then $R(G) = Z(G)^\circ$ is a torus which has only a finite intersection with (G, G).*

Proof. That $R(G)$ is a torus follows from (19.1). Rigidity (Corollary 16.3) implies that $G = N_G(R(G))^\circ = C_G(R(G))^\circ$, so $R(G)$ is central; conversely, of course, $Z = Z(G)^\circ$ always lies in $R(G)$. For the remaining assertion, embed G in some GL(V) and write $V = \coprod V_\alpha$ (α ranging over the *weights* of Z in V, cf. (16.4)). Evidently the centralizer of Z in GL(V) is the subgroup consisting of corresponding block diagonal matrices, which we write as $\prod$ GL(V_α). It includes G. Therefore (G, G) lies in $\prod$ SL(V_α), while each block of a matrix in Z is a scalar matrix. Since there are only finitely many scalar matrices of a given size having determinant 1, our assertion is proved. $\square$

Later on (27.5) we will be able to prove that a reductive group G is the product of its center and derived group (the intersection being finite, by the above lemma). The reader might try this now, in order to appreciate the difficulty of using the assumption $R_u(G) = e$.

Exercises

1. Show by examples that the assertions in Proposition 19.4 fail if G is not required to be solvable: H may fail to be commutative, may fail to lie in a torus even if it is commutative, etc.
2. Let G be a connected solvable algebraic group. Prove that $\mathscr{L}(G_u)$ is the set of all nilpotent elements of $\mathfrak{g}$. If G is nilpotent, $\mathscr{L}(G_s)$ is the set of all semisimple elements of $\mathfrak{g}$.
3. Let G be a connected solvable algebraic group. If T is a maximal torus of G, then $C_G(T)$ is nilpotent.
4. If G is a connected solvable algebraic group, $Z(G)_s$ is the intersection of all maximal tori of G.
5. If G is any connected algebraic group, T a torus in $Z(G)$, prove that $(G, G) \cap T$ is finite. [Cf. Lemma 19.5.]
6. A closed connected normal subgroup of positive dimension in a semisimple (resp. reductive) group is again semisimple (resp. reductive).
7. Let G be a connected algebraic group of positive dimension. Prove that G is semisimple if and only if G has no closed connected commutative normal subgroup except e.

Notes

Theorem 19.3 is due to Borel [1], cf. Borel [4, 10.6].

20. One Dimensional Groups

The aim of this section is to prove that $\mathbf{G}_m$ and $\mathbf{G}_a$ are the only connected one dimensional algebraic groups. We first show that such a group is commutative, hence consists entirely of semisimple elements or of unipotent elements. The only difficult case—that of a unipotent group in characteristic p—is treated in the more general context of "e-groups" (not needed elsewhere in the text), to bring out an analogy with d-groups. Other treatments of the one dimensional groups are indicated in the Notes.

20.1. Commutativity of G

Let G be a connected one dimensional algebraic group. Since (G, G) is connected, either G is commutative or else $G = (G, G)$. The powerful structure theory of the next chapter would easily rule out the second possibility;

but we can reason in an elementary way, as follows. The assumption that $G = (G, G)$ forces Ad $G = (\text{Ad } G, \text{Ad } G)$ to lie in $\text{SL}(\mathfrak{g})$ ($= 1$, $\mathfrak{g}$ being one dimensional). In particular, if $s \in G$ is semisimple, $\mathfrak{c}_{\mathfrak{g}}(s) = \mathfrak{g}$ implies that $C_G(s) = G$, since G is connected (Proposition 18.1). This means that all semisimple elements of G lie in $Z(G)$, which must of course be finite. Now $G/Z(G) = G'$ is again connected and one dimensional, but has no semisimple elements except e. Therefore G' is unipotent, hence *nilpotent* (Corollary 17.5). In turn G is nilpotent (Lemma 17.4 (b)), contradicting the assumption that $G = (G, G)$.

G being commutative, Theorem 15.5 applies: either $G = G_s$ or $G = G_u$. In the first case G is a connected d-group, hence a torus (Theorem 16.2), hence isomorphic to $\mathbf{G}_m$. In the second case, if char $\mathsf{K} = 0$, Lemma 15.1C and Exercise 15.11 show that G is isomorphic to $\mathbf{G}_a$. If char $\mathsf{K} = p > 0$, notice that $x \mapsto x^p$ is a morphism of algebraic groups $G \to G$, with image G^p connected and unequal to G. Therefore $G^p = e$ (we say G has **exponent p** in this case). It is not at all obvious that G is isomorphic to $\mathbf{G}_a$. In the remainder of this section we shall study closely the structure of commutative groups of exponent p, obtaining the desired theorem on one dimensional groups as a by-product.

20.2. Vector Groups and *e*-groups

Until further notice, char $\mathsf{K} = p > 0$. Let us call **e-group** any commutative algebraic group G of exponent p (G is then unipotent). One example is a finite elementary abelian p-group $\mathbb{Z}/p\mathbb{Z} \times \cdots \times \mathbb{Z}/p\mathbb{Z}$. Another relevant example is affine n-space K^n regarded as $\mathbf{G}_a \times \cdots \times \mathbf{G}_a$: any e-group isomorphic to one of the latter will be called a **vector group**. The group operation is often written additively.

There is a good analogy between tori and vector groups, or more generally, between d-groups and e-groups. Let us recall some salient properties of the former from §16: (a) Any d-group is isomorphic to a closed subgroup of a torus. (b) Any closed subgroup of a torus is the intersection of kernels of characters. (c) A closed connected d-group is a torus (and a direct factor of any larger torus). In this subsection we prove the analogue of the first of these (with a refinement not possible for d-groups, cf. Exercise 3).

Proposition. *Any e-group G is isomorphic to a closed subgroup of some vector group, which can be chosen to be of dimension 1 larger than G.*

Proof. We attempt to imitate the maps log, exp in characteristic 0 (15.1). Begin with $G \subset \text{GL}(n, \mathsf{K})$ for some n. Let $\mathfrak{N}$ be the subspace of $\text{M}(n, \mathsf{K})$ spanned by the matrices $x - 1$ ($x \in G$), $\mathfrak{A} = \mathsf{K} + \mathfrak{N} \supset G$. The equation $(x - 1)(y - 1) = (xy - 1) - (x - 1) - (y - 1)$ shows that $\mathfrak{N}$ is closed under multiplication, hence that $\mathfrak{A}$ is a commutative ring having $\mathfrak{N}$ as an ideal. Because G has exponent p, $(x - 1)^p = 0$ ($x \in G$) and therefore $y^p = 0$

for all $y \in \mathfrak{N}$. The commuting set $\mathfrak{N}$ of nilpotent matrices can be simultaneously trigonalized (Proposition 15.4). It follows that all k-fold products from $\mathfrak{N}$ are 0 if k is chosen large enough, i.e., $\mathfrak{N}$ is a nilpotent ideal of $\mathfrak{A}$ (cf. (17.3)).

Since $x^p = 0$ for all $x \in \mathfrak{N}$, we may define a truncated exponential $\exp'(x) = \sum_{k=0}^{p-1} x^k/k!$, which clearly lies in $1 + \mathfrak{N} \subset \mathfrak{A}$. It is easy to check (Exercise 4) that $\exp'(ax) \exp'(bx) = \exp'((a + b)x)$ for $a, b \in \mathsf{K}$. Therefore we can define, relative to an ordered basis $(x_1, \ldots, x_t)$ of $\mathfrak{N}$, specified below, a morphism $\varphi: \mathfrak{N} \to \mathfrak{A}$ by setting $\varphi(\sum a_i x_i) = \prod \exp'(a_i x_i)$. This is a morphism of algebraic groups, $\mathfrak{N}$ being given its natural additive structure of vector group and $\operatorname{Im} \varphi$ being a subgroup of the multiplicative group of $\mathfrak{A}$ in $\mathrm{G\,L}(n, \mathsf{K})$. Of course, G itself lies in the latter, though not obviously in $\operatorname{Im} \varphi$.

We want to show that φ has an inverse morphism (defined on the invertible elements of $\mathfrak{A}$). For this we use the nilpotence of $\mathfrak{N}$: Define φ relative to an ordered basis $(x_1, \ldots, x_t)$ of $\mathfrak{N}$ such that each $\mathfrak{N}^k$ is spanned by a final segment. Let T_i, U_i be two sets of t indeterminates, and write $\prod \exp'(\mathsf{T}_i x_i) = 1 + \sum \mathsf{U}_i x_i$. Expansion of the left side reveals at once that $\mathsf{U}_i = \mathsf{T}_i + f_i(\mathsf{T}_1, \ldots, \mathsf{T}_{i-1})$, f_i some polynomial in $i - 1$ variables. Invertibility of φ is then a matter of expressing the T_i in terms of the U_i. But $\mathsf{T}_1 = \mathsf{U}_1$, $\mathsf{T}_2 = \mathsf{U}_2 - f_2(\mathsf{T}_1) = \mathsf{U}_2 - f_2(\mathsf{U}_1), \ldots$. Therefore we can proceed recursively.

From this construction it follows that G does lie in $\operatorname{Im} \varphi$ and that φ^{-1} maps G isomorphically onto a closed subgroup of the vector group $\mathfrak{N}$.

Now let V be any vector group (written additively), G a closed subgroup of codimension $\geqslant 2$ in V. We will replace V by a vector group 1 dimension smaller, containing an isomorphic copy of G; induction then completes the argument.

It is an elementary geometric fact (Exercise 5) that the union of all lines through 0 passing through other points of G cannot exhaust V, since codim $G \geqslant 2$. Identify (canonically) V with $\mathscr{L}(V)$. Then the linear subspace $\mathscr{L}(G)$ of $\mathscr{L}(V) = V$ obviously cannot exhaust what remains of V, since it too has codimension at least 2. Therefore we can locate $x \in V$ such that no nonzero scalar multiple of x lies in G or $\mathscr{L}(G)$. Let $X = \mathsf{K}x$, and let $\pi: V \to W = V/X$ be the canonical morphism. Because of the identification $V = \mathscr{L}(V)$, it is clear that $\operatorname{Ker} d\pi = \mathscr{L}(X) = X$, X viewed as vector subgroup of V. By choice of x, $\operatorname{Ker} \pi \cap G = 0 = \operatorname{Ker} d\pi \cap \mathscr{L}(G)$, so $\pi|G$ is bijective and separable, hence an isomorphism of G onto $\pi(G)$, cf. remark (1) of (12.4). $\quad\square$

20.3. Properties of p-polynomials

A character of an algebraic group G is simply a morphism of algebraic groups $G \to \mathbf{G}_m$. The obvious unipotent counterpart would be a morphism $\varphi: G \to \mathbf{G}_a$. What form must φ take when K has prime characteristic p? Since $\mathbf{G}_a$ is isomorphic (as variety) to the affine line K, φ is given by a single polynomial. Consider first the case where G is a vector group K^n (written additively).

Lemma A.   *Let $f(T_1, \ldots, T_n)$ be an additive polynomial, i.e., $f(a_1 + b_1, \ldots, a_n + b_n) = f(a_1, \ldots, a_n) + f(b_1, \ldots, b_n)$ for all $(a), (b) \in K^n$. Then $f(T)$ is a* **p-polynomial**, *i.e., $f(T)$ is a linear combination of terms $T_i^{p^r}\ (r \geqslant 0)$.*

Proof.   Use induction on the degree of $f(T)$ and denote by $(D_i f)(T)$ the "partial derivative" of $f(T)$ with respect to T_i. From the assumed identity $f(T_1 + a_1, \ldots, T_n + a_n) = f(T) + f(a)$, $(a) \in K^n$, we obtain $(D_i f)(T + a) = (D_i f)(T)$. Therefore the polynomial $(D_i f)(T)$ takes constant value c_i on K^n, and all partial derivatives of $f(T) - (c_1 T_1 + \cdots + c_n T_n)$ are 0. It is easy to check that any polynomial with this property has the form $g(T_1^p, \ldots, T_n^p)$, $g(T) \in K[T]$. Since K is algebraically closed, arbitrary $(a), (b) \in K^n$ satisfy $a_i = d_i^p$, $b_i = e_i^p$ for some $d_i, e_i \in K$. So $g(a + b) = g((d_1 + e_1)^p, \ldots, (d_n + e_n)^p) = f(d + e) - \sum c_i(d_i + e_i) = g(a) + g(b)$. Now $g(T)$ is additive but of smaller degree than $f(T)$, so by induction $g(T)$ (hence also $f(T)$) is a p-polynomial. $\square$

Conversely, of course, any p-polynomial defines a morphism of algebraic groups $K^n \to G_a$. We shall abuse terminology by identifying the polynomial and the morphism it affords. Evidently the kernel of a nonzero p-polynomial (i.e., the set of zeros of $f(T)$) is a closed subgroup of K^n (of codimension one). This property actually characterizes p-polynomials:

Lemma B.   *Let $f(T_1, \ldots, T_n)$ be a squarefree polynomial, viewed as a morphism $K^n \to K$. If the set of zeros of $f(T)$ is a subgroup of K^n, then $f(T)$ is a p-polynomial.*

Proof.   We may assume that $f(T) \neq 0$. To prove that $f(T)$ is a p-polynomial, it will suffice (by Lemma A) to prove that it is additive. Let G be its set of zeros. If $(a) \in G$, $f(T + a)$ has the same set of zeros as $f(T)$, because G is a group. But $f(T)$ is squarefree, so this forces the former to be a (nonzero) scalar multiple of the latter, say $\gamma(a)f(T)$ ($K[T]$ is a UFD (0.2)). If $(a), (b) \in G$, then in turn $\gamma(a + b)f(T) = f(T + a + b) = \gamma(b)f(T + a) = \gamma(a)\gamma(b)f(T)$. The function $\gamma: G \to G_m$ is therefore a homomorphism, hence identically 1, because G is a p-group while G_m has no p-torsion. This says that $f(T + a) = f(T)$ for all $(a) \in G$.

Turning this around, for all $(a) \in K^n$, the polynomial $f(T + a) - f(a) \neq 0$ vanishes on G, and is clearly squarefree. This yields a function $\delta: K^n \to G_m$, satisfying $f(T + a) - f(a) = \delta(a)f(T)$. For any $(b) \in K^n$, direct calculation shows that $f(T + a + b) = f(a) + \delta(a)f(b) + \delta(a)\delta(b)f(T)$, while also $f(T + a + b) = f(a + b) + \delta(a + b)f(T)$. Comparing highest terms, we deduce that δ is a homomorphism, hence trivial, i.e., $f(T + a) = f(a) + f(T)$ for all $(a) \in K^n$. This proves that f is additive on K^n. $\square$

In (20.2) we proved that e-groups are the same thing as closed subgroups of vector groups, and that a given e-group can even be realized as such a subgroup of codimension 1. Now we can go a step further.

Proposition.   *Let V be a vector group (say $V = K^n$), G a closed subgroup of codimension 1. Then there exists a p-polynomial (in n variables) having G as its set of zeros.*

Proof.   By the results of §3 (cf. Exercise 3.6), G is the set of zeros of some polynomial $f(\mathsf{T}) \in \mathsf{K}[\mathsf{T}]$, which may be assumed to be squarefree. Then Lemma B says that $f(\mathsf{T})$ must be a p-polynomial. $\square$

20.4. Automorphisms of Vector Groups

To show that a connected d-group is a torus, we first embedded the given group in a torus, then found an automorphism of the torus which would take this subgroup onto a canonical subtorus. An analogous procedure works for e-groups. We shall construct automorphisms by using induction on degrees, so it is convenient to make a definition: Let $f(\mathsf{T}_1, \ldots, \mathsf{T}_n) = \sum_i \sum_r c_{ir}\mathsf{T}_i^{p^r}$ be a p-polynomial. If $r(i)$ is the largest exponent for which $c_{ir} \neq 0$, the **principal part** of $f(\mathsf{T})$ is the polynomial $c_{1r(1)}\mathsf{T}_1^{p^{r(1)}} + \cdots + c_{nr(n)}\mathsf{T}_n^{p^{r(n)}}$.

Proposition.   *Let $V = \mathsf{K}^n$, and let $f(\mathsf{T})$ be a p-polynomial, viewed as a morphism of algebraic groups $V \to \mathbf{G}_a$. Then there exists an automorphism φ of V (i.e., a change of coordinates $\mathsf{T}_i \mapsto \mathsf{U}_i$) such that the kernel of $f(\mathsf{U}) = f(\mathsf{T}) \circ \varphi$ contains all $(a, 0, \ldots, 0) \in V$.*

Proof.   Write the principal part of $f(\mathsf{T})$ as $\sum_i c_i \mathsf{T}_i^{p^{r(i)}}$, for brevity. Evidently a permutation of $\mathsf{T}_1, \ldots, \mathsf{T}_n$ induces an automorphism of V, so we may assume that all $c_i \neq 0$ (otherwise we are done) and that $r(1) \geq r(2) \geq \cdots \geq r(n)$. Proceed by induction on $\sum r(i)$. If this is 0, then $f(\mathsf{T})$ is a *linear* polynomial, and a linear change of variables $\mathsf{T}_i \mapsto \mathsf{U}_i$ does the trick. In general, find a zero $(a) \in V$ for the principal part of $f(\mathsf{T})$. Say $a_1 = a_2 = \cdots = a_{m-1} = 0 \neq a_m$. The following equations define an automorphism in V (Exercise 6):

$$\mathsf{U}_1 = \mathsf{T}_1$$
$$\vdots$$
$$\mathsf{U}_{m-1} = \mathsf{T}_{m-1}$$
$$\mathsf{U}_m = a_m\mathsf{T}_m$$
$$\mathsf{U}_{m+1} = \mathsf{T}_{m+1} + a_{m+1}\mathsf{T}_m^{p^{r(m)-r(m+1)}}$$
$$\vdots$$
$$\mathsf{U}_n = \mathsf{T}_n + a_n\mathsf{T}_m^{p^{r(m)-r(n)}}.$$

After substituting U_i for T_i in $f(\mathsf{T})$, we obtain a new p-polynomial: $g(\mathsf{T}) = \sum_{i<m} c_i\mathsf{T}_i^{p^{r(i)}} + \left(\sum_{i=m}^{n} c_i a_i^{p^{r(i)}}\right)\mathsf{T}_m^{p^{r(m)}} + \text{(lower degree terms)}$. By the choice of (a), the sum of degrees in the principal part of $g(\mathsf{T})$ is smaller than $\sum r(i)$. So the induction hypothesis applies. $\square$

If $V = \mathsf{K}^n$, let V_i be the set of vectors having all coordinates 0 except possibly the ith.

Corollary.   *With $V = \mathsf{K}^n = \prod_{i=1}^{n} V_i$ as above, let G be a closed connected subgroup of V. Then there exists an automorphism φ of V such that $\varphi(G)$ is* *the direct product of some of the canonical subgroups V_i of V. In particular, G is itself a vector group.*

Proof. Use induction on n (for $n = 1$, there is nothing to prove). Consider first the case: dim $G = n - 1$. By Proposition 20.3, G is the kernel of some p-polynomial $f(\mathsf{T})$ (viewed as a morphism $V \to \mathbf{G}_a$), Then the above proposition yields an automorphism φ of V such that $\varphi(G)$ includes V_1. If H is the projection of $\varphi(G)$ on $V' = \prod\limits_{i=2}^{n} V_i$, then clearly $\varphi(G) = V_1 \times H$. Since H has codimension 1 in V' (and is connected), induction yields an automorphism of V', which extends to an automorphism of V fixing V_1, mapping H onto a product of some of the V_i.

In case dim $G < n - 1$, define $V' = \prod\limits_{i=2}^{n} V_i$ as above, and project G to a closed connected subgroup G' of V'. By induction, there is an automorphism φ' of V' sending G' onto a product of some V_i. Extend φ' to $\varphi : V \to V$ ($\varphi|V_1 = 1_{V_1}$) as above. Then $\varphi(G)$ lies in the product of fewer than n of the V_i, so we can throw away the others and appeal again to induction. $\square$

20.5. The Main Theorem

From the preceding description of e-groups we obtain as a special case the fact that a connected 1-dimensional e-group G is isomorphic to $\mathbf{G}_a$. The reader interested only in this result can take a few (but very few) shortcuts. The essential line of argument is this: G is isomorphic to a closed subgroup of the vector group K^2 (20.2), hence is precisely the set of zeros of a p-polynomial in two variables (20.3). Using this p-polynomial, one constructs an automorphism of K^2 such that the image of G includes (hence equals) a canonical copy of $\mathbf{G}_a$ (20.4).

Since a connected 1-dimensional unipotent group is commutative (20.1), hence of exponent p, hence an e-group, this completes the proof of the main theorem:

Theorem. *Let* K *have arbitrary characteristic. Up to isomorphism, the only connected 1-dimensional algebraic groups are* $\mathbf{G}_a$ *and* $\mathbf{G}_m$. $\square$

Exercises

1. In (20.1), prove that G is commutative by using (18.2) in place of (18.1): semisimple conjugacy classes are closed (and connected), hence trivial.
2. In (20.1), prove that G is commutative without invoking §18, as follows: Let $x \notin Z(G)$. The class of x is infinite, hence dense (as well as locally closed); deduce that its complement is finite. Assume $G \subset \mathrm{GL}(n, \mathsf{K})$, $c_i(y) = i^{th}$ coefficient of characteristic polynomial of $y \in G$. Then $\varphi(y) = (c_0(y), \ldots, c_n(y))$ defines a morphism $G \to \mathsf{K}^{n+1}$ with finite image. Conclude that G is unipotent, hence nilpotent.

3. Show that a d-group cannot always be embedded in a torus as a closed subgroup of codimension 1.

4. Let char $K = p > 0$. If $x \in M(n, K)$, $x^p = 0$, show that $\exp'(ax)\exp'(bx) = \exp'((a + b)x)$ for $a, b \in K$, where $\exp'$ is the truncated exponential (20.2). On the other hand, if $x, y \in M(n, K)$ commute and if $x^p = 0 = y^p$, it need not be the case that $\exp'(x + y) = \exp'(x)\exp'(y)$: e.g., let $2 = $ char K, and take

$$x = \begin{pmatrix} 0 & 1 & 0 & 0 \\ 0 & 0 & 0 & 0 \\ 0 & 0 & 0 & 1 \\ 0 & 0 & 0 & 0 \end{pmatrix}, \quad y = \begin{pmatrix} 0 & 0 & 1 & 0 \\ 0 & 0 & 0 & 1 \\ 0 & 0 & 0 & 0 \\ 0 & 0 & 0 & 0 \end{pmatrix}.$$

5. Let V be an affine variety in $\mathbf{A}^n$ of codimension $\geqslant 2$. Prove that $\{ax | a \in K, x \in V\}$ is properly contained in $\mathbf{A}^n$. [Consider $K \times V$.]

6. Let char $K = p > 0$. Verify that the change of variable $T_i \mapsto U_i$ described in the proof of Proposition 20.4 induces an automorphism of V. Try to formulate necessary and sufficient conditions for a collection of n p-polynomials to define an automorphism of V.

7. Let char $K = p > 0$. Prove that each closed subgroup of a vector group is the intersection of kernels of p-polynomials.

Notes

The material in this section is drawn from Tits [9, III.3.3]. An earlier proof of Theorem 20.5, given by Grothendieck in Chevalley [8, exposé 7], is sketched in Borel [4, 11.6]; yet another proof appears in Borel [4, 10.7–10.9]. The study of commutative unipotent groups (and group schemes) can be pursued in Demazure, Gabriel [1], Oort [1], Seligman [4] [5], Serre [1, VII §2], Tits [9], Schoeller [1], Simon [1], Nakamura [1] [2]. The example in Exercise 4 was furnished by George Seligman.

Chapter VIII

Borel Subgroups

21. Fixed Point and Conjugacy Theorems

The structure of connected solvable groups is made clear enough by the results of the preceding chapter. What is still lacking is insight into the interaction between an arbitrary algebraic group and its subgroups of this type. Borel's fixed point theorem (21.2) provides this insight. Here, for the first time, we make essential use of homogeneous spaces G/H which are *projective* (or complete) varieties. G will denote an arbitrary algebraic group, assumed from (21.3) on to be connected.

21.1. Review of Complete Varieties

For the reader's convenience we shall list those properties of complete varieties which are needed in the sequel. By definition, a variety X is **complete** if, for all varieties Y, the projection map $X \times Y \to Y$ is closed. In Chapter I the following facts were proved:

(a) A closed subvariety of a complete (resp. projective) variety is complete (resp. projective), cf. (6.1) and (1.6).

(b) If $\varphi : X \to Y$ is a morphism of varieties, with X complete, then the image is closed in Y, and complete (6.1).

(c) A complete affine variety has dimension 0 (6.1).

(d) Projective varieties are complete (6.2); complete quasiprojective varieties are projective (6.1).

(e) The flag variety $\mathfrak{F}(V)$ of a finite dimensional vector space V is projective (1.8), hence complete.

Besides these facts, we need a lemma.

Lemma. *Let G act transitively on two irreducible varieties X, Y, and let $\varphi : X \to Y$ be a bijective, G-equivariant morphism. If Y is complete, then X is complete.*

Proof. Notice that the assertion is not obvious, though in the reverse direction it would be (cf. (b) above). Thanks to (6.1), it just has to be shown that $\mathrm{pr}_2 : X \times Z \to Z$ is closed for all affine varieties Z. Since Y is complete, it suffices to prove that $\varphi \times 1 : X \times Z \to Y \times Z$ is closed.

Corollary 4.3 yields affine open subsets $U \subset X$, $V \subset Y$ such that $\varphi(U) \subset V$ and $\varphi | U$ is a finite morphism. Let R, S, T be the respective affine algebras of U, V, Z. Since R is integral over S, it is trivial to check that $R \otimes T$ is integral over $S \otimes T$. Therefore $\varphi \times 1 : U \times Z \to V \times Z$ is also a finite morphism (cf. (2.4)); in particular, it is a closed map (4.2). Because G acts transitively on X, Y, and φ is G-equivariant, X (resp. Y) is covered by finitely many translates $x_i \cdot U$ (resp. $x_i \cdot V$) for some $x_i \in G$. It follows that $\varphi \times 1 : X \times Z \to Y \times Z$ is closed. $\square$

21.2. Fixed Point Theorem

Theorem. *Let G be a connected solvable algebraic group, and let X be a (nonempty) complete variety on which G acts. Then G has a fixed point in X.*

Proof. If $\dim G = 0$, $G = e$ and there is nothing to prove. Proceed by induction on $\dim G$. $H = (G, G)$ is connected (17.2), solvable, and of lower dimension (G being solvable), so by induction the set Y of fixed points of H in X is nonempty. Y is closed (Proposition 8.2), hence complete (21.1)(a). G keeps Y stable, because H is normal in G, so we may as well replace X by Y.

This leaves us with the situation: $H \subset G_x$ for all $x \in X (= Y)$. In particular, all isotropy groups are *normal* in G, so G/G_x is an *affine* variety. Choose $x \in X$ whose orbit $G \cdot x$ is closed, hence again complete: this is possible, by Proposition 8.3. Now the canonical morphism $G/G_x \to G \cdot x$ is bijective, with the left side affine and the right side complete. According to Lemma 21.1, G/G_x is complete. By (21.1)(c), $G = G_x$, so x is the desired fixed point. $\square$

We use the theorem first to obtain a new proof of the Lie-Kolchin Theorem (17.6). Let G be a closed connected solvable subgroup of $\mathrm{GL}(V)$. Then G acts on the flag variety $\mathfrak{F}(V)$, which is complete (21.1)(e), so G fixes a flag $0 = V_0 \subset V_1 \subset \cdots \subset V_n = V$. In other words, G is triangular for suitable choice of basis in V.

Next we shall apply the fixed point theorem in a more essential way.

21.3. Conjugacy of Borel Subgroups and Maximal Tori

A **Borel subgroup** of G is a closed connected solvable subgroup properly included in no other (the word "closed" being redundant). Since Borel subgroups of G and of G° coincide, *we shall always assume in what follows that G is connected*. A connected solvable subgroup of largest possible dimension in G is evidently a Borel subgroup; but it is not obvious that every Borel subgroup has the same dimension. This emerges from a stronger result:

Theorem. *Let B be any Borel subgroup of G. Then G/B is a projective variety, and all other Borel subgroups are conjugate to B.*

Proof. Begin with a Borel subgroup S of largest possible dimension. Represent G in some $\mathrm{GL}(V)$ with a 1-dimensional subspace V_1 whose stabilizer in G is precisely S (Theorem 11.2). The induced action of S on

V/V_1 is trigonalizable, thanks to Lie-Kolchin (17.6), so there is a full flag $0 \subset V_1 \subset \cdots \subset V$ stabilized by S, call it f. In fact, S is the isotropy group of f in G, by choice of V_1. It follows that the induced morphism of G/S onto the orbit of f in $\mathfrak{F}(V)$ is bijective. On the other hand, the stabilizer of any flag is solvable and therefore has dimension no bigger than dim S. It follows that the orbit of f has smallest possible dimension, hence is *closed* (8.3). So the orbit is complete (21.1)(a)(e). This forces G/S to be complete (Lemma 21.1), hence projective (21.1)(d).

Now the given Borel subgroup B acts by left multiplication on the complete variety G/S. By Theorem 21.2, it fixes a point xS, i.e., $BxS = xS$, or $x^{-1}Bx \subset S$. Each of these being a Borel subgroup, we conclude that $x^{-1}Bx = S$. From this both assertions of the theorem follow. $\square$

Corollary A. *The maximal tori (resp. maximal connected unipotent subgroups) of G are those of the Borel subgroups of G, and are all conjugate.*

Proof.   Let T be a maximal torus of G, U a maximal connected unipotent subgroup. Evidently T is included in some Borel subgroup B, so it is a maximal torus of B, to which all other maximal tori of B are conjugate (Theorem 19.3). Similarly, U lies in some Borel subgroup B', with $U = B'_u$ by maximality. Since all Borel subgroups of G are conjugate, the corollary follows. $\square$

Call the common dimension of the maximal tori of G the **rank** of G. For example, SL(n, K) has rank $n - 1$.

The fact that G/B is a projective variety may appear at this point to be merely incidental, but in fact the geometry of this variety is of central importance in what follows. Let us just observe now that *G/B is the largest homogeneous space for G having the structure of projective variety*. Indeed, if H is any closed subgroup of G such that G/H is projective, then B fixes a point and therefore has a conjugate in H, forcing dim $G/H \leqslant$ dim G/B. Conversely, if H is a closed subgroup including B, then $G/B \to G/H$ is a surjective morphism from a complete variety, forcing G/H to be complete (21.1)(b). But then G/H is projective (21.1)(d), since all homogeneous spaces are quasi-projective by construction (11.3). This discussion leads us to define a special class of closed subgroups P of G: P is **parabolic** if and only if G/P is projective (equivalently, complete). We have shown:

Corollary B.   *A closed subgroup of G is parabolic if and only if it includes a Borel subgroup. In particular, a connected subgroup H of G is a Borel subgroup if and only if H is solvable and G/H projective.* $\square$

For example, let $G = $ GL(n, K), $B = $ T(n, K). The Lie-Kolchin Theorem implies that B is a Borel subgroup of G. Of course, G/B is just the orbit in $\mathfrak{F}(\mathsf{K}^n)$ of the standard flag (i.e., $\mathfrak{F}(\mathsf{K}^n)$ itself). Which parabolic subgroups of G include B? If $(e_1, \ldots, e_n)$ is the standard basis of K^n, then for each partial flag $(e_1, \ldots, e_{i(1)}) \subset (e_1, \ldots, e_{i(2)}) \subset \cdots$, the stabilizer in G is evidently a closed subgroup including B. In $GL(3, K)$ the two parabolic subgroups other than B and G which arise this way consist of matrices in one of the forms

$$\begin{pmatrix} * & * \\ \hline 0\;0 & * \end{pmatrix}, \quad \begin{pmatrix} * & * \\ \hline \begin{matrix}0\\0\end{matrix} & * \end{pmatrix}.$$

The characterization of Borel subgroups given in Corollary B can be used to good advantage to describe their behavior under homomorphisms.

Corollary C. *Let $\varphi : G \to G'$ be an epimorphism of (connected) algebraic groups. Let H be a Borel subgroup (resp. parabolic subgroup, maximal torus, maximal connected unipotent subgroup) of G. Then $\varphi(H)$ is a subgroup of the same type in G' and all such subgroups of G' are obtained in this way.*

Proof. In view of Corollaries A, B, it suffices to prove this when $H = B$ is a Borel subgroup. Evidently $B' = \varphi(B)$ is connected and solvable. But the natural map $G \to G' \to G'/B'$ induces a surjective morphism $G/B \to G'/B'$, forcing the latter to be complete (21.1)(b), i.e., B' is a parabolic subgroup of G'. Corollary B then says it is a Borel subgroup. Since some Borel subgroup of G' has the form $\varphi(B)$, it follows from the conjugacy theorem (for G') that all do. $\square$

21.4. Further Consequences

At this point it is easy to carry over to an arbitrary connected group G some facts proved in the preceding chapter for solvable groups. The results we obtain are still a bit crude, but will be refined in subsequent sections.

Proposition A. *An automorphism σ of G which fixes all elements of a Borel subgroup B must be the identity.*

Proof. The morphism $\varphi : G \to G$, $\varphi(x) = \sigma(x) x^{-1}$, sends B to e, hence factors through $G \to G/B$. By Theorem 21.3 and (21.1)(b), $\varphi(G)$ is closed (therefore affine) and complete. By (21.1)(c), $\varphi(G) = e$. $\square$

Corollary. $Z(G)^\circ \subset Z(B) \subset C_G(B) = Z(G)$.

Proof. $Z(G)^\circ$ is connected and solvable, hence lies in some Borel subgroup of G. Conjugating the latter onto B (Theorem 21.3) leaves $Z(G)^\circ$ unmoved, so the first inclusion follows. The second inclusion is obvious, as is the inclusion $Z(G) \subset C_G(B)$. Finally, if $x \in C_G(B)$, then $\sigma = \mathrm{Int}\, x$ satisfies the hypothesis of the proposition, forcing $\mathrm{Int}\, x = 1$, $x \in Z(G)$. $\square$

Later on (Corollary 22.2B) we shall see that $Z(B) = Z(G)$, but this is not yet obvious.

Proposition B.   (a) *If some Borel subgroup B is nilpotent (in particular, if $B = B_s$ or $B = B_u$), then $G = B$.*
   (b) *G is nilpotent if and only if G has just one maximal torus.*

Proof.   (a) Note first that if $B = B_s$ or $B = B_u$, then B is a torus or a unipotent group (Theorem 19.3), hence nilpotent in either case. Whenever B is nilpotent, of positive dimension, its center also has positive dimension (Proposition 17.4(a)). But $Z(G)^\circ \subset Z(B) \subset Z(G)$, by the preceding corollary. So we can pass to the lower dimensional group $G/Z(G)^\circ$, in which $B/Z(G)^\circ$ is a nilpotent Borel subgroup (this is obvious even without Corollary C of (21.3)). By induction, these groups are equal, so $G = B$ follows. (The induction starts with dim $G = 0$.)
   (b) We know from (19.2) that a nilpotent group has a unique maximal torus. Conversely, if T is the only maximal torus of G and B some Borel group containing it, then B must be nilpotent (19.2) (19.3), forcing $B = G$ by part (a). $\square$

Corollary.   *Let T be a maximal torus of G, $C = C_G(T)^\circ$. Then C is nilpotent and $C = N_G(C)^\circ$.*

Proof. T is the unique maximal torus of C, thanks to the conjugacy theorem (Corollary 21.3A), hence C is nilpotent, by the proposition just proved. Evidently T is normal in $N_G(C)^\circ$, hence also central, by rigidity of tori (Corollary 16.3). $\square$
   The connected centralizer C of a maximal torus T of G is often called a **Cartan subgroup** because of the analogy with Cartan subalgebras (nilpotent and self-normalizing) in Lie algebra theory. It will be shown in §22 that $C_G(T)$ is already connected (if G is); but in general $N_G(C)$ need not be.

Exercises

G denotes a connected algebraic group.
1. Prove that rank $G = 0$ if and only if G is unipotent.
2. Prove that a torus may be characterized as a connected algebraic group consisting of semisimple elements.
3. Let T be a torus which is normal in G, and suppose G/T is also a torus. Prove that G is a torus.
4. If dim $G \leqslant 2$, then G is solvable.
5. Let H be a closed subgroup of G. Denote by L a Borel subgroup (resp. maximal torus, maximal connected unipotent subgroup) of H. Prove that $L = (H \cap M)^\circ$ for a suitable subgroup M of G of the same type. If H is *normal* in G, show that any such $(H \cap M)^\circ$ is a Borel subgroup (resp., ...) of H. What can you say about parabolic subgroups of H?
6. The radical (resp. unipotent radical) of G is the identity component in the intersection of all Borel subgroups (resp., of their unipotent parts).

7. Verify that every subgroup of GL(3, K) including T(3, K) is the stabilizer of some partial flag (hence is automatically closed), and is connected and self-normalizing.

8. Describe Borel subgroups of the various classical groups, as well as some parabolic subgroups.

9. Let P be a parabolic subgroup of G. Then the Borel subgroups of P are also Borel subgroups of G.

10. If B is a Borel subgroup of G, then $B = N_G(B)^\circ$ (and similarly for parabolic subgroups).

11. Let $\varphi: G \to G'$ be an epimorphism of connected algebraic groups. If T is a maximal torus of G, $T' = \varphi(T)$, prove that the Cartan subgroup $C_G(T)^\circ$ is mapped by φ onto the Cartan subgroup $C_{G'}(T')^\circ$.

12. Let T be a maximal torus of G, $C = C_G(T)^\circ$. Show that any Borel subgroup of G which includes T must also include C.

Notes

The theorems of this section originate with Borel [1], cf. Borel [4, §10–11]. The proof of Lemma 21.1 is taken from Steinberg [13, appendix to 2.11]. Sweedler [1] has another proof of the conjugacy of Borel subgroups, which does not require the construction of homogeneous spaces.

22. Density and Connectedness Theorems

If G is a connected solvable group, then each semisimple (resp. unipotent) element of G lies in a maximal torus (resp. maximal connected unipotent subgroup), and the centralizer of a torus in G is always connected (cf. Proposition 19.4). In this section we shall extend these results to an arbitrary connected algebraic group, via its Borel subgroups. Unless otherwise specified, G always denotes a *connected* group.

22.1. The Main Lemma

First we ask what can be said about the union of all conjugates of a closed subgroup H of G. The special assumptions made about H in parts (a), (b) of the following lemma will turn out to be true when H is (respectively) a Borel subgroup or a Cartan subgroup.

Lemma. *Let H be a closed connected subgroup of G, and set $X = \bigcup_{x \in G} xHx^{-1}$.*

(a) If G/H is complete, then X is closed.

(b) If some element of H fixes only finitely many points of G/H, then X contains an open subset of G (in particular, X is dense).

Proof. Since X is rather hard to study directly, we make use of some auxiliary morphisms:

$$G \times G \xrightarrow{\varphi} G \times G \xrightarrow{\psi} (G/H) \times G \underset{\text{pr}_2}{\overset{\text{pr}_1}{\rightrightarrows}} {}^{G/H}_{G}.$$

Here $\varphi(x, y) = (x, xyx^{-1})$, so φ is clearly an isomorphism of varieties, while $\psi(x, y) = (xH, y)$. Notice that ψ may be viewed as the canonical morphism $G \times G \to (G \times G)/(H \times e)$, so ψ is an *open* map (12.2). Let $M = \psi \circ \varphi(G \times H)$; evidently, $X = \text{pr}_2(M)$.

To prove (a), it suffices (by the definition of complete variety) to prove that M *is closed*. In turn it suffices (ψ being an open map) to prove that $\psi^{-1}(M)$ is closed. Now $\varphi(G \times H) \subset \psi^{-1}(M)$, and the former is closed because φ is an isomorphism of varieties, so we need only show that equality holds. Let $(x, y) \in G$ with $(xH, y) \in M$, i.e., $y = xzx^{-1}$ for some $z \in H$. Then $(x, y) = \varphi(x, z) \in \varphi(G \times H)$, as desired.

Next consider (b). Since X is constructible (as the image of the morphism $\text{pr}_2 \circ \psi \circ \varphi$ (4.4)), it is enough to show that X is dense in G, or that $\dim \bar{X} = \dim G$. We showed in general that M is closed, and we can easily compute its dimension: pr_1 maps M *onto* G/H, and the fibre over xH of this morphism is clearly isomorphic to xHx^{-1}, so $\dim M = \dim G/H + \dim H = \dim G$. On the other hand, the assumption of (b) says that the fibre of $\text{pr}_2 : M \to X \subset G$ over at least one element of H is 0-dimensional (and, of course, nonempty). Moreover, M is connected because H is. It follows that the fibres of pr_2 in M are "generically" finite (Theorems 4.1, 4.3). Since $\dim M = \dim G$, we conclude that pr_2 is dominant and $\bar{X} = G$. $\square$

22.2. Density Theorem

Let T be a maximal torus of G, $C = C_G(T)^{\circ}$. We know that C is nilpotent (Corollary of Proposition 21.4B), hence $C = T \times C_u$ (Proposition 19.2). Furthermore, there exists a semisimple element $t \in T$ such that $C = C_G(t)^{\circ}$ (Proposition 16.4). We claim that the group $H = C$ and the element t meet the requirement of Lemma 22.1(b). Indeed, t fixes a point $xC \in G/C$ precisely when $x^{-1}tx \in C$. Then $x^{-1}tx \in C_s = T$, so $T \subset C_G(x^{-1}tx)^{\circ} = x^{-1}C_G(t)^{\circ}x = x^{-1}Cx$, or $xTx^{-1} \subset C$. This forces $xTx^{-1} = T$, $xCx^{-1} = C$. But $C = N_G(C)^{\circ}$ (Corollary of Proposition 21.4B), so this allows only finitely many choices for xC.

Thanks to (21.3), the choice $H = B$ (any Borel subgroup) meets the requirement of Lemma 22.1(a), so we can state:

Theorem. *Let B be a Borel subgroup of G, T a maximal torus of G, $C = C_G(T)^{\circ}$ a Cartan subgroup. Then the union of all conjugates of B (resp. T, resp. B_u) is G (resp. G_s, resp. G_u). Moreover, the union of the conjugates of C contains a (dense) open subset of G.*

Proof. The statement about C follows from Lemma 22.1(b) and the preceding remarks. C lies in some conjugate of B, so it follows that the union of all conjugates of B is also dense in G; by Lemma 22.1(a), it is closed, hence equal to G. The statement about T now follows from Proposition 19.4, while that about B_u is obvious. $\square$

Corollary A. *Each semisimple (resp. unipotent) element of G lies in a connected nilpotent subgroup of G.* $\square$

Corollary B. *If B is any Borel group of G, then $Z(G) = Z(B)$.*

Proof. As a consequence of the conjugacy of Borel subgroups, we already showed (Corollary of Proposition 21.4A) that $Z(G)° \subset Z(B) \subset Z(G)$. If $x \in Z(G)$, the theorem above shows that x lies in some Borel subgroup, hence (by the conjugacy theorem) in all B. $\square$

There is a Lie algebra analogue of the theorem, which asserts that $\mathfrak{g}$ is the union of its Borel subalgebras. The proof is based on a straightforward analogue of Lemma 22.1, but the final step requires more detailed information than we have yet, cf. Borel [4, 14.16].

22.3. Connectedness Theorem

The following result is very useful (and somewhat surprising).

Theorem. *Let S be a torus in G. Then $C_G(S)$ is connected.*

Proof. If G is solvable, this is already known (Proposition 19.4). To reduce the general case to this one, it suffices to show that there exists a Borel subgroup which includes both S and a given element $x \in C_G(S)$: then x must lie in $C_G(S)°$, as desired. According to Theorem 22.2, x does lie in some conjugate of a given Borel subgroup B, i.e., the fixed point set X of x in G/B is nonempty. X is closed in G/B, hence complete. Since S commutes with x, S leaves X stable and therefore has a fixed point (21.2). This yields the required Borel group. $\square$

Corollary A. *The Cartan subgroups of G are precisely the centralizers of maximal tori.* $\square$

Corollary B. *If $x \in G$ is semisimple and centralizes a torus S of G, then $\{x\} \cup S$ lies in a torus of G.*

Proof. The proof of the theorem allows us to place x, S in a Borel subgroup, so we may assume that G is solvable. But then $C_G(S)$ is connected and solvable; S lies in some (hence every) maximal torus of $C_G(S)$, and one of these contains x (Corollary 19.3). $\square$

22.4. Borel Subgroups of $C_G(S)$

If B is a Borel subgroup and H an arbitrary closed subgroup of G, it is hardly to be expected that $B \cap H$ will be a Borel subgroup of H, or even be connected. But when H is the centralizer of a torus, this does turn out to be the case. (Since the Borel subgroups of H are all conjugate, it will follow that all of them are gotten in this fashion.)

Let B be a fixed Borel subgroup of G, containing the torus S, $C = C_G(S)$. Consider how S acts on the variety G/B. Its fixed point set X is closed (Proposition 8.2(c)), hence complete (21.1)(a), and C evidently stabilizes X. Since C is *connected* (22.3), it must in fact stabilize each connected component of X (Proposition 8.2(d)). These components are also complete, being closed in X.

Theorem. *Let S be a torus in G, $C = C_G(S)$, B a Borel subgroup of G containing S, X the set of fixed points of S in G/B, Y any connected component of X. Then C acts transitively on Y.*

Proof. We noted already that C stabilizes Y. Let us assume (without loss of generality) that B is the isotropy group of some $y \in Y$. It has to be shown that $C \cdot y = Y$. Equivalently, if Z is the inverse image of Y under the canonical map $G \to G/B$, it has to be shown that $CB = Z$.

Notice that Z is a *connected* variety, since Y is connected and the canonical map has connected fibres. Moreover, $z^{-1}Sz \subset B$ for $z \in Z$, by construction. So the morphism $Z \times S \to B$ defined by $(z, s) \mapsto z^{-1}sz$ makes sense. Let φ be the composite of this with the canonical map $B \to B/B_u$. For fixed $z \in Z$, the resulting morphism $S \to B/B_u$ is clearly a homomorphism of tori. Z being connected, it follows from rigidity (16.3) that this map is independent of z. But $e \in Z$, so we conclude: $z^{-1}sz \equiv s \pmod{B_u}$ for all $z \in Z$, $s \in S$. In particular, $z^{-1}Sz$ lies in the connected solvable group SB_u, of which both it and S are maximal tori. Thanks to (19.3), there exists $u \in B_u$ such that $u^{-1}z^{-1}Szu = S$, or $zu \in N = N_G(S)$. This proves that $CB \subset Z \subset NB$. But $[N:C]$ is finite (16.3), forcing $\dim CB = \dim Z = \dim NB$. Finally, $CB = Z$, since both are connected. $\square$

Corollary. $C_B(S) = C \cap B$ is a Borel subgroup of C (and all Borel subgroups of C are obtained in this way).

Proof. Let X and Y be as before, with B the isotropy group of a point $y \in Y$. Y is a complete variety and, according to the theorem, is the orbit of y under C. The isotropy group of y in C is just $C \cap B = C_B(S)$. Therefore, $C/C \cap B$ is complete (Lemma 21.1). Now the criterion of Corollary 21.3B is applicable: $C \cap B$ is solvable, and connected (19.4), so it must be a Borel subgroup of C. $\square$

22.5. Cartan Subgroups: Summary

If S is any torus of G, we have seen that $C_G(S)$ is connected (22.3). In particular, a Cartan subgroup may now be described simply as the centralizer of a maximal torus (Corollary 22.3A). Since all maximal tori are conjugate (21.3), so are all Cartan subgroups. Let $C = C_G(T)$ be one of them. We showed in (21.4) that C is nilpotent and that $C = N_G(C)^\circ$. Being nilpotent, $C = T \times C_u$ (T the unique maximal torus of C). Moreover, the union of the conjugates of C includes an open subset of G (and is therefore dense), thanks to (22.2).

In case G is *reductive*, it will turn out later on (§26) that $C_u = e$, $C = T$. (For semisimple or reductive Lie algebras in characteristic 0, the analogous result is proved fairly soon, by a Killing form argument. Here it is not so easy.) The emphasis therefore will be placed on maximal tori rather than on Cartan subgroups. This is consistent also with the fact that semisimple elements (and tori) behave rather uniformly in characteristics 0 and p, whereas unipotent elements do not.

Exercises

G denotes a connected algebraic group.

1. $Z(G)_s$ is the intersection of all maximal tori in G.
2. If $x \in G$ has semisimple part x_s, then $x \in C_G(x_s)^\circ$. Must x lie in $C_G(x_u)^\circ$?
3. Let B be a Borel subgroup of G, $U = B_u$. Prove that $B = N_G(U)^\circ$. [Show that $U = H_u$, where $H = N_G(U)^\circ$; deduce that H/U is a torus.]
4. Deduce from Exercise 3 that if P is any parabolic subgroup of G (e.g., $P = B$), then $P = N_G(P)^\circ$ (cf. Exercise 21.10). [When char $\mathsf{K} = p > 0$, U may be thought of as a "p-sylow subgroup" of G; this result then resembles a well-known theorem about finite groups.]
5. If $\varphi : G \to G'$ is an epimorphism of algebraic groups, the Cartan subgroups of G' are precisely the images of the Cartan subgroups of G. [Cf. Exercise 21.11.]
6. A Cartan subgroup of G is maximal among the (not necessarily closed) nilpotent subgroups of G, but not every maximal nilpotent subgroup need be a Cartan subgroup.
7. Suppose some (hence every) maximal torus T of G is equal to $C_G(T)$. Then the set of semisimple elements $s \in G$ for which $C_G(s)$ is a torus contains a nonempty open set. (Such elements are called **regular semisimple**.)
8. For SL(n, K) and other classical groups, prove directly that each element of G lies in some Borel subgroup and that the centralizer of a maximal torus is itself.
9. Let C be a closed connected nilpotent subgroup of G, with $C = N_G(C)^\circ$. Prove that C is a Cartan subgroup of G.
10. If $x \in G$ is semisimple, must $C_G(x)$ be connected?

Notes

This material is based on Borel [4, §10–11], which in turn is based on Borel [1].

23. Normalizer Theorem

As in the previous section, G will denote a *connected* algebraic group. To make effective use of the Borel subgroups of G, we have to prove one more rather subtle fact, that $B = N_G(B)$. That B has finite index in its normalizer is not hard to see (cf. Exercises 21.10, 22.4), so this may be viewed as another connectedness theorem.

23.1. Statement of the Theorem

First, we prove a fact just alluded to.

Lemma. *Let B be a Borel subgroup of G, $N = N_G(B)$. Then $B = N^\circ$.*

Proof. Evidently B is a Borel subgroup of N°. Thanks to the conjugacy theorem (21.3) and the fact that B is normal in N°, B is the sole Borel subgroup of N°. The density theorem (22.2) then forces $B = N^\circ$. $\square$

Next we state the main theorem, to be proved in (23.2), and derive some corollaries.

Theorem. *Let B be a Borel subgroup of G. Then $B = N_G(B)$.*

Corollary A. *B is maximal in the collection of solvable (not necessarily connected or closed) subgroups of G.*

Proof. If S is a maximal solvable subgroup of G, then evidently S is closed. If $S \supset B$, then $S^\circ = B$ (by definition of Borel subgroup), whence $S \subset N_G(B) = B$. $\square$

In general, however, there exist maximal solvable subgroups of G which are *not* Borel subgroups (see Notes). They are necessarily disconnected and of lower dimension than B (Exercise 1). Notice that Corollary A is actually equivalent to the theorem (Exercise 2).

Corollary B. *Let P be a parabolic subgroup of G. Then $P = N_G(P)$. In particular, P is connected.*

Proof. By definition, P includes some Borel subgroup B of G. Let $x \in N_G(P)$. Then both B and xBx^{-1} are Borel subgroups of P°, so they are conjugate by some $y \in P^\circ$ (21.3). In turn, $xy \in N_G(B) = B$ (thanks to the theorem). But then $xy, y \in P^\circ$ together force $x \in P^\circ$, i.e., $P^\circ = P = N_G(P)$. $\square$

Corollary C.   *Let P, Q be parabolic subgroups of G, both of which include a Borel subgroup B. If P, Q are conjugate in G, then $P = Q$.*

Proof.   Let $x^{-1}Px = Q$. Then B and $x^{-1}Bx$ are two Borel subgroups of the connected group Q, so there exists $y \in Q$ such that $B = y^{-1}x^{-1}Bxy$. Therefore $xy \in N_G(B) = B$, forcing $x \in Q$ and $P = Q$. $\square$

This shows that the number of conjugacy classes of parabolic subgroups of G is precisely the number of parabolic subgroups containing a given B.

Corollary D.   *Let B be a Borel subgroup of G, $U = B_u$. Then $B = N_G(U)$.*

Proof.   Set $N = N_G(U)$. Since U is a maximal connected unipotent subgroup of N°, it contains a conjugate of every unipotent element of N° (Theorem 22.2). But U is normal in N°, so we conclude that N°/U consists of semisimple elements and hence is a torus (cf. Exercise 21.2). In particular, N° is solvable. Since $B \subset N^{\circ}$, we get $B = N^{\circ}$. But $B = N_G(B)$, so finally $N^{\circ} = N$. $\square$

23.2. Proof of the Theorem

We are given a Borel subgroup B, with normalizer N (where $B = N^{\circ}$, thanks to Lemma 23.1). To show that $N = B$, we proceed by induction on $\dim G$. It is clear that $R(G)$ lies in all Borel subgroups of G (cf. Exercise 21.6), so we may assume without loss of generality that G is positive dimensional and *semisimple*: otherwise the induction hypothesis applies to $G/R(G)$.

Let $x \in N$, and let T be any maximal torus of G contained in B. Then xTx^{-1} is another such torus, so there exists $y \in B$ for which $yxTx^{-1}y^{-1} = T$ (Corollary 21.3A). Evidently x belongs to B if and only if yx does, so we may as well assume at the outset that $x \in N_G(T)$. Let $S = C_T(x)^{\circ}$, a subtorus of T. Two cases are possible:

Case 1. $S \neq e$.
Then $C = C_G(S)$ has nontrivial radical, so C is a proper subgroup of G. According to (22.4), $B' = B \cap C$ is a Borel subgroup of C. Moreover, C is connected (22.3). The induction hypothesis says that $N_C(B') = B'$. But x lies in C and normalizes B, so $x \in N_C(B') = B' \subset B$.

Case 2. $S = e$.
Since x normalizes T, while T is commutative, a quick calculation shows that the commutator morphism $\gamma_x : T \to T$ (sending t to $txt^{-1}x^{-1}$) is actually a group homomorphism with kernel $C_T(x)$. This is finite, because $S = e$, so γ_x must be surjective, for reasons of dimension (Proposition 7.4B). Therefore T lies in (M, M), where M is the subgroup of N generated by x and B. As a result, $B = TB_u$ lies in the subgroup of M generated by B_u and (M, M).

Now choose a rational representation $\rho : G \to \mathrm{GL}(V)$, where V contains a line D whose isotropy group in G is precisely M (Theorem 11.2). Let $\chi: M \to \mathbf{G}_m$ be the associated character. Then χ is trivial on (M, M) as well as on the unipotent group B_u; so χ is trivial on B. If $0 \neq v \in V$ spans D, this implies that ρ induces a morphism $G/B \to Y =$ orbit of v under $\rho(G)$. Since G/B is complete, its image Y is closed in V (hence affine) as well as complete, by (21.1)(b). But then Y is a point (21.1)(c). Therefore $G = M$. Since G is connected and $[M:B] < \infty$ (Lemma 23.1), this forces $G = B$, which is absurd. $\quad\square$

23.3. The Variety G/B

Let B be any Borel subgroup of G (thanks to the conjugacy theorem, it makes no real difference which B we choose). The normalizer theorem enables us to identify the collection $\mathfrak{B}$ of all Borel subgroups of G with the projective variety G/B. Indeed, if $B' \in \mathfrak{B}$, then B' has a fixed point xB on G/B (i.e., $x^{-1}B'x = B$). If yB is any fixed point of B', then $xBx^{-1} = B' = yBy^{-1}$, or $y^{-1}x \in N_G(B) = B$, whence $xB = yB$. This shows that the assignment $B' \mapsto xB$ is unambiguous. It is *surjective*, since $xBx^{-1} \mapsto xB$ for arbitrary $x \in G$. Finally, it is *injective*, again by the normalizer theorem.

Under the 1–1 correspondence just described, the natural action of G on $\mathfrak{B}$ ($B' \mapsto xB'x^{-1}$) evidently goes over into the natural action of G on G/B ($yB \mapsto xyB$). In particular, for any subgroup H of G, the set of all Borel groups (if any) containing H corresponds to the set of fixed points of H on G/B. We shall denote these sets respectively by $\mathfrak{B}^H$ and $(G/B)^H$. The first is of course intrinsically defined, while the second is dependent on the choice of B.

23.4. Summary

The results of this chapter form the essential foundation for the study of reductive groups in subsequent chapters. To make reference easier and to etch these results more sharply in the reader's mind, we summarize briefly the main ones (G is a connected group):

(a) All Borel subgroups (resp. all maximal tori) of G are conjugate (21.3).

(b) If B is a Borel subgroup, then G/B is a projective variety (21.3).

(c) A closed subgroup P of G includes a Borel subgroup if and only if G/P is a complete variety (then P is called parabolic) (21.3).

(d) The union of all Borel subgroups of G is G itself (22.2).

(e) If S is any torus of G, $C_G(S)$ is connected; in case S is maximal, $C_G(S)$ is nilpotent and of finite index in its normalizer, and lies in every Borel subgroup which includes S (22.3). For any torus S, any Borel subgroup B of G, $B \cap C_G(S)$ is a Borel subgroup of $C_G(S)$ (22.4).

(f) If P is a parabolic subgroup of G, $P = N_G(P)$ (23.1). In particular, if B is a Borel subgroup, G/B may be identified with the collection $\mathfrak{B}$ of all Borel subgroups of G (23.3).

All of these results depend in turn on Borel's fixed point theorem (21.2):
If X is a nonempty complete variety on which a connected solvable algebraic
group G acts, then there exists a point of X fixed by G.

Exercises

G denotes a connected algebraic group.

1. Let S be a maximal solvable subgroup of G, S not a Borel subgroup.
   Prove that S is closed, disconnected, and of lower dimension than any
   Borel subgroup of G.
2. Prove that Theorem 23.1 is equivalent to its Corollary A.
3. Let B be a Borel subgroup of G. If P, Q are parabolic subgroups of G
   which include B, prove that $P \subset Q$ implies $R_u(P) \supset R_u(Q)$.
4. For $G = \mathrm{SL}(n, \mathsf{K})$, describe the parabolic subgroups including $B =
   G \cap \mathsf{T}(n, \mathsf{K})$, and deduce that the number of conjugacy classes of para-
   bolics in G is 2^{n-1} ($n - 1 = \mathrm{rank}\ G$). Treat the other classical groups
   similarly.
5. If T is a maximal torus of G, the number of distinct Borel subgroups
   including T is finite, equal to $[N_G(T){:}C_G(T)]$.
6. If char $\mathsf{K} = 0$, B a Borel subgroup of G, then $\mathfrak{b}$ is a maximal solvable
   subalgebra of $\mathfrak{g}$, $\mathfrak{n}_\mathfrak{g}(\mathfrak{b}) = \mathfrak{b}$, and $N_G(\mathfrak{b}) = B$.
7. If char $\mathsf{K} = 0$, why is it obvious that every unipotent element normalizing
   a Borel subgroup B of G must belong to B?
8. Prove that a closed subgroup H of G is a Borel subgroup if and only
   if H is solvable and G/H is complete.

Notes

Theorem 23.1 was first proved by Chevalley [8, exposé 9], thereby paving
the way for the further structure and classification theory. The elegant proof
given here is due to Borel (1973 Buenos Aires lectures). Maximal solvable
subgroups of algebraic groups are investigated by Platonov [4], who shows
that they form only finitely many conjugacy classes.

Chapter IX

Centralizers of Tori

Throughout this chapter G denotes a *connected* algebraic group.

The collection $\mathfrak{B}$ of all Borel subgroups of G may be identified with the variety G/B, for any fixed B in $\mathfrak{B}$ (23.3). The action of G on $\mathfrak{B}$ induced by conjugation corresponds to left multiplication on G/B. Here we shall examine how various tori S and their centralizers act on $\mathfrak{B}$. Since $C_G(S)$ is itself connected (22.3), and often of smaller dimension than G, it is possible to reduce many questions to the case of "semisimple rank 1", where the action on $\mathfrak{B}$ yields decisive information about the structure of G.

24. Regular and Singular Tori

It is useful to categorize a torus S in G by its number of fixed points on $\mathfrak{B}$ (i.e., by the number of Borel subgroups it lies in). If $\mathfrak{B}^S$ is finite, call S **regular**; otherwise call S **singular**. (In Chevalley [8], Borel [4], the term "semiregular" is used in place of "regular".) For example, we shall soon see that a maximal torus is regular, while in cases like $SL(n, K)$, the identity component in the kernel of a root turns out to be a singular torus.

24.1. Weyl groups

Let S be any torus in G. Because of the rigidity of tori (16.3), $N_G(S)/C_G(S)$ is a finite group, called the **Weyl group of G relative to S** and denoted $W(G, S)$. Since all maximal tori are conjugate, all their Weyl groups are isomorphic; such a group will be called simply the **Weyl group of G**.

Lemma. *Let T be a maximal torus of G, $C = C_G(T)$. Then C lies in every Borel subgroup which contains T.*

Proof. Since C is connected (22.3) and nilpotent (21.4), at least one Borel subgroup B includes C. If $B' = xBx^{-1}$ is also in $\mathfrak{B}^T$, then T and xTx^{-1} are maximal tori of B', hence are conjugate by some $y \in B'$. So $y(xCx^{-1})y^{-1} = C \subset B'$, as required. $\square$

The lemma implies that C acts trivially on the set $\mathfrak{B}^T$. On the other hand, $N = N_G(T)$ obviously permutes $\mathfrak{B}^T$, so we obtain an action of $W = N/C$ on $\mathfrak{B}^T$.

Proposition A. *Let T be a maximal torus of G, $W = W(G, T)$. Then W permutes the set $\mathfrak{B}^T$ simply transitively. In particular,* Card $\mathfrak{B}^T = |W|$ *is finite, so T is regular.*

Proof.   First we show that W acts transitively. Let $B_1, B_2 \in \mathfrak{B}^T$. The conjugacy theorem shows that $xB_2x^{-1} = B_1$, for some $x \in G$. Since xTx^{-1} and T are maximal tori of B_1, there also exists $y \in B_1$, such that $yxTx^{-1}y^{-1} = T$, so $z = yx \in N_G(T)$. But $zB_2z^{-1} = B_1$, so the coset of z in W sends B_2 to B_1.

To say that W acts simply transitively is just to say that the isotropy group in W of $B \in \mathfrak{B}^T$ is trivial, i.e., that if $x \in N_G(T)$ satisfies $xBx^{-1} = B$, then $x \in C_G(T)$. But $N_G(B) = B$ (23.1), while $N_B(T) = C_B(T)$ (19.4), so this follows. $\square$

To prepare for the semisimple (or reductive) case, we ask how passage to $G/R(G)$ affects W. In general, if T is a maximal torus of G and $\varphi: G \to G'$ an epimorphism, we know that $T' = \varphi(T)$ is a maximal torus of G' (Corollary 21.3C).

Proposition B.   *Let $\varphi: G \to G'$ be an epimorphism of algebraic groups, with T and $T' = \varphi(T)$ respective maximal tori. Then φ induces surjective maps $\mathfrak{B}^T \to \mathfrak{B}'^T$ and $W(G, T) \to W(G', T')$, which are also injective in case $\mathrm{Ker}\ \varphi$ lies in all Borel subgroups of G.*

Proof.   If B is a Borel subgroup of G containing T, then $B' = \varphi(B)$ is a Borel subgroup of G' containing T' (Corollary 21.3C), so $\mathfrak{B}^T \to \mathfrak{B}'^T$ is defined. Since $\varphi(N_G(T)) \subset N_{G'}(T')$, $\varphi(C_G(T)) \subset C_{G'}(T')$, it is clear that $W = W(G, T) \to W(G', T') = W'$ is also defined.

φ induces a surjection $\mathfrak{B} \to \mathfrak{B}'$ (again by Corollary 21.3C). If $B' \in \mathfrak{B}'^T$, then each Borel subgroup of $H = \varphi^{-1}(B')^\circ$ is a Borel subgroup of G mapped by φ onto B'; but T is a maximal torus of H, so it lies in one of these. This shows that $\mathfrak{B}^T \to \mathfrak{B}'^T$ is surjective. In case $\mathrm{Ker}\ \varphi$ lies in all Borel subgroups of G, $\mathfrak{B} \to \mathfrak{B}'$ becomes injective as well.

For the Weyl groups, we exploit their simple transitivity (Proposition A). Choose $B \in \mathfrak{B}^T$, set $B' = \varphi(B)$, and use the vertical (bijective) orbit maps to construct a commutative diagram:

$$\begin{array}{ccc} W & \to & W' \\ \downarrow & & \downarrow \\ \mathfrak{B}^T & \to & \mathfrak{B}'^{T'} \end{array}$$

Surjectivity of the top map comes from the just proved surjectivity of the bottom map; injectivity likewise follows in case $\mathfrak{B} \to \mathfrak{B}'$ is injective. $\square$

Recall that $R(G)$ is the identity component in the intersection of all Borel subgroups of G (Exercise 21.6), so the proposition yields bijections in the important special cases: $G' = G/R(G)$, $G' = G/R_u(G)$.

Before discussing arbitrary subtori of T, let us note that the Weyl group W acts naturally on the root system $\Phi = \Phi(G, T)$ as a finite group of permutations. Indeed, for any torus T the natural action of the normalizer on $X(T)$, $Y(T)$ induces an action of $W = N_G(T)/C_G(T)$ (Exercise 8). Correspondingly, $N = N_G(T)$ (and hence W) permutes the eigenspaces of T in $\mathfrak{g}$

along with the roots. Explicitly, $(\sigma\alpha)(t) = \alpha(n^{-1}tn)$ if $n \in N$ represents $\sigma \in W$. In turn, W permutes the root spaces $\mathfrak{g}_\alpha$ (via Ad N): Ad $n(\mathfrak{g}_\alpha) = \mathfrak{g}_{\sigma(\alpha)}$ (cf. (16.4)).

24.2. Regular Tori

Let S be an arbitrary torus, $C = C_G(S)$. In (22.4) we studied the set $\mathfrak{B}^S$ of Borel groups containing S, which may be identified with a closed subset X of some G/B, $B \supset S$. The connected group C stabilizes, and in fact acts transitively on, each connected component Y of X. Therefore, the complete variety Y looks just like $C/C_B(S)$. From this we deduced in particular that $C_B(S)$ is a Borel subgroup of C.

Recall that S is called *regular* if $\mathfrak{B}^S$ is finite. For example, maximal tori are regular (Proposition 24.1A).

Proposition.   *Let S be a torus in G, $C = C_G(S)$. Then S is regular if and only if C is solvable. In that case C lies in every Borel subgroup of G containing S, and for each maximal torus T containing S, $\mathfrak{B}^T = \mathfrak{B}^S$.*

Proof.   As above, identify $\mathfrak{B}^S$ with a closed subset X of some G/B, $B \supset S$. The connected components Y of X all have dimension equal to the codimension in C of a Borel subgroup of C. As a result, S is regular $\Leftrightarrow \mathfrak{B}^S$ is finite $\Leftrightarrow C$ is a Borel subgroup of itself (i.e., C is solvable). The argument shows also that for regular S, C lies in each Borel subgroup of G containing S (i.e., C fixes each point of $\mathfrak{B}^S$). In particular, since any maximal torus T containing S lies in C, $\mathfrak{B}^T = \mathfrak{B}^S$. $\square$

24.3. Singular Tori and Roots

A torus $S \subset G$ is *singular* if $\mathfrak{B}^S$ is infinite, or equivalently (by Proposition 24.2), if $C = C_G(S)$ is nonsolvable. The largest tori of this kind arise in connection with roots, as follows. Recall that if T is a maximal torus of G, the *roots* of G relative to T are the nontrivial weights of Ad T in $\mathfrak{g}$:

$$\mathfrak{g} = \mathfrak{c}_\mathfrak{g}(T) \oplus \coprod_{\alpha \in \Phi} \mathfrak{g}_\alpha,$$

where $\mathfrak{g}_\alpha = \{x \in \mathfrak{g} \mid \text{Ad } t(x) = \alpha(t)x, t \in T\}$, $\alpha \in X(T)$. (Cf. (16.4).)

Since Ad T is diagonalizable, we can split off a complement to $\mathscr{L}(H)$ in $\mathfrak{g}$ stable under Ad T for any closed subgroup H of G normalized by T. In particular, take H to be $I(T)$, the identity component in the intersection of all Borel subgroups of G containing T, and write

$$\mathfrak{g} = \mathscr{L}(I(T)) \oplus \coprod_{\alpha \in \Psi} \mathfrak{g}'_\alpha.$$

Notice that $C_G(T) \subset I(T)$ (Lemma 24.1), so $\Psi \subset \Phi$. It is clear that Ψ is independent of the choice of complement. (Eventually we shall show that $I(T) = T \cdot R_u(G)$.)

With this set-up, define $T_\alpha = (\text{Ker } \alpha)^\circ \subset T$ for $\alpha \in \Psi$. Evidently T_α is a torus of codimension 1 in T. We intend to show that T_α is singular.

Proposition. *Let S be a torus in G, T a maximal torus containing it. Then S is singular if and only if $S \subset T_\alpha$ for some $\alpha \in \Psi$.*

Proof. Let S be singular. Then $C = C_G(S)$ is nonsolvable (24.2); in particular, C has larger dimension than the solvable group $C \cap I(T)$, so the space of fixed points of S in $\mathfrak{g}$ is not wholly included in $\mathscr{L}(I(T))$. If S fixes $\mathsf{x} \in \mathfrak{g}$, it also fixes the various $\mathfrak{g}'_\alpha$ components of x, so we conclude that $S \subset \text{Ker } \alpha$ for at least one $\alpha \in \Psi$. But then $S \subset T_\alpha$.

Conversely, let $S \subset T_\alpha$ for some $\alpha \in \Psi$. If S is not singular, then C must be solvable and lie in all Borel subgroups of G including S; in particular, $C \subset I(T)$. Since the global and infinitesimal centralizers of S correspond (Proposition 18.4A), all the fixed points of S in $\mathfrak{g}$ must lie in $\mathscr{L}(I(T))$, contrary to the definition of Ψ. $\square$

Corollary. *For $\alpha \in \Psi$, set $Z_\alpha = C_G(T_\alpha)$. Then $G_\alpha = Z_\alpha/R(Z_\alpha)$ is a semisimple group of rank 1.*

Proof. Since T_α is singular, the connected group Z_α is not solvable (24.3). Therefore G_α is semisimple. But $T_\alpha \subset Z(Z_\alpha)^\circ \subset R(Z_\alpha)$, so rank $G_\alpha = 1$. $\square$

In §25 we shall prove that G_α is "locally isomorphic" to $\text{PGL}(2, \mathsf{K})$; in particular, its dimension is 3. (This fact will be a first step in showing that Ψ is an abstract root system in the sense of the Appendix.) The proof is closely tied up with showing that the Weyl group of G is not "too small". In case it has not already occurred to the reader, we point out that we have yet to prove the existence of more than one Borel subgroup containing a given maximal torus (in case G is nonsolvable).

24.4. Regular 1-parameter Subgroups

Fix a maximal torus T of G. At the opposite extreme from the singular tori T_α, $\alpha \in \Psi$ (24.3), which have codimension 1 in T, are certain regular subtori of dimension 1. Call a 1-parameter subgroup $\lambda \in \mathsf{Y}(T)$ **regular** if the torus $\lambda(\mathbf{G}_m) \subset T$ is regular. Write $\mathsf{Y}(T)_{reg}$ for the set of all regular 1-psg's. Provided dim $T > 0$, we can show that such 1-psg's abound. They will play a vital role in §25.

Proposition. *$\lambda \in \mathsf{Y}(T)$ is regular $\Leftrightarrow \langle \alpha, \lambda \rangle \neq 0$ for all $\alpha \in \Psi$.*

Proof. Thanks to Proposition 24.3, $S = \lambda(\mathbf{G}_m)$ is singular if and only if $S \subset T_\alpha$ for some $\alpha \in \Psi$. But this happens if and only if $\langle \alpha, \lambda \rangle = 0$ for some $\alpha \in \Psi$, by the way the dual pairing of $\mathsf{X}(T)$ and $\mathsf{Y}(T)$ is defined. $\square$

Exercises

1. If S is a torus in G, $X = \mathfrak{B}^S$, all connected components of X have dimension equal to the dimension of the variety of Borel subgroups of $C_G(S)$. Show by example that these components need not be permuted transitively by $W(G, S)$.

2. Let T be a maximal torus of G, S a subtorus of T. Prove that T fixes a point in each connected component of $\mathfrak{B}^S$, and deduce that the number of such components is less than or equal to $|W(G, T)|$.

3. For $G = \mathrm{SL}(n, \mathsf{K})$ (or other classical group), prove that the group $I(T)$ in (24.3) is equal to T, so that $\Psi = \Phi(G, T)$.

4. Prove that Z_α (24.3) has Weyl group of order 2. [Use Proposition 24.1B.] Describe the Z_α occurring in $\mathrm{SL}(n, \mathsf{K})$.

5. Prove that a subtorus of the diagonal torus in $\mathrm{SL}(n, \mathsf{K})$ is regular if and only if it contains a regular semisimple element (Exercise 22.7).

6. Let T be a maximal torus of G, S a subtorus of T. Then S is regular $\Leftrightarrow$ $C_G(S) \subset I(T) \Leftrightarrow \mathfrak{B}^S = \mathfrak{B}^T$.

7. If S is a singular torus of G, $\dim C_G(S) \geqslant \mathrm{rank}\ G + 2$.

8. Let T be a torus in G, $W = W(G, T)$, $\mathrm{X} = \mathrm{X}(T)$, $\mathrm{Y} = \mathrm{Y}(T)$, If $\sigma \in W$ has coset representative $n \in N_G(T)$, show that the following formulas yield an action of W on X, Y independent of the choice of n:

$$(\sigma\chi)(t) = \chi(n^{-1}tn) \qquad (t \in T,\ \chi \in \mathrm{X}),$$
$$(\sigma\lambda)(a) = n\lambda(a)n^{-1} \qquad (a \in \mathbf{G}_m,\ \lambda \in \mathrm{Y}).$$

Verify that $\langle \sigma\chi, \sigma\lambda \rangle = \langle \chi, \lambda \rangle$ for $\sigma \in W$, $\chi \in \mathrm{X}$, $\lambda \in \mathrm{Y}$.

Notes

This chapter follows, with minor expository changes, Borel [4, Chapter IV].

25. Action of a Maximal Torus on G/B

Any maximal torus T of G fixes at least one point of G/B, but at most finitely many (24.1). In particular, the action of T on G/B is highly nontrivial when $G \neq B$. We shall prove that T fixes precisely two points of G/B when $\dim G/B = 1$, and fixes at least three points when $\dim G/B > 1$ (25.2). At the same time we shall obtain a good description of $G/R(G)$ in case this group has rank 1 (to be applied when G is the centralizer Z_α of a singular torus T_α in a larger connected group).

For all of this, we use 1-psg's $\lambda: \mathbf{G}_m \to T$ and extend the resulting action of $\mathbf{G}_m$ on G/B to $\mathbf{P}^1 = \mathbf{G}_m \cup \{0\} \cup \{\infty\}$. In order to study G/B concretely, we can embed G in some $\mathrm{GL}(V)$ and find a point in $\mathbf{P}(V)$ whose precise stabilizer in G is B, thereby identifying G/B with the G-orbit of this point in $\mathbf{P}(V)$ (11.3). This orbit is of course closed, because G/B is complete. We can even arrange matters so that the image of G/B in $\mathbf{P}(V)$ lies in no hyperplane, simply by replacing V by the subspace spanned by the preimage of G/B under $\pi: V - \{0\} \to \mathbf{P}(V)$. (Similarly, we could realize G/P concretely, P any parabolic subgroup of G.)

25.1. Action of a 1-parameter Subgroup

(In this subsection G plays no role.)

Let T be a torus in $\mathrm{GL}(V)$ (maximal or otherwise), with 1-psg $\lambda: \mathbf{G}_m \to T$. Then $\mathbf{G}_m$ acts (via λ) on $\mathbf{P}(V)$. To make this action explicit, choose a basis $(v_1, \ldots, v_n)$ of V consisting of eigenvectors for T. The weights of T in V are the restrictions $\gamma_1, \ldots, \gamma_n$ of the corresponding coordinate functions; they generate $X(T)$. Set $m_i = \langle \gamma_i, \lambda \rangle$. If $v = \sum a_i v_i \in V$, $t \in \mathbf{G}_m$, we have by definition: $\lambda(t) . v = \sum a_i t^{m_i} v_i$. The action of $\mathbf{G}_m$ on $\mathbf{P}(V)$ is gotten by passing from $v \neq 0$ to $\pi(v) = [v]$.

Fix an arbitrary nonzero vector $v \in V$, $v = \sum a_i v_i$. For reasons which will soon become clear, we wish to extend the orbit map $t \xmapsto{\varphi} \lambda(t)[v]$ to a morphism $\mathbf{P}^1 \to \mathbf{P}(V)$. Here we view $\mathbf{P}^1$ as covered by the two affine open subsets $\mathsf{K}^* \cup \{0\}$, $\mathsf{K}^* \cup \{\infty\}$, so it suffices to extend φ to each of these. (Actually, the *existence* of an extension of φ to $\mathbf{P}^1$ can be proved quite generally when the target variety is complete; but we need an explicit description.)

With $m_i = \langle \gamma_i, \lambda \rangle$ as above, let m_0 (resp. m^0) be the minimum (resp. maximum) value of m_i for $i \in I = \{i | a_i \neq 0\}$. Set $I_0 = \{i \in I | m_i = m_0\}$, $I^0 = \{i \in I | m_i = m^0\}$. Notice that for $t \in \mathsf{K}^*$, $[v] = [t^{-m_0} v] = [t^{-m^0} v]$. Accordingly, we can express φ in two ways: $\varphi_0(t) = [\sum a_i t^{m_i - m_0} v_i]$ makes sense even when $t = 0$, while $\varphi^0(t) = [\sum a_i t^{m_i - m^0} v_i]$ makes sense when $t = \infty$. (In each case the vector in brackets is nonzero in V.) Evidently φ_0, φ^0 define morphisms of the respective affine lines $\mathsf{K}^* \cup \{0\}$, $\mathsf{K}^* \cup \{\infty\}$ into $\mathbf{P}(V)$ which agree with φ on K^*, so they yield the desired morphism $\mathbf{P}^1 \to \mathbf{P}(V)$ (2.3). The images of 0, ∞ will be denoted $\lambda(0)[v]$, $\lambda(\infty)[v]$:

$$\lambda(0)[v] = \left[\sum_{i \in I_0} a_i v_i \right],$$
$$\lambda(\infty)[v] = \left[\sum_{i \in I^0} a_i v_i \right].$$

These explicit formulas show that $\lambda(\mathbf{G}_m)$ *fixes each of the two points* $\lambda(0)[v]$ and $\lambda(\infty)[v]$ in $\mathbf{P}(V)$. (This was our motive for constructing them.) It still has to be decided when these two points are distinct. But this is easy: $\lambda(0)[v] = \lambda(\infty)[v] \Leftrightarrow I_0 = I^0 \Leftrightarrow m_0 = m^0 \Leftrightarrow v$ is an eigenvector of the torus $\lambda(\mathbf{G}_m) \Leftrightarrow [v]$ is fixed by $\lambda(\mathbf{G}_m)$. In the contrary case, it is easy to see that $\lambda(t)[v]$ is fixed *only* for $t = 0$, ∞ (Exercise 1).

One further observation: If λ had been chosen so that the $m_i = \langle \gamma_i, \lambda \rangle$ for distinct γ_i were all distinct, then the eigenvectors of T and of $\lambda(\mathbf{G}_m)$ would coincide. In this case, we would have: $\lambda(0)[v] = \lambda(\infty)[v] \Leftrightarrow T$ fixes $[v]$. Such a choice of λ is always possible, since $X(T) \times Y(T) \to \mathbb{Z}$ is a dual pairing and since the γ_i generate $X(T)$.

25.2. Existence of Enough Fixed Points

First we prove an auxiliary geometric lemma.

Lemma. *Let W be a subspace of codimension 1 in a vector space V, X an irreducible closed subvariety of $\mathbf{P}(V)$ not wholly contained in $\mathbf{P}(W)$. If dim $X > 0$, then X meets $\mathbf{P}(W)$, and each irreducible component of their intersection has codimension 1 in X.*

Proof. If $X \cap \mathbf{P}(W)$ were empty, then X would be a complete closed subset of the affine variety $\mathbf{P}(V) - \mathbf{P}(W)$, forcing dim $X = 0$ (21.1)(c). Now the remaining assertion follows from Theorem 3.3, combined with Corollary 3.2. $\square$

Theorem. *Let $T \subset \mathrm{GL}(V)$ be a torus, acting naturally on $\mathbf{P}(V)$. Let X be an irreducible closed set in $\mathbf{P}(V)$ stabilized by T. Assume that X lies in no hyperplane $\mathbf{P}(W)$ of $\mathbf{P}(V)$.*
   (a) *If* dim $X \geqslant 1$, *T fixes at least two points of X.*
   (b) *If* dim $X \geqslant 2$, *T fixes at least three points of X.*

Proof. Let $(v_1, \ldots, v_n)$ be a basis of V consisting of eigenvectors for T, with corresponding weights $\gamma_i \in \mathrm{X}(T)$. Choose $\lambda \in \mathrm{Y}(T)$ so that $m_i = \langle \gamma_i, \lambda \rangle$ are distinct when the γ_i are distinct. As remarked at the end of (25.1), T and $\lambda(\mathbf{G}_m)$ then have precisely the same fixed points in $\mathbf{P}(V)$. So we may as well assume that $T = \lambda(\mathbf{G}_m)$.

(a) Since dim $X \geqslant 1$, X is infinite. In case T fixes all points of X, we therefore have nothing to prove. Otherwise let $[v] \in X$ be not fixed by T. Following the recipe in (25.1) for this choice of v, we get two *distinct* fixed points $\lambda(0)[v]$ and $\lambda(\infty)[v]$ in $\mathbf{P}(V)$. But the associated morphism $\mathbf{P}^1 \to \mathbf{P}(V)$ has closed image (21.1)(b), and X is closed, so these points must lie in X.

(b) As in the proof of (a), we may assume that some $[v]$ in X is not fixed by T. Using the notation of (25.1) (for this choice of v), order the basis of V so that $m_1 = m^0$ has the maximum value for basis vectors involved in v. Let W be the subspace of V spanned by $v_2, \ldots, v_n$. Then the lemma shows that the irreducible components of $X \cap \mathbf{P}(W)$ have codimension 1 in X. They are obviously stable under T (since X and W are), and of dimension at least 1 (since dim $X \geqslant 2$). Let Y be one of these components. Then part (a) implies that T fixes at least two points of Y. It remains to find a third fixed point. By choice of v, $\lambda(\infty)[v] = [\sum_{i \in I^0} a_i v_i]$ does not lie in $\mathbf{P}(W)$, but is fixed by T, so this does the trick. $\square$

We now deduce from the theorem that G has sufficiently many Borel subgroups containing a given maximal torus T.

Corollary A. *Let P be a proper parabolic subgroup of G, T any torus in G. Then T fixes at least two points of G/P (resp. at least three, when* dim $G/P > 1$). *In particular, a maximal torus lies in at least two Borel sub-groups (resp. at least three, when* dim $G/B > 1$).

Proof. As remarked at the beginning of this section, G may be embedded into some $GL(V)$ in such a way that G/P is isomorphic to the orbit X of some point in $\mathbf{P}(V)$; we can even guarantee that X lies in no hyperplane. Since G/P is complete and irreducible, X is closed and irreducible (21.1)(a). T stabilizes X, so the theorem applies. $\square$

This corollary can be formulated also as an assertion that the Weyl group of G is not "too small", in view of Proposition 24.1A:

Corollary B. *Let B be a Borel subgroup of G, T a maximal torus, $W = N_G(T)/C_G(T)$ the Weyl group of G relative to T. If* dim $G/B \geqslant 1$ *(resp.* $\geqslant 2$), *then* $|W| \geqslant 2$ *(resp.* $\geqslant 3$). $\square$

Corollary C. *Let T be a maximal torus of G. Then G is generated by the set $\mathfrak{B}^T$ of Borel subgroups containing T.*

Proof. Use induction on dim G. The subgroup P of G generated by $\mathfrak{B}^T$ is evidently parabolic. Suppose $P \neq G$. By Corollary A, there is at least one conjugate P' of P containing T, with $P' \neq P$. But T is a maximal torus of P', and the Borel subgroups of P' containing T are also Borel subgroups of G (Exercise 21.9). The latter generate P', by induction, forcing $P' \subset P$, which is absurd. $\square$

Later on it will be shown that *two* suitably chosen elements of $\mathfrak{B}^T$ already suffice to generate G. For example, when $G = \mathrm{SL}(n, \mathsf{K})$, take the upper and the lower triangular groups (Exercise 2).

25.3. Groups of Semisimple Rank 1

We can use (25.2) along with (6.3), (6.4), to obtain a description of semi-simple groups of rank 1. The **semisimple rank**, $\mathrm{rank}_{ss}G$ (resp. **reductive rank**, $\mathrm{rank}_{red}G$) is the rank of $G/R(G)$ (resp. $G/R_u(G)$). For example, SL(2, K), GL(2, K), PGL(2, K), and the groups Z_α (24.3) all have semisimple rank 1, while $G\,L(2, \mathsf{K})$ has reductive rank 2.

Theorem. *Let T be a maximal torus of G, $W = W(G, T)$. The following statements are equivalent:*
   (a) $\mathrm{Rank}_{ss}G = 1$.
   (b) $|W| = 2$.
   (c) Card $\mathfrak{B}^T = 2$.
   (d) Dim $G/B = 1$.
   (e) $G/B \cong \mathbf{P}^1$.
   (f) *There exists an epimorphism* $\varphi : G \to \mathrm{PGL}(2, \mathsf{K})$, *with* $(\mathrm{Ker}\ \varphi)^\circ = R(G)$.

Proof. (a) $\Rightarrow$ (b) Since every Borel subgroup includes $R(G)$, the Borel subgroups of G including T are in 1–1 correspondence with those of $G' = G/R(G)$ including $T' = T/T \cap R(G)$, and the number in each case is the order of the respective Weyl group W, W' (Proposition 24.1B). But dim $T' = 1$, by assumption, and Aut $\mathbf{G}_m = \mathbb{Z}/2\mathbb{Z}$ (Exercise 7.4), so $|W| = |W'| \leqslant 2$. On the other hand, G is nonsolvable by assumption, so $|W| \geqslant 2$ by Corollary 25.2B.

(b) $\Rightarrow$ (c) The equivalence of (b) and (c) (24.1) was just noted.

(c) $\Rightarrow$ (d) Since T lies in two distinct Borel subgroups, G/B has dimension at least 1. If dim $G/B \geqslant 2$, then Corollary 25.2A says that Card $\mathfrak{B}^T \geqslant 3$, contrary to (c).

(d) $\Rightarrow$ (e) Let dim $G/B = 1$. As in (25.1), we may construct a nonconstant morphism $\mathbf{P}^1 \to G/B \subset \mathbf{P}(V)$ relative to any regular 1-psg $\lambda \in \mathrm{Y}(T)$, T a maximal torus of G. Its image is closed, because $\mathbf{P}^1$ is complete (21.1)(b), hence equal to G/B, because G/B is irreducible and of dimension 1. This implies that G/B is isomorphic to $\mathbf{P}^1$ (6.3).

(e) $\Rightarrow$ (f) G acts as a group of automorphisms of $G/B \cong \mathbf{P}^1$. But we know that Aut $\mathbf{P}^1 = \mathrm{PGL}(2, \mathsf{K})$, whence a morphism of algebraic groups $\varphi : G \to \mathrm{PGL}(2, \mathsf{K})$ (6.4). Evidently Ker φ is the intersection of all Borel subgroups of G, so $(\mathrm{Ker}\, \varphi)^\circ = R(G)$ (cf. Exercise 21.6). Since $G' = G/R(G)$ is nonsolvable, its dimension is at least 3 (Exercise 21.4), forcing φ to be an epimorphism.

(f) $\Rightarrow$ (a) This is clear. $\square$

The theorem tells us a considerable amount about the structure of a semisimple group of rank 1, notably that its dimension must be 3. More generally, we get information about reductive groups of semisimple rank 1, which will turn out to be applicable to the groups Z_α when G is reductive:

Corollary. *Let G be reductive, $\mathrm{rank}_{ss}G = 1$, T a maximal torus of G, $Z = Z(G)^\circ$. Then:*

(a) *(G, G) is semisimple, of dimension 3.*

(b) *$G = (G, G) \cdot Z$, the intersection of (G, G) with Z being finite.*

(c) *$C_G(T) = T$, and in particular, $Z(G) \subset T$.*

(d) *If $\varphi : G \to \mathrm{PGL}(2, \mathsf{K})$ is an epimorphism, Ker $\varphi = Z(G)$.*

Proof. We get from part (f) of the theorem an epimorphism $\varphi : G \to \mathrm{PGL}(2, \mathsf{K})$, with $(\mathrm{Ker}\, \varphi)^\circ = R(G)$. Since G is reductive, $R(G) = Z$ and is a torus (Lemma 19.5). On the other hand, $\mathrm{PGL}(2, \mathsf{K})$ is its own derived group: for example, otherwise the derived group would be solvable (Exercise 21.4), contrary to the semisimplicity of $\mathrm{PGL}(2, \mathsf{K})$. It follows that φ maps (G, G) onto $\mathrm{PGL}(2, \mathsf{K})$. Combined with the fact that (G, G) is connected (17.2) and the fact that $(G, G) \cap Z$ is finite (19.5), this proves both (a) and (b). It is easy to see that a maximal torus in $\mathrm{PGL}(2, \mathsf{K})$ is its own centralizer, by computing in $\mathrm{SL}(2, \mathsf{K})$, so $C_G(T)$ and T have the same image under φ, forcing $T \subset C_G(T) \subset T \cdot \mathrm{Ker}\, \varphi$. But T has finite index in the right side, while $C_G(T)$ is connected (22.3). So (c) and (d) follow. $\square$

25.4. Weyl Chambers

Let us look more closely at Borel subgroups of the centralizer of a singular torus.

Fix a maximal torus T of G. For each $\alpha \in \Psi$ (24.3), $T_\alpha = (\mathrm{Ker}\ \alpha)^\circ$ is a singular subtorus of T, whose centralizer Z_α has semisimple rank 1. From Theorem 25.3 (in fact, already from Exercise 24.4) it follows that $W_\alpha = W(Z_\alpha, T)$ has order 2, so the maximal torus T of Z_α lies in precisely two Borel subgroups (say B_α and B'_α) of Z_α. Since Z_α includes $C_G(T)$, W_α may be identified canonically with a subgroup of $W = W(G, T)$; denote a generator by σ_α, and choose some representative n_α of σ_α in $N_{Z_\alpha}(T)$. Then $n_\alpha B_\alpha n_\alpha^{-1} = B'_\alpha$. For each $B \in \mathfrak{B}^T$, we know (22.4) that $B \cap Z_\alpha$ is a Borel subgroup of Z_α, hence is either B_α or B'_α. Say $B \cap Z_\alpha = B_\alpha$, so $n_\alpha B n_\alpha^{-1} \cap Z_\alpha = B'_\alpha$. Then it is clear that precisely half the elements of W (whose order is even!) send B to other Borel subgroups with the same property; thus α induces a natural partition of $\mathfrak{B}^T$. We can make this partition more recognizable if we formulate it in terms of 1-psg's.

The idea is to assign to each $\lambda \in Y(T)_{reg}$ a Borel subgroup $B(\lambda)$ containing T, in such a way that $B(\lambda) \cap Z_\alpha = B_\alpha$ precisely when $\langle \alpha, \lambda \rangle > 0$. Since λ is regular if and only if $\langle \alpha, \lambda \rangle \neq 0$ for all $\alpha \in \Psi$ (24.4), this idea is certainly reasonable. Of course, the correspondence $\lambda \mapsto B(\lambda)$ will be many-to-one, since Card $\mathfrak{B}^T = |W|$ is finite.

Choose an embedding $G \subset GL(V)$, so that G/B is identified with the orbit X of a point of $\mathbf{P}(V)$ (X being contained in no hyperplane $\mathbf{P}(W)$). As in the set-up of (25.1), let $m_i = \langle \gamma_i, \lambda \rangle$ for $\lambda \in Y(T)_{reg}$. Order the basis $(v_1, \ldots, v_n)$ of V so that $m^0 = m_1 = \cdots = m_r > m_{r+1} \geqslant \cdots \geqslant m_n = m_0$. Let W be the span of $(v_2, \ldots, v_n)$. By construction, X does not lie wholly within $\mathbf{P}(W)$, so we may choose $[v] = [\sum a_i v_i] \in X$ with $a_1 \neq 0$. The discussion in (25.1) shows that $\lambda(\mathbf{G}_m)$ fixes the point $\lambda(\infty)[v] = [a_1 v_1 + \cdots + a_r v_r]$. Because λ is regular, we can argue that $r = 1$: Otherwise we can find infinitely many distinct points in X of the type $[v'] = [v_1 + bv_2 + \cdots]$ for distinct b, since X cannot lie in the union of $\mathbf{P}(W)$ and only finitely many other hyperplanes defined by conditions "b = ratio of second to first coordinate". So we can find infinitely many distinct fixed points $\lambda(\infty)[v']$ of $\lambda(\mathbf{G}_m)$ in X, contradicting the regularity of λ. Therefore, $r = 1$, $[v] = [v_1]$.

Consider the neighborhood $U = X - (X \cap \mathbf{P}(W))$ of $[v_1]$. The preceding argument shows that $\lambda(\infty)[v'] = [v_1]$ for all $[v'] \in U$. This singles out the fixed point $[v_1]$ of $\lambda(\mathbf{G}_m)$ for special attention: the corresponding Borel group is denoted $B(\lambda)$. Notice that T fixes $[v_1]$ (λ being regular), so $B(\lambda) \in \mathfrak{B}^T$, as desired.

The Weyl group W permutes $\mathfrak{B}^T$ simply transitively (24.1). On the other hand, W acts naturally on $Y(T)$ via $(\sigma\lambda)(a) = n\lambda(a)n^{-1}$ (n any representative of σ in $N_G(T)$); this action satisfies $\langle \sigma\chi, \sigma\lambda \rangle = \langle \chi, \lambda \rangle$ for $\chi \in X(T)$, $\lambda \in Y(T)$ (Exercise 24.8). Since λ is regular $\Leftrightarrow \langle \alpha, \lambda \rangle \neq 0$ for all $\alpha \in \Psi$, this shows that W permutes $Y(T)_{reg}$. We claim that the two actions of W respect the map $\lambda \mapsto B(\lambda)$. Let $\sigma \in W$ be represented by $n \in N_G(T)$. Concretely, n permutes

the eigenspaces of T in V, hence permutes the weights γ_i and the integers
$m_i = \langle \gamma_i, \lambda \rangle$. But then the above recipe singles out the fixed point $[n \cdot v_1]$
of $\sigma\lambda(\mathbf{G}_m)$ and labels the corresponding Borel group $nB(\lambda)n^{-1}$ as $B(\sigma\lambda)$.
This proves our claim. In particular, the correspondence $\lambda \mapsto B(\lambda)$ is *onto*
$\mathfrak{B}^T$. Write $\mathfrak{C}(B) = \{\lambda \in \mathrm{Y}(T)_{reg} | B = B(\lambda)\}$ for each $B \in \mathfrak{B}^T$, and call this
the **Weyl chamber** of B (relative to T). The (nonempty) sets $\mathfrak{C}(B)$ therefore
form a disjoint partition of $\mathrm{Y}(T)_{reg}$.

Return now to the discussion of Z_α ($\alpha \in \Psi$) which began this subsection,
with σ_α, n_α, etc. Choose $\lambda_0 \in \mathrm{Y}(T)_{reg}$ so that $\langle \alpha, \lambda_0 \rangle > 0$, set $B = B(\lambda_0)$, and
define $B_\alpha = B \cap Z_\alpha$, $B'_\alpha = n_\alpha B_\alpha n_\alpha^{-1}$. Then B_α, B'_α are the two Borel subgroups
of Z_α containing T. Moreover, since $\sigma_\alpha \alpha = -\alpha$ (in additive notation!),
we see from the W-equivariance of the map $\lambda \mapsto B(\lambda)$ that $\langle \alpha, \lambda \rangle > 0$
precisely when $B(\lambda) \cap Z_\alpha = B_\alpha$. To summarize our findings:

Proposition. *Fix a maximal torus T of G, and assign to each $\lambda \in \mathrm{Y}(T)_{reg}$
a Borel subgroup $B(\lambda)$ containing T, as above. Let $\alpha \in \Psi$, $Z_\alpha = C_G(T_\alpha)$. Then
the two Borel subgroups B_α, B'_α of Z_α containing T may be labelled so that for all
$\lambda \in \mathrm{Y}(T)_{reg}$, $\langle \alpha, \lambda \rangle > 0$ precisely when $B(\lambda) \cap Z_\alpha = B_\alpha$.* $\quad\square$

Exercises

1. In the notation of (25.1), suppose that $\lambda(0)[v] \neq \lambda(\infty)[v]$. Prove that
   no other point $\lambda(t)[v]$ can be fixed by $\lambda(\mathbf{G}_m)$.
2. Let $G = \mathrm{SL}(n, \mathsf{K})$ (or other classical group). Show that G is generated
   by two suitably chosen Borel subgroups.
3. Fix a maximal torus T of G. Then G is generated by all $C_B(S)$, B ranging
   over $\mathfrak{B}^T$ and S ranging over the subtori of codimension 1 in T.
4. Let G be a reductive group of semisimple rank 1, T a maximal torus,
   $\Phi = \Phi(G, T)$, $W = W(G, T)$. Prove that:
     (a) Ker Ad $= Z(G)$.
     (b) $\Phi = \{\alpha, -\alpha\}$ for some $\alpha \in \mathrm{X}(T)$, with $\sigma\alpha = -\alpha$ if σ generates W.
     (c) If B, B' are the two Borel subgroups containing T, then $B \cap B' =$
   T, while B_u and B'_u are isomorphic to $\mathbf{G}_a$.
     (d) $\mathfrak{g} = \mathfrak{t} \oplus \mathfrak{g}_\alpha \oplus \mathfrak{g}_{-\alpha}$, with $\mathfrak{g}_\alpha = \mathscr{L}(B_u)$, $\mathfrak{g}_{-\alpha} = \mathscr{L}(B'_u)$ (for suitable
   labelling of $\alpha, -\alpha$).
5. Exhibit three reductive groups of dimension 4 and semisimple rank 1
   which are pairwise nonisomorphic (as algebraic groups). [To show non-
   isomorphism when char $\mathsf{K} = 2$, compare Lie algebras.]

26. The Unipotent Radical

Let T be a fixed maximal torus of G. We are going to prove that $R_u(G)$
coincides with the unipotent part of the intersection of all Borel subgroups

containing T. Coupled with the earlier description of the groups Z_α, this will yield a reasonably clear picture of the root system of a reductive group.

26.1. Characterization of $R_u(G)$

First we prove a lemma which insures the existence of enough regular 1-psg's. Recall from (24.3) the set Ψ of roots "outside" $I(T)_u$.

Lemma. *Let α, $\beta \in \Psi$ be nonproportional. Then there exists $\lambda \in Y(T)_{reg}$ for which $\langle \alpha, \lambda \rangle > 0$ and $\langle \beta, \lambda \rangle < 0$.*

Proof. The criterion for $\lambda \in Y(T)$ to be regular is that $\langle \gamma, \lambda \rangle \neq 0$ for all $\gamma \in \Psi$ (24.4). The free abelian groups $X(T)$ and $Y(T)$, of rank $n = \dim T$, become dual vector spaces over Q if we tensor with Q over Z and extend the pairing $\langle \gamma, \lambda \rangle$. In this setting α, β are linearly independent vectors, while the condition for regularity is that λ avoid finitely many hyperplanes (including those "orthogonal" to α, β). So it is geometrically obvious that the conditions of the lemma can be met by some Q-linear combination of 1-psg's: α and β simply have to lie on different sides of the hyperplane orthogonal to this vector. (The reader should supply an algebraic proof.) Multiplying by a common denominator of the coefficients gets us back into $Y(T)$, where the conditions of the lemma are still satisfied. $\square$

As in (24.3), $I(T)$ denotes the identity component in the intersection of the (finitely many) Borel subgroups containing T.

Theorem. $R_u(G) = I(T)_u$.

Proof. Abbreviate $I(T)_u$ by U. Since $R_u(G)$ lies in B_u for all Borel subgroups B, it is clear that $R_u(G) \subset U$. To get the reverse inclusion, it will be enough to show that $U \lhd G$. But G is generated by the Borel subgroups containing T (Corollary 25.2C). In turn we claim that $B \in \mathfrak{B}^T$ is generated by its subgroups $C_B(S)$, where S ranges over the subtori of codimension 1 in T: Let A be the subgroup of B so generated. Since $C_B(T) \subset C_B(S)$, all occurrences of the weight 0 in $\mathfrak{b}$ occur in $\mathfrak{a}$ (global and infinitesimal centralizers of tori correspond: Proposition 18.4A). If α is a nontrivial weight, $S = (\operatorname{Ker} \alpha)^\circ$ has codimension 1 in T, so all occurrences of α in $\mathfrak{b}$ already occur in $\mathfrak{a}$ (for the same reason). Conclusion: $\mathfrak{a} = \mathfrak{b}$, hence $A = B$ (since $A \subset B$ and B is connected).

This reduces matters to showing that U is normalized by all $C_B(S)$, S as above. If S is *regular*, then $C_B(S) = C_G(S)$ lies in all Borel subgroups containing T (24.2), hence in $I(T)$, so there is nothing to prove. Next let S be *singular*, so $S = (\operatorname{Ker} \alpha)^\circ = T_\alpha$ for some $\alpha \in \Psi$ (24.3). Here $C_B(T_\alpha) = Z_\alpha \cap B$ is one of the two Borel subgroups B_α, B'_α of Z_α (25.4). We can even say which it is if we write $B = B(\lambda)$ for some (nonunique) $\lambda \in Y(T)_{reg}$: $Z_\alpha \cap B = B_\alpha \Leftrightarrow \langle \alpha, \lambda \rangle > 0$ (25.4).

It is awkward to prove directly that B_α (or B'_α) normalizes U. Instead let us look at the root structure as a whole for some guidance. This is shown schematically in Figure 1.

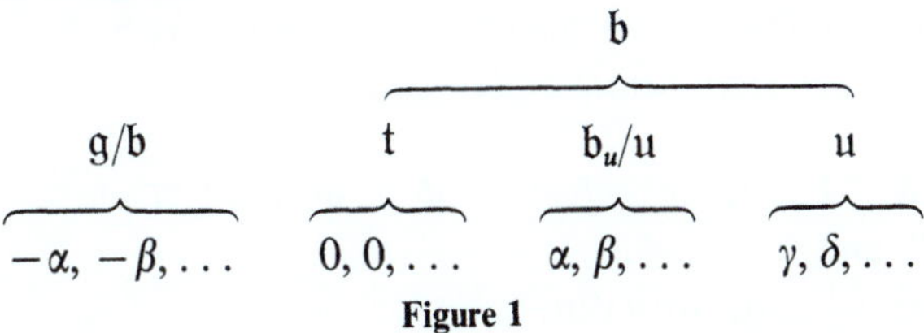

Figure 1

Here $\Psi = \{\pm\alpha, \pm\beta, \ldots\}$, while various $\gamma, \delta, \ldots$ over which we have no control (some may be 0) occur in u. Varying the choice of $B \in \mathfrak{B}^T$ just replaces some of the roots in Ψ by their negatives. B_α involves (besides 0 weights corresponding to t) α along with some unknown weights occurring in u.

Our strategy is to find a closed subgroup H of B_u containing both U and $(B_\alpha)_u$, with U of *codimension one* in H. Then Proposition 17.4(b) will force U to be *normal* in H.

Define H to be the unipotent part of the intersection of all $B \in \mathfrak{B}^T$ for which $B \cap Z_\alpha = B_\alpha$. This amounts (25.4) to intersecting those $B(\lambda)_u$, $\lambda \in Y(T)_{reg}$, for which $\langle \alpha, \lambda \rangle > 0$. It is clear that U and $(B_\alpha)_u$ lie in H, so we just have to compare dimensions. For this it suffices to show that u has dimension one less than $\mathfrak{h}$. Now H is stable under conjugation by T, so $\mathfrak{h}$ is the sum of u and certain $\mathfrak{g}'_\beta$ for $\beta \in \Psi$ (notation of (24.3)), α included. Among the latter roots none can appear along with its negative (cf. Figure 1), so for $\beta \neq \alpha$, we deduce from (25.3) that β and α are nonproportional. The above lemma furnishes a regular 1-psg λ satisfying: $\langle \alpha, \lambda \rangle > 0$, $\langle \beta, \lambda \rangle < 0$. Thus $B(\lambda) \cap Z_\beta = B'_\beta$, while $H \subset B(\lambda)$. Assuming that β (but not $-\beta$) occurs in $\mathfrak{h}$, this forces $C_H(T_\beta) = e$. But global and infinitesimal centralizers of tori correspond (18.4A), so β cannot occur in $\mathfrak{h}$ after all. We conclude that only α occurs in $\mathfrak{h}$ outside u. Since dim $\mathfrak{g}'_\alpha = 1$ (25.3), the proof is complete. $\square$

26.2. Some Consequences

With Theorem 26.1 in hand, it is possible (for the first time) to exploit fully the assumption that a group is reductive.

Corollary A. *Let G be reductive, S any subtorus of T. Then:*
(a) *$C_G(S)$ is reductive.*
(b) *If S is regular, then $C_G(S) = T$. In particular, the Cartan subgroups of G are just the maximal tori, and $Z(G) \subset T$.*

Proof. Let $C = C_G(S)$. Theorem 26.1, applied to C, shows that $R_u(C)$ is the unipotent part of the intersection of all Borel subgroups of C containing the maximal torus T of C. This in turn lies in $I(T)_u = e$, proving (a). In case

S is regular, C is solvable (24.2) as well as reductive, i.e., C is a torus. So (b) follows. $\square$

The fact that centralizers of tori in reductive groups are again reductive often makes it possible to use inductive arguments (which is why we emphasize these rather than semisimple groups).

Corollary B. *Let G be reductive, $\Phi = \Phi(G, T)$. Then:*
(a) $\Phi = \Psi$, *and* $\Phi = -\Phi$.
(b) $\mathfrak{g} = \mathfrak{t} \oplus \coprod_{\alpha \in \Phi} \mathfrak{g}_\alpha$, *with* $\dim \mathfrak{g}_\alpha = 1$.
(c) *The singular tori of codimension 1 in G are the* $T_\alpha = (\mathrm{Ker}\ \alpha)^\circ$, $\alpha \in \Phi$.
(d) *Let* $Z_\alpha = C_G(T_\alpha)$ $(\alpha \in \Phi)$. *Then* Z_α *is a reductive group of semisimple rank* 1, *with* $\mathfrak{z}_\alpha = \mathfrak{t} \oplus \mathfrak{g}_\alpha \oplus \mathfrak{g}_{-\alpha}$. *Moreover, the* Z_α *generate G.*
(e) $Z(G) = \bigcap_{\alpha \in \Phi} T_\alpha$.
(f) *The rank of the subgroup R of $X(T)$ generated by Φ is equal to* $\mathrm{rank}_{ss} G$.

Proof. Assertions (a), (b), (c), and the first part of (d) are all immediate, in view of Corollary A and (25.3). That the Z_α generate G follows from the fact that they generate a closed subgroup of G having Lie algebra $\mathfrak{g}$ (cf. (b)). Consider (e). According to Corollary A, $Z(G) \subset T$, forcing $Z(G) \subset T_\alpha$ for all α. Conversely, if $t \in T_\alpha$ for all α, then t centralizes every Z_α; these generate G (by (d)), so $t \in Z(G)$. As to (f), we need only prove this when G is semisimple. Let R' be the annihilator of R in $Y(T)$, relative to the dual pairing $X(T) \times Y(T) \to \mathbb{Z}$. It just has to be shown that $R' = 0$. Now $\langle \alpha, \lambda \rangle = 0$ for all $\alpha \in \Phi$ means that $\lambda(\mathbf{G}_m) \subset \mathrm{Ker}\ \alpha$, hence that $\lambda(\mathbf{G}_m) \subset T_\alpha$, for all α. By (e), $\lambda(\mathbf{G}_m) \subset Z(G)$, which is finite because G is semisimple. But $\lambda(\mathbf{G}_m)$ is connected, forcing $\lambda = 0$. $\square$

Corollary C. *Let G be reductive. For each Borel subgroup B containing T, there exists $B^- \in \mathfrak{B}^T$ such that $B \cap B^- = T$. Then* $\mathfrak{g} = \mathfrak{b}^- + \mathfrak{b} = \mathfrak{u}^- \oplus \mathfrak{t} \oplus \mathfrak{u}$, *where* $B = TU$, $B^- = TU^-$.

Proof. We know that $B = B(\lambda)$ for some regular 1-psg λ, and that each $\alpha \in \Phi$ satisfies $\langle \lambda, \alpha \rangle \neq 0$ (cf. Corollary B). Set $B^- = B(-\lambda)$. For each $\alpha \in \Phi$, $Z_\alpha \cap B$ and $Z_\alpha \cap B^-$ are therefore the two Borel subgroups of Z_α containing T (25.4). In view of part (b) of Corollary B, $B \cap B^- = T$, while $\mathfrak{g}$ decomposes as indicated. $\square$

B^- is in fact unique (Exercise 6); it is called the Borel subgroup **opposite** B (relative to T).

26.3. The Groups U_α

Let G henceforth be reductive, $\Phi = \Phi(G, T)$, $W = W(G, T)$. So far we have studied roots as characters of T, permuted by W (or N). The decomposition $\mathfrak{g} = \mathfrak{t} \oplus \coprod_{\alpha \in \Phi} \mathfrak{g}_\alpha$ offers a more concrete portrayal of Φ, making it clear (for example) that $\dim G = \mathrm{rank}\ G + \mathrm{Card}\ \Phi$.

The root system may even be viewed as living within G itself, in the following sense. For each $\alpha \in \Phi$, $\mathfrak{g}_\alpha$ is the Lie algebra of the unipotent part U_α of one of the two Borel subgroups of Z_α containing T (part (d) of Corollary 26.2B). In fact, U_α *is the unique connected T-stable subgroup of G having Lie algebra* $\mathfrak{g}_\alpha$. (This is clear if char $\mathsf{K} = 0$, cf. (13.1)). For suppose V_α satisfies these conditions. Since its Lie algebra consists of nilpotent elements, V_α must be unipotent rather than toral (cf. (20.5)). Moreover, $H = TV_\alpha$ is a connected solvable subgroup of G, so it lies in some $B \in \mathfrak{B}^T$. Since T_α centralizes $\mathfrak{h} = \mathfrak{t} + \mathfrak{g}_\alpha$, T_α must also centralize H: global and infinitesimal centralizers of tori correspond (Proposition 18.4A). Therefore, $V_\alpha \subset B \cap Z_\alpha$ (one of the two Borel subgroups of Z_α containing T), forcing $V_\alpha = U_\alpha$.

Theorem. *Let G be reductive, $\alpha \in \Phi$.*

(a) *There exists a unique connected T-stable subgroup U_α of G having Lie algebra $\mathfrak{g}_\alpha$ (and $U_\alpha \subset Z_\alpha$).*

(b) *If $n \in N$ represents $\sigma \in W$, then $nU_\alpha n^{-1} = U_{\sigma(\alpha)}$.*

(c) *There exists an isomorphism $\varepsilon_\alpha : \mathbf{G}_a \to U_\alpha$ such that for all $t \in T$, $x \in \mathbf{G}_a$, $t\varepsilon_\alpha(x)t^{-1} = \varepsilon_\alpha(\alpha(t)x)$.*

(d) *G is generated by the groups U_α ($\alpha \in \Phi$), along with T.*

Proof. We just verified (a). Since $\mathscr{L}(nU_\alpha n^{-1}) = \mathrm{Ad}\ n(\mathscr{L}(U_\alpha)) = \mathrm{Ad}\ n(\mathfrak{g}_\alpha) = \mathfrak{g}_{\sigma(\alpha)}$ (24.1), while $nU_\alpha n^{-1}$ is obviously connected and T-stable, (b) follows from (a). For (d), use part (d) of Corollary 26.2B.

It remains to prove (c). The fact that U_α is a connected 1-dimensional unipotent group insures the existence of some isomorphism $\varepsilon : \mathbf{G}_a \to U_\alpha$ (20.5). The action of T on U_α by inner automorphisms therefore yields an action on $\mathbf{G}_a$, hence a morphism of algebraic groups $T \to \mathrm{Aut}\ \mathbf{G}_a \cong \mathbf{G}_m$ (cf. Exercise 7.4), i.e., a character γ of T. Thus $t\varepsilon(x)t^{-1} = \varepsilon(\gamma(t)x)$. For each $t \in T$, there is a corresponding commutative diagram, the left vertical arrow being multiplication by $\gamma(t)$:

$$
\begin{array}{ccc}
\mathbf{G}_a & \xrightarrow{\varepsilon} & U_\alpha \\
\downarrow & & \downarrow \quad \mathrm{Int}\ t \\
\mathbf{G}_a & \xrightarrow{\varepsilon} & U_\alpha
\end{array}
$$

If we identify the Lie algebra of $\mathbf{G}_a$ with K, the differential of multiplication by $\gamma(t)$ is again multiplication by $\gamma(t)$, while $d(\mathrm{Int}\ t) = \mathrm{Ad}\ t$. It follows that $\mathrm{Ad}\ t(\mathsf{x}) = \gamma(t)\mathsf{x}$ for $\mathsf{x} \in \mathfrak{g}_\alpha$, which forces $\gamma = \alpha$. So we can set $\varepsilon_\alpha = \varepsilon$. $\square$

The proof of (c) shows that *any* isomorphism $\mathbf{G}_a \to U_\alpha$ will be compatible with the action of T; but for emphasis we call such an isomorphism **admissible**. It is unique only up to a scalar factor (i.e., an automorphism of $\mathbf{G}_a$), which amounts to the choice of a basis for $\mathfrak{g}_\alpha$.

Exercises

G denotes a reductive group.

1. Let S be any torus in a connected group H. Prove that $C_{R_u(H)}(S) = R_u(C_H(S))$.

2. $Z(G)$ is the intersection of all maximal tori of G.
3. A subtorus of G is regular if and only if it contains a regular semisimple
   . element (Exercise 22.7).
4. $Z(G) = \bigcap_{\alpha \in \Phi} \operatorname{Ker} \alpha$.
5. If $\mathsf{x} \in \mathfrak{g}$ is semisimple, then $C_G(\mathsf{x})^\circ$ is reductive.
6. Prove that B^- in Corollary 26.2C is unique.
7. All elements of U_α (other than e) are conjugate in G.
8. The character group of a maximal torus in Ad G is generated by roots.
9. Find the index in $X(T)$ of the subgroup R generated by Φ when $G = \mathrm{SL}(n, \mathsf{K})$ (or other classical group). What is the structure of $X(T)/R$?
10. The groups U_α generate (G, G).
11. If B is a Borel subgroup containing the maximal torus T, then $N_G(T) \cap B = T$.
12. G is generated by its semisimple elements.

Chapter X

Structure of Reductive Groups

By studying the actions of tori and their centralizers on G/B, we showed in Chapter IX that a reductive group is generated by the centralizers of singular tori (the latter being precisely the connected kernels of roots). Moreover, we showed that the quotient of such a centralizer by its center is essentially PGL(2, K). The goal of this chapter is a more detailed description of G: properties of the root system, structure of normal subgroups of G, "normal form" for elements of G, structure of parabolic subgroups.

Except in §29, G will denote a reductive group, T a fixed maximal torus, $\Phi = \Phi(G, T)$, $W = W(G, T)$, X = X(T), Y = Y(T).

27. The Root System

Our first objective is to prove that Φ is an abstract root system, in the sense of the Appendix. This will lead in turn to a determination of the closed normal subgroups of G.

27.1. Abstract Root Systems

Given a finite dimensional vector space E over R, an **abstract root system** in E consists of a subset Ψ of E satisfying the following axioms:

(R1) Ψ is finite, spans E, and does not contain 0.

(R2) If $\alpha \in \Psi$, the only multiples of α in Ψ are $\pm\alpha$.

(R3) If $\alpha \in \Psi$, there exists a reflection τ_α relative to α which leaves Ψ stable.

(R4) If $\alpha, \beta \in \Psi$, then $\tau_\alpha(\beta) - \beta$ is an integral multiple of α.

Recall (A.1) that τ_α in (R3) is uniquely determined by α, so (R4) is unambiguous. Moreover, $\tau_\alpha = \tau_{-\alpha}$, and $\tau_\alpha(\beta) - \beta$ is in any case a scalar multiple of α. The **rank** of Ψ is dim E. The **abstract Weyl group** $W(\Psi)$ is the (finite) subgroup of GL(E) generated by all τ_α $(\alpha \in \Psi)$.

With these notions in hand, we can formulate the main result of this section.

Theorem. *Let G be semisimple,* $E = R \otimes_Z X$. *Then* Φ *is an abstract root system in* E, *whose rank is* rank G *and whose abstract Weyl group is isomorphic to* W.

The proof of (R4) will be given in (27.2), but the other points can be settled right away. (R1) follows from part (f) of Corollary 26.2B, while (R2) is a consequence of the fact (25.3) that $\pm\alpha$ are the only roots of Z_α relative to T. The study of Z_α also yields a unique nontrivial element $\sigma_\alpha \in W(Z_\alpha, T) \subset W$

163

(cf. Proposition 24.1B), which sends α to $-\alpha$ (and has order 2). This is our candidate for reflection relative to α. For σ_α to be a reflection, it should fix pointwise a subgroup of corank 1 in X. Select a 1-psg λ for which $\sigma_\alpha\lambda = -\lambda$; then the subgroup of X consisting of characters χ for which $\langle\chi, \lambda\rangle = 0$ does the trick. (R3) requires that σ_α stabilize Φ, but this was observed earlier for arbitrary elements of W (24.1) (26.3).

Once (R4) has been verified, the abstract Weyl group $W(\Phi)$ may be identified with the subgroup of W generated by all σ_α ($\alpha \in \Phi$). We can then conclude that $W(\Phi) = W$ by noting that each acts simply transitively on a set of the same cardinality: $W(\Phi)$ does so on the collection of abstract Weyl chambers in E (A.4), while W does so on the set $\mathfrak{B}^T$(24.1), or equivalently, on the set of Weyl chambers in Y. To see that the sets in question have the same number of elements, it is enough to observe that after extending scalars to $\mathbb{R}$ and identifying E with $\mathrm{E}^* = \mathbb{R} \otimes_\mathbb{Z} \mathrm{Y}$, the two notions of Weyl chamber coincide. Indeed, one looks in each case at the connected components in the complement of the union of hyperplanes "orthogonal" to roots.

While the theorem applies directly only to *semisimple* groups, there is no difficulty about treating an arbitrary reductive group. The roots of G and of (G, G) (or $G/Z(G)^\circ$) are in natural 1–1 correspondence, and the Weyl groups may be identified (24.1).

In case $G = \mathrm{SL}(n, \mathbf{K})$, it is easy to see that Φ is the abstract root system whose Dynkin diagram (A.7) has type A_{n-1} (Exercise 1); other classical groups lead to types B, C, D.

27.2. The Integrality Axiom

In order to verify (R4) we have to look concretely at the way in which W permutes Φ. If $\sigma \in W$ is represented by $n \in N_G(T)$, then Ad n maps $\mathfrak{g}_\beta$ onto $\mathfrak{g}_{\sigma(\beta)}$ (24.1). Proposition 26.3 shows that this infinitesimal action actually derives from the action of Int n on the collection of subgroups $\{U_\alpha, \alpha \in \Phi\}$, viz., $nU_\beta n^{-1} = U_{\sigma(\beta)}$. Recall further that for each $\beta \in \Phi$, there exists a (nonunique) isomorphism $\varepsilon_\beta : \mathbf{G}_a \to U_\beta$ such that for all $t \in T$, $t\varepsilon_\beta(x)t^{-1} = \varepsilon_\beta(\beta(t)x)$. We have to deal with the case $\sigma = \sigma_\alpha$ ($\alpha \in \Phi$), β arbitrary.

The strategy is to show that Ad Z_α stabilizes the subspace $\mathfrak{m} = \sum \mathfrak{g}_{\beta+k\alpha}$ ($k \in \mathbb{Z}$); since σ_α has a representative in $N_{Z_\alpha}(T)$, this will imply (R4). Now Z_α is generated by T, U_α, $U_{-\alpha}$ (Theorem 26.3), and Ad T certainly stabilizes $\mathfrak{m}$. It is therefore enough (by symmetry) to show that Ad U_α also does.

Notice that our problem has been reduced to one concerning the reductive group Z_α, acting on $\mathfrak{g}$ via the adjoint representation of G. There is really nothing special here about Ad. Rather, the essential situation is this: A reductive group G is represented in some $\mathrm{GL}(V)$, where a weight χ relative to T is singled out (16.4). (For example, take Ad: $Z_\alpha \to \mathrm{GL}(\mathfrak{g})$ and the weight β.) It just has to be shown that U_α stabilizes $\sum_k V_{\chi+k\alpha}$, or merely that U_α sends V_χ into $\sum V_{\chi+k\alpha}$ (where we can specify that $k \in \mathbb{Z}^+$.) This more general formulation will be applied in §31 to the study of representations of G.

It is technically convenient to consider only those representations $\rho: G \to GL(V)$ for which $\mathrm{Ker}\ \rho \subset Z(G)$. Actually, by the end of this section it will be clear that $\mathrm{Ker\ Ad} = Z(G)$ for any reductive group G. In the case of the restricted $\mathrm{Ad}: Z_\alpha \to GL(\mathfrak{g})$, comparison with the homomorphic image $PGL(2, \mathsf{K})$ of Z_α implies directly that the kernel in question is central. Indeed, $PGL(2, \mathsf{K})$ is nonsolvable and therefore cannot have a proper closed normal subgroup of positive dimension (cf. Exercise 21.4). In turn, since $T \not\subset \mathrm{Ker\ Ad}$, the restriction of Ad to Z_α cannot be trivial, so its kernel must lie in the center of Z_α.

Proposition. *Let $\rho: G \to GL(V)$ be a rational representation, with $\mathrm{Ker}\ \rho \subset Z(G)$. Let V_χ be a weight space relative to $\rho(T)$. If $\alpha \in \Phi$, then $\rho(U_\alpha)$ sends V_χ into $\sum V_{\chi + k\alpha}$ $(k \in \mathbb{Z}^+)$.*

Proof. Since $\mathrm{Ker}\ \rho \subset Z(G) \subset T$, there is no loss of generality in replacing G by the linear group $\rho(G)$, which can be viewed as a matrix group if we choose a basis for V, say consisting of weight vectors relative to T. Since TU_α lies in a Borel group, the Lie-Kolchin Theorem even allows us to take U_α in upper triangular (unipotent) form, while T consists of diagonal matrices $t = \mathrm{diag}\ (t_1, \ldots, t_n)$. Notice that if $u \in U$, $t \in T$, the matrix $t\,u\,t^{-1}$ has (i, j) entry $t_i t_j^{-1} u_{ij}$.

Now we exploit the isomorphism $\varepsilon_\alpha: \mathbf{G}_a \to U_\alpha$. Concretely, $\varepsilon_\alpha(x)_{ij}$ must be a polynomial in x, say $c_0 + c_1 x + \cdots + c_m x^m$ $(c_k \in \mathsf{K}$ independent of $x)$. Therefore $\varepsilon_\alpha(\alpha(t)x)_{ij} = c_0 + c_1(\alpha(t)x) + \cdots + c_m(\alpha(t)^m x^m)$. But this is also equal to $t_i t_j^{-1} \varepsilon_\alpha(x)_{ij}$, since $t\varepsilon_\alpha(x)t^{-1} = \varepsilon_\alpha(\alpha(t)x)$. Equating these two results, we have a polynomial equation valid for all $x \in \mathsf{K}$:

$$c_0(1 - t_i t_j^{-1}) + c_1(\alpha(t) - t_i t_j^{-1})x + \cdots + \cdots + c_m(\alpha(t)^m - t_i t_j^{-1})x^m = 0.$$

K being an infinite field, all coefficients must vanish. Whenever $c_k \neq 0$, we must therefore have $\alpha^k = \gamma$, where γ is the character $t \mapsto t_i t_j^{-1}$ of T. Since α is not trivial, this implies that at most one c_k can be nonzero, and then $\varepsilon_\alpha(x)_{ij} = c_k x^k$.

Let us see that this means in terms of weights. Say the jth basis vector is of weight χ, i.e., $\chi(t) = t_j$. Then U_α sends this vector to a sum of vectors of weights $\eta(t) = t_i$. Our calculation shows that those η which actually occur must satisfy: $\eta \chi^{-1} = $ nonnegative power of α. In additive notation, this says that $\eta = \chi + k\alpha$ (for some $k \in \mathbb{Z}^+$). $\square$

As remarked earlier, this proposition completes the proof of Theorem 27.1.

27.3. Simple Roots

Now that Φ is known to be an abstract root system, all the results summarized in the Appendix are available. Recall (A.4) that a **base** of Φ is a subset $\Delta = \{\alpha_1, \ldots, \alpha_\ell\}$, $\ell = \mathrm{rank}\ \Phi$, which spans E (hence is a basis of E) and relative to which each root α has a (unique) expression $\alpha = \sum c_i \alpha_i$, where the c_i are integers of like sign. The elements of Δ are called **simple roots**. Bases do exist; in fact, $W = W(\Phi)$ permutes them simply transitively, and each root lies in at least one base. Moreover, W is generated by the reflections σ_α ($\alpha \in \Delta$), for any base Δ (A. 5).

In view of the discussion in (27.1), the following objects correspond 1–1 in a natural way and are permuted simply transitively by W: bases, Weyl chambers in E, Weyl chambers in Y, Borel subgroups containing T. In particular, each choice of $B \in \mathfrak{B}^T$ amounts to a choice of Δ, or to a choice of **positive** roots Φ^+ (those for which all $c_i \geqslant 0$ above). If $B = B(\lambda)$, $\lambda \in Y_{reg}$, then α is positive precisely when $\langle \alpha, \lambda \rangle > 0$.

The generation of W by $\{\sigma_\alpha, \alpha \in \Delta\}$ leads to an analogous statement about G.

Theorem.   *Let Δ be a base of Φ. Then G is generated by the Z_α ($\alpha \in \Delta$), or equivalently, by T along with all U_α ($\pm\alpha \in \Delta$).*

Proof.   The subgroup H of G generated by the Z_α ($\alpha \in \Delta$) includes T as well as representatives of all $\sigma \in W$ (since the σ_α, $\alpha \in \Delta$, generate W). So the Lie algebra $\mathfrak{h}$ includes $\mathfrak{t}$ along with all root spaces $\mathfrak{g}_{\sigma(\alpha)}$ ($\alpha \in \Delta$, $\sigma \in W$). As recalled above, these $\sigma(\alpha)$ exhaust Φ, forcing $\mathfrak{h} = \mathfrak{g}$ and hence $H = G$. Theorem 26.3 (d) completes the proof. $\square$

27.4. The Automorphism Group of a Semisimple Group

In (27.5) we shall show that a semisimple group is the product of its "simple" subgroups, which correspond to the irreducible components of the root system. But first we have to get some information about automorphisms. Assume that G is semisimple, and denote by Aut G the (abstract) group consisting of all algebraic group automorphisms of G. Int G is a normal subgroup. Another subgroup of Aut G, denoted D, consists of those automorphisms which leave stable a given maximal torus T and a Borel subgroup B containing it.

We fix the choice of B, which amounts to fixing a base Δ of Φ (27.3). Each $\sigma \in D$ induces in an obvious way an automorphism $\hat{\sigma}$ of Φ, since $\sigma(T) = T$. Since moreover $\sigma(B) = B$, $\hat{\sigma}$ stabilizes Δ. Therefore $\hat{\sigma}$ belongs to the group Γ of **diagram** (or **graph**) **automorphisms** of Φ (A.8). The correspondence $\sigma \mapsto \hat{\sigma}$ is evidently a group homomorphism $D \to \Gamma$.

Theorem.   *Let G be semisimple.*
(a) Aut $G = (\text{Int } G)D$.
(b) *The natural map $D \to \Gamma$ induces a monomorphism* Aut $G/\text{Int } G \to \Gamma$; *in particular,* Int G *has finite index in* Aut G.

Proof.   (a) If $\sigma \in$ Aut G, the conjugacy of Borel groups yields $x \in G$ such that Int $x(\sigma(B)) = B$. In turn, the maximal tori T and $x\sigma(T)x^{-1}$ of B are conjugate by some $y \in B$, so that Int $y \circ$ Int $x \circ \sigma = \text{Int}(yx) \circ \sigma$ belongs to D.

(b) It suffices to show that $(\text{Int } G) \cap D$ is the precise kernel of the homomorphism $D \to \Gamma$, thanks to part (a). In one direction, this is easy: If Int x lies in D, then x normalizes B (hence lies in B by Theorem 23.1) and x normalizes T (i.e., $x \in N = N_G(T)$). But $N \cap B = T$ (cf. Proposition 19.4, Corollary 26.2A). So Int x induces the identity automorphism of Φ.

In the other direction, let $\sigma \in D$ induce the identity on Φ. We have to show that σ is inner. For each $\alpha \in \Delta$, select as in (26.3) an isomorphism $\varepsilon_\alpha : \mathbf{G}_a \to U_\alpha$. By assumption, $\sigma(U_\alpha) = U_\alpha$, so there exists $c_\alpha \in \mathsf{K}^*$ such that $\sigma(\varepsilon_\alpha(x)) = \varepsilon_\alpha(c_\alpha x)$ for all $x \in \mathsf{K}$. (Recall that Aut $\mathbf{G}_a = \mathsf{K}^*$, cf. Exercise 7.4.) In turn, the linear independence of the set Δ enables us to select $t \in T$ for which $\alpha(t) = c_\alpha (\alpha \in \Delta)$ (Lemma 16.2C). After replacing σ by $(\text{Int } t^{-1})\sigma$, we may assume that each $c_\alpha = 1$. The new σ does nothing to $U_\alpha (\alpha \in \Delta)$. In turn, if $t \in T$ is arbitrary, $\alpha(\sigma(t)) = \alpha(t) (\alpha \in \Delta)$. G being semisimple, Δ spans a subgroup of finite index in $X(T)$ (part (f) of Corollary 26.3B), from which it follows that $\sigma(t) = t$ for all $t \in T$. Now σ fixes T_α and therefore stabilizes Z_α $(\alpha \in \Delta)$, acting on its Borel subgroup TU_α as the identity; so σ is the identity on Z_α (Proposition 21.4A). But the $Z_\alpha (\alpha \in \Delta)$ generate G (Theorem 27.3), forcing $\sigma = 1$. This means that the original σ was inner. $\quad\square$

27.5. Simple Components

As in (27.4), we assume that G is semisimple. Any closed connected normal subgroup $\neq e$ is also semisimple: its radical is a characteristic subgroup, hence normal in G. Now we are in a position to describe these subgroups.

Theorem. *Let G be semisimple, and let $\{G_i | i \in I\}$ be the minimal closed connected normal subgroups of positive dimension. Then:*

(a) If $i \neq j$, $(G_i, G_j) = e$.

(b) I is finite, say $I = \{1, \ldots, n\}$, and the product morphism $G_1 \times \cdots \times G_n \to G$ is surjective, with finite kernel.

(c) An arbitrary closed connected normal subgroup of G is the product of those G_i which it contains, and is centralized by the remaining ones.

(d) $G = (G, G)$.

(e) G is generated by the groups U_α $(\pm \alpha \in \Delta)$.

Proof. (a) (G_i, G_j) is a closed connected normal subgroup of G contained in both G_i and G_j, so it must be e, by minimality.

(b) Let $J = \{i(1), \ldots, i(r)\} \subset I$. It follows from (a) that the product map $\pi : G_{i(1)} \times \cdots \times G_{i(r)} \to G$ is a morphism of algebraic groups. In particular, $G_{i(1)} \cdots G_{i(r)}$ is a closed connected normal subgroup of G, semisimple in its own right. It also follows from (a) that any other G_i $(i \notin J)$ centralizes this group, so $G_i \cap G_{i(1)} \cdots G_{i(r)}$ is finite. As a result, Ker π must be finite, and for large enough J the collection must be exhausted, i.e., I is finite. Say $I = \{1, \ldots, n\}$, $H = G_1 \cdots G_n$. It remains to show that $H = G$.

The action of G on H by inner automorphisms yields a group homomorphism $\psi : G \to \text{Aut } H$, with $\psi(H) = \text{Int } H$. Since H is semisimple, Int H

has finite index in Aut H (Theorem 27.4). On the other hand, Ker $\psi = C_G(H)$; call its identity component H'. Then $H\,H'$ has finite index in G, hence equals G by connectedness. Since $H' \lhd G$, H' is also semisimple. Its minimal closed connected normal subgroups $\neq e$ centralize H and are therefore also normal in G; but then they are among the G_i and lie in H. This is absurd unless $H' = e, H = G$.

(c) Let H be an arbitrary closed connected normal subgroup of G, $H \neq e$. The argument in (b) shows that each $G_i\,(i \in I)$ either lies in H or centralizes H. From this we deduce that the minimal closed connected normal subgroups of H are certain of the G_i, and then (as in (b)) that H is their product.

(d) The derived group of G_i is a characteristic subgroup, hence normal in G; since G_i is not commutative, minimality forces $(G_i, G_i) = G_i$. Thanks to parts (a) and (b), $(G, G) = (G_1, G_1) \cdots (G_n, G_n) = G_1 \cdots G_n = G$.

(e) Since G is generated by T along with the $U_\alpha\,(\pm\alpha \in \varDelta)$ (Theorem 27.3), the group H generated by the U_α alone is closed, connected, and normal in G, while G/H is a torus. But then $H \supset (G, G)$, so (d) forces $H = G$. $\square$

It is an immediate consequence of part (d) that a *reductive* group G is the product of its center and derived group, the intersection being finite (cf. Lemma 19.5). For the reductive groups Z_α this was already observed in Corollary 25.3.

Recall (A.7) that $\varPhi$ is called **irreducible** if it (or equivalently, $\varDelta$) cannot be partitioned into two disjoint "orthogonal" subsets, and that $\varPhi$ decomposes uniquely into the disjoint union $\varPhi_1 \cup \cdots \cup \varPhi_t$ of irreducible root systems (in subspaces of E). Consider what happens when α, β lie in different components. First of all, no $\beta + k\alpha\,(k \in \mathbb{Z})$ can be a root except β itself. So Proposition 27.2 implies that Ad Z_α stabilizes the 1-dimensional space $\mathfrak{g}_\beta$; in particular, Ad U_α acts trivially (being unipotent). If $x \in U_\alpha$, $\mathscr{L}(xU_\beta x^{-1}) =$ Ad $x(\mathfrak{g}_\beta) = \mathfrak{g}_\beta$, so from Proposition 26.3 (a) we conclude that $xU_\beta x^{-1} = U_\beta$, i.e., U_α normalizes U_β. By symmetry, U_β normalizes U_α. It follows that $(U_\alpha, U_\beta) = e$. This argument shows that the subgroups H_i of G generated by the U_α for $\alpha \in \varPhi_i$ centralize each other, and together generate G (part (e) of theorem). Therefore the H_i are closed connected normal subgroups, whose product is G. According to the theorem, each H_i is a product of certain G_j. But it cannot be a product of several, since the roots occurring in distinct G_j are "orthogonal". To summarize:

Corollary.   *The decomposition* $G = G_1 \cdots G_n$ *corresponds precisely to the decomposition of* $\varPhi$ *into its irreducible components.* $\square$

As a byproduct of the preceding arguments, we see that each G_i is noncommutative and has no closed connected normal subgroups other than itself and e. Such an algebraic group is called **simple** (or **almost simple**, if we wish to emphasize that the group need not be simple as an abstract group). Example: SL(n, K). It will be shown in §29 that when G is a simple algebraic group, the abstract group $G/Z(G)$ is simple in the usual sense of having no proper normal subgroup $\neq e$.

Exercises

1. Verify that $SL(n, K)$ has root system of type A_{n-1}.
2. Let $\alpha, \beta \in \Delta$ (a base of Φ). Prove that Z_α, Z_β generate a reductive subgroup of semisimple rank 2 in G (if $\alpha \neq \beta$).
3. What can you say about Aut G (G reductive)?
4. Let G be semisimple. If Φ has t irreducible components, G has precisely 2^t closed connected normal subgroups.
5. Prove that Ker Ad $= Z(G)$.
6. A proper closed normal subgroup of an almost simple group G lies in $Z(G)$.
7. If H is a closed normal subgroup of G, T a maximal torus of G, then $T \cap H$ is a maximal torus of H.

Notes

The treatment here differs somewhat from that in Borel [4, §14]; the use of Proposition 27.2 was suggested by Steinberg, cf. Steinberg [10, Lemma 72] or [13, 3.3], Humphreys [1, Lemma 2.1].

28. Bruhat Decomposition

As in §27, G will be a reductive group, T a maximal torus. We fix a Borel subgroup B of G containing T (i.e., we fix a base Δ of Φ) and set $U = B_u$, $N = N_G(T)$.

The object now is to develop a normal form for elements of G, parametrized by B and the Weyl group W. In case $G = GL(n, K)$, the reader should already be familiar with the underlying idea. A given matrix can be multiplied on the left and on the right by diagonal or upper triangular unipotent matrices (corresponding to elementary row and column operations) until it becomes just a permutation matrix (a representative of W in N). With a little more care, this decomposition $G = BNB$ can be made unique (the representatives of W in N having been fixed).

28.1. *T*-Stable Subgroups of B_u

The Lie algebra of $U = B_u$ is just the direct sum of 1-dimensional root spaces $\mathfrak{g}_\alpha$ ($\alpha > 0$), with $[\mathfrak{g}_\alpha, \mathfrak{g}_\beta] \subset \mathfrak{g}_{\alpha+\beta}$. (A precise description of structure constants is, however, difficult to obtain at this point.) We have to develop analogous information about U itself, or more generally, about its T-stable subgroups, e.g., (U_α, U_β) or $U \cap nUn^{-1}$ ($n \in N$). If H is a T-stable closed subgroup of U, $\mathfrak{h}$ is stable under Ad T and is therefore just the sum of certain $\mathfrak{g}_\alpha$. We want to show that H is the product of the corresponding groups U_α. More precisely, we say that an algebraic group H is **directly spanned** by its closed subgroups $H_1, \ldots, H_n$ in the given order if the product morphism $H_1 \times \cdots \times H_n \to H$ is bijective. (If we know that $\mathfrak{h} = \mathfrak{h}_1 \oplus \cdots \oplus \mathfrak{h}_n$, this morphism will also be separable; since the varieties involved are smooth, it can then be shown to be an isomorphism of varieties, by invoking a special case of Zariski's Main Theorem. We shall obtain this more directly in cases of interest to us, cf. (28.5) below.)

It is important to notice that T acts on U without fixing any element other than e (26.2).

Proposition. *Let H be a closed, T-stable subgroup of U. Then H is connected and is directly spanned by those U_α for which $u_\alpha = g_\alpha$ lies in $\mathfrak{h}$ (taken in any order).*

Proof. Let $\Psi = \Phi(H, T)$, so $\mathfrak{h} = \coprod_{\alpha \in \Psi} g_\alpha$. Order Ψ in any way as $(\alpha_1, \ldots, \alpha_n)$, and let $\pi : U_{\alpha_1} \times \cdots \times U_{\alpha_n} \to H$ be the product morphism. This makes sense, because for $\alpha \in \Psi$, $U_\alpha = C_U(T_\alpha)$ does lie in H (apply Proposition 18.4A to the action of T_α on H). We have to show that π is bijective (which will imply that H is connected). We use induction on dim H; but for this purpose we assume only that U is a connected unipotent group on which T acts without nontrivial fixed points, while u is the direct sum of 1-dimensional eigenspaces for T corresponding to characters α with distinct connected kernels T_α.

First we treat the case in which H is *connected*. If H is commutative, then π is actually a homomorphism, with Ker π finite and T-stable. The connected group T must permute the finite set Ker π trivially (Proposition 8.2(d)), forcing Ker $\pi \subset$ fixed point set of T on $H = e$. On the other hand, Im π is a closed subgroup of the same dimension as H, so it coincides with H (H being connected).

Even if H is not commutative, it is nilpotent and therefore has a center of positive dimension (Proposition 17.4(a)), which is evidently T-stable. The commutative case shows that $Z = Z(H)^\circ$ is directly spanned by those U_α it contains (in any order!). Notice that it is enough to prove that π is bijective when the central U_α are pushed to the end in the chosen ordering; say these are $U_{\alpha_{k+1}}, \ldots, U_{\alpha_n}$. If $\varphi : H \to H/Z$ is the canonical map, then T acts in a natural way on H/Z. By induction on dimension, $\varphi(H)$ is directly spanned by $\varphi(U_{\alpha_1}), \ldots, \varphi(U_{\alpha_k})$ in the given order. This implies that H is directly spanned by $U_{\alpha_1}, \ldots, U_{\alpha_n}$.

Finally, we drop the assumption that H is connected. The preceding argument gives the structure of both U and H°, allowing us in particular to write $U = H^\circ V$, where V is the product of those U_α not in H°. So H is the set-theoretic product of H° and $H \cap V$. But $H \cap V$ is finite, T-stable, hence trivial (as in the argument above). $\quad\square$

Notice that the proposition leaves unsettled the question: Which subsets Ψ of Φ^+ actually belong to T-stable subgroups of U? But we can single out some interesting examples. Let $\sigma \in W$ be represented by $n \in N$. Then nUn^{-1} is independent of the choice of coset representative, since T normalizes U, so we may allow ourselves to write $\sigma U \sigma^{-1}$. Similarly, we write $\sigma U_\alpha \sigma^{-1}$

$(= U_{\sigma(\alpha)})$. If $B^- = TU^-$ is the Borel subgroup containing T opposite to B (26.2), we have T-stable subgroups of $U: U_\sigma = U \cap \sigma U \sigma^{-1}$, $U'_\sigma = U \cap \sigma U^- \sigma^{-1}$. Their respective sets of roots partition Φ^+:

$$\Phi_\sigma^+ = \{\alpha > 0 | \sigma(\alpha) > 0\},$$
$$\Phi_\sigma^- = \{\alpha > 0 | \sigma(\alpha) < 0\}.$$

The proposition shows that $U = U_\sigma U'_\sigma = U'_\sigma U_\sigma$ ($\sigma \in W$); but in general this direct span is *not* a semidirect product.

Consider what happens when $\alpha \in \Delta$, $\sigma = \sigma_\alpha$. Since α is simple, σ_α permutes the positive roots other than α (A.5). Thus U_{σ_α} has codimension 1 in U (and $U = U_\alpha U_{\sigma_\alpha}$). From Proposition 17.4 we conclude that U_{σ_α} is normal in U. Even more is true: Since $\sigma_\alpha^2 = e$, U_{σ_α} is also normalized by σ_α (i.e., by $N_{Z_\alpha}(T)$). It follows that U_{σ_α} is normalized by Z_α. This will be exploited presently.

28.2. Groups of Semisimple Rank 1

First, a couple of general observations. Since T lies in B, the double coset BnB is independent of the choice of n representing a given $\sigma \in W = N/T$. It is convenient to write $B\sigma B$ in this case (as we wrote $\sigma U \sigma^{-1}$ in (28.1)). The mixture of roman and greek letters should suffice to warn the reader not to take this notation too literally. Observe next that if we wish to verify the equality $G = BNB$, it is enough to do so for the image of G under a morphism of algebraic groups whose kernel lies in $Z(G)$. More precisely, given $\varphi: G \to G'$, with $\mathrm{Ker}\ \varphi \subset Z(G) \subset T$, $G' = \varphi(G)$ is again reductive (or trivial), with Borel subgroup $B' = \varphi(B)$, maximal torus $T' = \varphi(T)$, normalizer $N' = \varphi(N)$. So $G' = B'N'B'$ will imply $G = BNB$ (or vice versa).

Proposition. *Suppose* $\mathrm{rank}_{ss}(G) = 1$. *Then* $G = B \cup B\sigma B$ *(disjoint union), where* σ *is the nontrivial element of* W. *Moreover,* $B\sigma B = U\sigma TU$, *and if* $n \in N$ *is a fixed representative of* σ, *each element of this double coset has a unique expression of the form* $untu'$ ($u, u' \in U$, $t \in T$).

Proof. Recall (Corollary 25.3) that there exists an epimorphism $G \to$ PGL(2, K) having $Z(G)$ as kernel. In view of the preceding remarks, it is enough to prove all assertions when $G = \mathrm{SL}(2, \mathsf{K})$. That the union is disjoint is clear, since two double cosets are either equal or disjoint, while $N \cap B = T$. It is also evident that $B\sigma B = U\sigma TU$. We just have to show that each element of G not in B has a unique expression of the indicated form. By direct calculation (when $c \neq 0$):

$$\begin{pmatrix} a & b \\ c & d \end{pmatrix} = \begin{pmatrix} 1 & ac^{-1} \\ 0 & 1 \end{pmatrix} \begin{pmatrix} 0 & -c^{-1} \\ c & 0 \end{pmatrix} \begin{pmatrix} 1 & -d \\ 0 & 1 \end{pmatrix}.$$

When a representative of σ is fixed, e.g.,

$$\begin{pmatrix} 0 & 1 \\ -1 & 0 \end{pmatrix} = \begin{pmatrix} 1 & 1 \\ 0 & 1 \end{pmatrix} \begin{pmatrix} 1 & 0 \\ -1 & 1 \end{pmatrix} \begin{pmatrix} 1 & 1 \\ 0 & 1 \end{pmatrix},$$

the diagonal part is clearly determined by c. The reader can easily check the other uniqueness assertions. $\square$

This result can be given a geometric interpretation. With G still of semisimple rank 1, consider how U acts on the (1-dimensional) variety G/B, which is essentially $\mathbf{P}^1$. U of course fixes the point corresponding to B (and no other). The second Borel subgroup containing T corresponds to the point σB (i.e., nB for n representing σ), and according to the proposition its U-orbit exhausts what is left of G/B.

28.3. The Bruhat Decomposition

Now we can derive the Bruhat decomposition of G, to be refined to a normal form in (28.4).

Theorem. $G = \bigcup_{\sigma \in W} B\sigma B$ *(disjoint union), with* $B\sigma B = B\tau B$ *if and only if* $\sigma = \tau$ *in* W.

Proof.   According to Theorem 27.3, G is generated by the groups Z_α for α *simple*. Proposition 28.2 shows that $Z_\alpha \subset BWB$. To show that $G = BWB$, it will therefore suffice to show that BWB is closed under (left) multiplication by Z_α, $\alpha \in \Delta$. At the end of (28.1) we observed that $U = U_{\sigma_\alpha} U_\alpha$ and that Z_α normalizes U_{σ_α}. So any element of Z_α (notably a representative of σ_α) can be moved past U_{σ_α}.

Let $\sigma B = x_0$ $(\sigma \in W)$ be a point of G/B corresponding to one of the $|W|$ Borel subgroups which contain T. The stabilizer of x_0 in Z_α is one of its two Borel subgroups (label it B_α) containing T (25.4). Therefore, Z_α/B_α maps 1–1 onto the orbit of x_0 under Z_α, the map being equivariant with respect to Z_α. The geometric interpretation following Proposition 28.2 shows that this orbit can be written as $U_\alpha \cdot x_0 \cup U_\alpha \sigma_\alpha \cdot x_0$ (regardless of whether U_α lies in B_α or in its opposite).

Combining these two paragraphs, we obtain: $Z_\alpha B\sigma B = Z_\alpha U_{\sigma_\alpha} U_\alpha \sigma B = U_{\sigma_\alpha} Z_\alpha \sigma B = U_{\sigma_\alpha} U_\alpha \sigma B \cup U_{\sigma_\alpha} U_\alpha \sigma_\alpha \sigma B = B\sigma B \cup B\sigma_\alpha \sigma B$. In particular, $Z_\alpha BWB \subset BWB$, as required.

Two double cosets $B\sigma B$, $B\tau B$ are either disjoint or identical, so it remains to prove that equality can hold only if $\sigma = \tau$. For this we use the fact that $N \cap B = T$, along with the preceding calculation in the form: $\sigma_\alpha B\sigma \subset B\sigma B \cup B\sigma_\alpha \sigma B$ $(\sigma \in W, \alpha \in \Delta)$. Since the σ_α $(\alpha \in \Delta)$ generate W, it is natural to exploit the length function (A.5) on W: $\ell(\sigma) = $ minimum t such that $\sigma = \sigma_{\alpha(1)} \cdots \sigma_{\alpha(t)}$, $\alpha(i) \in \Delta$. Clearly $\ell(\sigma) = 1$ precisely when $\sigma = \sigma_\alpha$ $(\alpha \in \Delta)$, while $\ell(e) = 0$.

Assume $B\sigma B = B\tau B$, with (say) $\ell(\sigma) \leqslant \ell(\tau)$, and use induction on $\ell(\sigma)$. If $\ell(\sigma) = 0, \sigma = e$, so $B = B\tau B$ forces τ to normalize B; but $B \cap N = T$, so $\tau = e$. In general, write $\sigma = \sigma_\alpha \sigma^*$ $(\alpha \in \Delta)$, with $\ell(\sigma^*) = \ell(\sigma) - 1$. From $B\sigma B = B\tau B$ we get $\sigma_\alpha \sigma^* B\tau \subset B\tau B$, hence (because $\sigma_\alpha^2 = e$) $\sigma^* B \subset \sigma_\alpha B\tau B \subset B\tau B \cup B_{\sigma_\alpha}\tau B$. Two cases are possible:

(a) $\sigma^* B \subset B\tau B$. Then $B\sigma^* B = B\tau B$, so by induction $\sigma^* = \tau$, contradicting $\ell(\sigma^*) < \ell(\sigma) \leqslant \ell(\tau)$.

(b) $\sigma^*B \subset B\sigma_\alpha\tau B$. Then $B\sigma^*B = B\sigma_\alpha\tau B$. Evidently $\ell(\sigma^*) \leq \ell(\sigma_\alpha\tau)$, so by induction $\sigma^* = \sigma_\alpha\tau$, whence $\sigma = \sigma_\alpha\sigma^* = \sigma_\alpha^2\tau = \tau$, as desired. $\square$

An axiomatic version of the Bruhat decomposition (devised by J. Tits) will be presented in §29. The key axiom, contained in the above proof that $G = BWB$, is as follows.

Corollary (of proof).   *If $\alpha \in \Delta$, $\sigma \in W$, then $\sigma_\alpha B\sigma \subset B\sigma B \cup B\sigma_\alpha\sigma B$.* $\square$

It is also worthwhile to note an easy corollary of the theorem itself, which is by no means obvious at an earlier stage.

Corollary.   *Let B' be any Borel subgroup of G. Then $B \cap B'$ includes a maximal torus of G.*

Proof.   We know that B' is conjugate to B (21.3). The theorem allows us to write a conjugating element in the form $u\sigma b$ ($u \in U$, $\sigma \in W$, $b \in B$); so $B' = u\sigma B\sigma^{-1}u^{-1}$. But $\sigma T\sigma^{-1} = T$, while $uTu^{-1} \subset B$, so the maximal torus $u\sigma T\sigma^{-1}u^{-1}$ of B' is also included in B. $\square$

28.4. Normal Form in G

The double coset $B\sigma B$ can also be written as $U\sigma B$ or as $U\sigma TU$, but the two U-components of an element are *not* uniquely determined, except in the rank 1 case treated in (28.2). In general, a portion of U can be moved past σ (and T) and combined with the other copy of U. With this refinement, the decomposition does turn out to be unique.

Recall the discussion at the end of (28.1), where for each $\sigma \in W$ we partitioned the positive roots and wrote $U = U_\sigma U'_\sigma = U'_\sigma U_\sigma$ (direct span). This implies that $B\sigma B = U\sigma B = U'_\sigma U_\sigma \sigma B = U'_\sigma \sigma B$, since $\sigma^{-1}U_\sigma\sigma \subset U$.

Theorem.   *For each $\sigma \in W$, fix a coset representative $n(\sigma) \in N$. Then each element $x \in G$ can be written in the form $x = u'n(\sigma)tu$, where $\sigma \in W$, $t \in T$, $u \in U$, $u' \in U'_\sigma$ are all determined uniquely by x.*

Proof.   The existence of such a decomposition, along with the fact that x determines the choice of σ, follows from Theorem 28.3 and the preceding remarks. Suppose we also have $x = v'n(\sigma)t_0v$ ($v \in U$, $v' \in U'_\sigma$, $t_0 \in T$). Then $tu = (n(\sigma)^{-1}(u'^{-1}v')n(\sigma))t_0v$. But the expression in parentheses belongs to $\sigma^{-1}U'_\sigma\sigma \subset U^-$, while $U^- \cap B = e$, so this expression is e and $u' = v'$. Moreover, $tu = t_0v$, forcing $t = t_0$ and $u = v$ (since $T \cap U = e$). $\square$

28.5. Complements

So far we have ignored the topological aspect of the Bruhat decomposition. Let $B^- = TU^-$ be as before , and let σ_0 be the (unique) element of W sending B to B^-.

Proposition. *The product map $\pi: U^- \times B \to G$ defines a bijection of $U^- \times B$ onto an open subset Ω of G (called the **big cell**).*

Proof. Since $U^- \cap B = e$, π is certainly injective. Moreover, dim Ω = dim $\mathfrak{u}^-$ + dim $\mathfrak{b}$ = dim $\mathfrak{g}$ = dim G, so the constructible set Ω contains a dense open subset of G. But Ω is the translate by σ_0 of the double coset $B\sigma_0 B$. Thus it suffices to show that $B\sigma_0 B$ is open, or that $B\sigma_0 . x_0$ is open in G/B (x_0 the point corresponding to B). Being a full B-orbit as well as a constructible set, $B\sigma_0 . x_0$ is actually open in its closure (8.3); but we observed that the latter is G/B. $\square$

The fact that Ω is open in G can be made more precise: Take a basis of $\mathfrak{g}$ including an $\mathsf{x}_\alpha \in \mathfrak{g}_\alpha$ for each $\alpha \in \Phi$ along with a basis of $\mathfrak{t}$, and form the mth exterior power of $\mathfrak{g}$, on which G acts via the adjoint representation ($m = \text{Card } \Phi^+$). Let f be the coordinate function on G corresponding to the exterior product of those x_α for which $\alpha > 0$. Then f is clearly a polynomial function on G, and it can be shown that Ω is precisely the principal open set consisting of nonzeros of f in G. (See Notes below.)

In the following chapter it will be important to know that $\pi: U^- \times B \to \Omega$ is actually an *isomorphism of varieties*. Indeed, π is bijective and separable, the latter because its differential maps $\mathfrak{u}^- \oplus \mathfrak{b}$ isomorphically onto $\mathfrak{g}$ ($= $ tangent space to Ω at e), cf. (5.5). This implies that π is birational (4.6), hence that its restriction to some dense open subset of $U^- \times B$ is an isomorphism (4.7). The collection of all such open sets has a maximal element (thanks to ACC), call it O. But π is (left) U^--equivariant and (right) B-equivariant, so O must be stable under left multiplication by U^- and right multiplication by B (thanks to maximality). It follows that $O = U^- \times B$.

This argument exploits (as in §12) the homogeneity resulting from group actions. In fact, a version of Zariski's Main Theorem would show more generally that a separable bijective morphism of varieties is an isomorphism provided that the target variety is smooth (or just "normal").

The product map $U_\sigma \times U_\sigma' \to U$ in (28.1) can be shown to be an isomorphism of varieties by imitating the preceding argument. In Chapter XI we shall also need this sort of result for more than two factors, e.g., we shall need to know that the product map $U_{\alpha_1} \times \cdots \times U_{\alpha_m} \to U$ is an isomorphism of varieties when $\alpha_1, \ldots, \alpha_m$ are the positive roots in any order. This fact may be proved inductively, using the existence of a series of T-stable connected normal subgroups of U each of codimension 1 in the next (Theorem 19.3), along with Proposition 28.1.

Exercises

1. The assignment $H \mapsto \mathfrak{h}$ is an injection of the collection of all T-stable subgroups of U into the collection of all T-stable subalgebras of $\mathfrak{u}$.
2. Let H, xHx^{-1} be T-stable closed subgroups of U, where $x \in U$. Prove that $H = xHx^{-1}$. [Use Proposition 28.1 and induction on dim U, passing to U/U_β for some $U_\beta \subset Z(U)$.]

3. Let $\alpha, \beta \in \Phi^+$. Prove that (U_α, U_β) is included in the product (in any order) of those $U_{i\alpha + j\beta}$ for which $i, j > 0$ and $i\alpha + j\beta \in \Phi$.

4. Write down some explicit formulas for commutators in U involving U_α, U_β when $G = SL(3, K), Sp(4, K)$.

5. Complete the proof of uniqueness in (28.2).

6. Call two Borel subgroups **distant** if their intersection is a torus (necessarily maximal, thanks to Corollary 28.3). If B', B'' are both distant from B, prove that B' is conjugate to B'' by some element of B.

7. Prove that dim $U'_\sigma = \ell(\sigma), \sigma \in W$. [Use (A.5).]

8. Let $G = SL(2, K), \Omega$ as in (28.5). Prove that $\begin{pmatrix} a & b \\ c & d \end{pmatrix}$ belongs to Ω if and only if $a \neq 0$. (More generally, when $G = SL(n, K)$, the criterion for $x \in G$ to lie in Ω is that all diagonal minors of x be nonzero.)

9. Let H be an algebraic group, $\varphi: G \to H$ a homomorphism of abstract groups. If $\varphi|B$ is a morphism, prove that φ is a morphism. [Use $\Omega \cong U^- \times B$.]

10. Assuming the truth of Corollary 28.3, deduce the existence half of Theorem 28.3, i.e., the assertion that $G = BNB$.

11. Give another proof of the uniqueness assertion in Theorem 28.3, along the following lines: If $B\sigma B = B\tau B$, then for chosen representatives $n(\sigma), n(\tau) \in N$, there exist $b, b' \in B$ such that $bn(\sigma) = n(\tau)b'$. For arbitrary $t \in T$, $bn(\sigma)tn(\sigma)^{-1}b^{-1} = n(\tau)b'tb'^{-1}n(\tau)^{-1} \in B$. Rewrite the left side as $n(\sigma)tn(\sigma)^{-1}v$ and the right side as $(n(\tau)tn(\tau)^{-1})(n(\tau)v'n(\tau)^{-1})$ for some $v, v' \in B_u$, and conclude that $\sigma = \tau$.

Notes

The Bruhat decomposition was first observed in the framework of classical linear Lie groups; it was then exploited in the study of finite simple groups and algebraic groups by Chevalley [7] [8], and subsequently axiomatized by Tits [3]. In Steinberg [13, 2.14], the Bruhat decomposition is obtained directly for some classical groups. For a more precise description of the big cell (28.5) and the role it plays, cf. Chevalley [11], Borel [6, 4.5].

29. Tits Systems

Inspired by Chevalley's Tôhoku paper [7], Tits devised an efficient set of axioms to describe the structure of Chevalley's simple groups (and the variations of them constructed afterwards), thereby obtaining a unified simplicity proof for all types. The resulting "Tits systems" are to be found in all reductive algebraic groups, as well as in their subgroups (perhaps finite) consisting of matrices over certain subfields of K. They also occur, in a rather different way, in simple algebraic groups over local fields (cf. (35.4)). Because the axiomatic development is quite useful and uncluttered, we shall present it here, even though our only applications are to reductive groups over K. Besides obtaining a more precise description of parabolic subgroups, we shall deduce from the axioms a useful presentation of the Weyl group, to be applied in an essential way in §32, and a simplicity criterion which implies that a simple algebraic group over K has no proper normal subgroups outside its center. (This latter result is not used elsewhere in the text).

To avoid proliferation of notation and to render the proofs somewhat more intuitive, we retain in the axioms much of the notation already introduced : $G, B, T, N, W, \ldots$. However, the *roots* of G relative to T do not appear explicitly. Implicitly, the choice of B fixes a set of simple roots, hence a set S of *simple reflections* (which generate W); we shall reserve the letter ρ for these reflections, but allow σ to denote any element of W.

29.1. Axioms

Let G be a group generated by two subgroups B and N, where $T = B \cap N$ is *normal* in N. Denote the quotient group N/T by W, and suppose that W is generated by a subset S consisting of *involutions* (elements of order 2). Notation such as $B\sigma B$ or σB or $B\sigma$ ($\sigma \in W$) is permitted, since two representatives of σ in N differ only by an element of T, and $T \subset B$. We call (G, B, N, S) a **Tits system** provided the following axioms are satisfied:

(T1) If $\rho \in S$, $\sigma \in W$, then $\rho B\sigma \subset B\sigma B \cup B\rho\sigma B$.

(T2) If $\rho \in S$, $\rho B\rho \neq B$.

W is called the **Weyl group** of the Tits system, while Card S is the **rank**. In most applications, the rank is finite, but W can be either finite or infinite. (In any case, no restrictions on either cardinality need be made in advance.) We call B, or any of its conjugates in G, a **Borel subgroup** of G.

Axiom (T1) just expresses the fact that the product of the two double cosets $B\rho B$ and $B\sigma B$ (with $\rho \in S$) lies in the union of the *two* double cosets indicated. The apparent asymmetry can be removed if we take inverses and use the assumption that $\rho^2 = e$:

(T1′) If $\rho \in S$, $\sigma \in W$, then $\sigma B\rho \subset B\sigma B \cup B\sigma\rho B$.

The example of a Tits system which ought to come to mind is that in which G is a reductive algebraic group, B a Borel subgroup including a maximal torus T, $N = N_G(T)$, $W = N/T$ the Weyl group, and S the set of simple reflections corresponding to the base of the root system determined by B. We know that G is generated by the Borel subgroups containing T (Corollary 25.2C), hence by B and N, while S consists of involutions and generates W (27.3). Axiom (T1) is just Corollary 28.2, while axiom (T2) follows from the fact that W acts simply transitively on $\mathfrak{B}^T$ (24.1). One cautionary word: In this example, Card S is the rank of the root system, i.e., the *semisimple rank* of G. So the rank of the Tits system is not necessarily the full rank of G. (The point is that the center of G plays no role here.)

29.2. Bruhat Decomposition

Henceforth (G, B, N, S) denotes a Tits system, with Weyl group W. Call an expression $\sigma = \rho_1 \cdots \rho_k$ ($\rho_i \in S$) **reduced** if k is as small as possible, and write $\ell(\sigma) = k$ (this is the **length** of σ, relative to S). By convention, $\ell(\sigma) = 0$ if and only if $\sigma = e$. Evidently $\ell(\sigma) = 1$ if and only if $\sigma \in S$.

For each subset $I \subset S$, let W_I be the subgroup of W generated by I. Set $P_I = BW_I B$ (product set), so $P_\emptyset = B$.

Theorem. (a) *If $I \subset S$, P_I is a subgroup of G (in particular, $P_S = BWB$ coincides with G, since B and N generate G).*

(b) *For $\sigma, \sigma' \in W$, $B\sigma B = B\sigma' B$ if and only if $\sigma = \sigma'$.*

Proof. (a) P_I is closed under taking inverses and under left multiplication by B, so it suffices to check closure under left multiplication by $\rho \in I$. But (T1) implies that $\rho BW_I B \subset BW_I B \cup B\rho W_I B \subset P_I$.

(c) The "only if" part needs proof. For this we repeat the argument in (28.3). Proceed by induction on $\ell(\sigma)$, which we may assume $\leqslant \ell(\sigma')$. If $\ell(\sigma) = 0$, $\sigma = e$ and the assumption reads: $B = B\sigma' B$. So a representative of σ' in N lies in B. But $B \cap N = T$, so $\sigma' = e$. For the induction step, let $B\sigma B = B\sigma' B$, with $\ell(\sigma) \geqslant 1$. Write $\sigma = \rho\sigma^*(\rho \in S)$ with $\ell(\sigma^*) = \ell(\sigma) - 1$. Therefore, $\sigma B = \rho\sigma^* B \subset B\sigma' B$, or $\sigma^* B \subset \rho B\sigma' B$ (because $\rho^2 = e$). Using (T1), $\sigma^* B \subset B\sigma' B \cup B\rho\sigma' B$. In particular, $B\sigma^* B$ coincides with one of these double cosets.

Suppose $B\sigma^* B = B\sigma' B$. By induction, $\sigma^* = \sigma'$, contrary to the fact that $\ell(\sigma^*) < \ell(\sigma) \leqslant \ell(\sigma')$. Therefore, $B\sigma^* B = B\rho\sigma' B$. Clearly $\ell(\sigma^*) \leqslant \ell(\rho\sigma')$, so induction implies that $\sigma^* = \rho\sigma'$, or $\rho\sigma^* = \sigma'$, or $\sigma = \sigma'$. $\square$

Notice that axiom (T2) has not yet been invoked.

29.3. Parabolic Subgroups

We are going to make axiom (T1) more precise by relating the inclusion there to the lengths of elements of W. This information will then be used to show that the *only* subgroups of G containing B are the groups P_I of (29.2). Notice that (T1), along with part (b) of Theorem 29.2, allows us to state (for $\rho \in S$):

$$(*) \qquad \rho B\sigma \subset B\rho\sigma B \Leftrightarrow \rho B\sigma \cap B\sigma B = \emptyset.$$

Lemma A. *Let $\rho \in S, \sigma \in W$.*

(a) $\ell(\rho\sigma) \geqslant \ell(\sigma)$ *implies* $\rho B\sigma \subset B\rho\sigma B$.

(b) $\ell(\rho\sigma) \leqslant \ell(\sigma)$ *implies* $\rho B\sigma \cap B\sigma B \neq \emptyset$.

(c) $\ell(\rho\sigma) = \ell(\sigma) \pm 1$.

Proof. The conclusions of (a) and (b) are incompatible, thanks to (*), so $\ell(\rho\sigma) \neq \ell(\sigma)$. On the other hand, $\ell(\rho\sigma)$ obviously cannot differ by more than 1 from $\ell(\sigma)$. So (c) will follow from (a), (b).

(a) Use induction on $\ell(\sigma)$, the case $\ell(\sigma) = 0$ being obvious. Write $\sigma = \sigma'\rho'$, where $\rho' \in S$ and $\ell(\sigma') = \ell(\sigma) - 1$. Suppose the conclusion is false, i.e., $\rho B\sigma$ meets $B\sigma B$. Right multiplication by ρ' yields: $\rho B\sigma' \cap B\sigma B\rho' \neq \emptyset$. But $\ell(\rho\sigma') \geqslant \ell(\rho\sigma'\rho) - 1 = \ell(\rho\sigma) - 1 \geqslant \ell(\sigma) - 1 = \ell(\sigma')$. By induction, $\rho B\sigma' \subset B\rho\sigma'B$, forcing $B\sigma B\rho' \cap B\rho\sigma'B \neq \emptyset$. In turn, (T1') says that $B\sigma B\rho' \subset B\sigma B \cup B\sigma\rho'B = B\sigma B \cup B\sigma'B$. Therefore, $B\rho\sigma'B$ intersects (hence equals) one of these double cosets. By Theorem 29.2(b), either $\rho\sigma' = \sigma$ (absurd, since $\ell(\sigma') < \ell(\sigma) \leqslant \ell(\rho\sigma)$ by assumption), or else $\rho\sigma' = \sigma'$, $\rho = e$ (absurd, since ρ has order 2). This contradiction shows that (a) is true.

(b) Here we use both axioms. (T1) implies that $\rho B\rho \subset B\rho B \cup B$; therefore, (T2) says in effect that $\rho B\rho \cap B\rho B \neq \emptyset$. Multiplying on the right by $\rho\sigma$, we get $\rho B\sigma \cap B\rho B\rho\sigma \neq \emptyset$. But by assumption, $\ell(\rho^2\sigma) = \ell(\sigma) \geqslant \ell(\rho\sigma)$, so part (a) forces $B\rho B\rho\sigma \subset B\sigma B$, whence $\rho B\sigma \cap B\sigma B \neq \emptyset$. $\square$

As a consequence of (c), the number of ρ's in *any* expression (reduced or not) for $\sigma \in W$ has the same parity as $\ell(\sigma)$.

Lemma B. *Let* $\sigma = \rho_1 \cdots \rho_k$ *(reduced form)*, $I = \{\rho_1, \ldots, \rho_k\}$. *Then the group* P_I *is generated by* B *and* $\sigma B\sigma^{-1}$, *or by* B *and* σ.

Proof. Set $\sigma' = \rho_1\sigma$, so $\ell(\sigma') < \ell(\sigma)$. Part (b) of Lemma A says that $\rho_1 B\sigma$ meets $B\sigma B$; in particular, $\rho_1 \in B\sigma B\sigma^{-1}B$, which lies in the group generated by B and $\sigma B\sigma^{-1}$. Using induction on $\ell(\sigma)$, we may assume already that $\rho_2, \ldots, \rho_k$ lie in the group generated by B and $\sigma'B\sigma'^{-1} = \rho_1\sigma B\sigma^{-1}\rho_1$. Combining these two steps, we conclude that $P_I = BW_I B$ is generated by B and $\sigma B\sigma^{-1}$. But $\langle B, \sigma B\sigma^{-1}\rangle \subset \langle B, \sigma\rangle \subset P_I$, so equality holds throughout. $\square$

It follows from Lemma B that the set of $\rho \in S$ occurring in a reduced expression for σ is determined uniquely by σ (Exercise 4). Some other consequences of the lemma are of more immediate interest:

Lemma C. *S is precisely the set of those* $\sigma \in W$ *for which* $B \cup B\sigma B$ *is a group (so* G, B, N *determine* S *uniquely), and* S *is a minimal generating set for* W.

Proof. If $I = \{\rho\}$, $\rho \in S$, then $W_I = \{e, \rho\}$ and $P_I = B \cup B\rho B$ is a group, thanks to Theorem 29.2(a). Conversely, if $B \cup B\sigma B$ is a group, and $\sigma = \rho_1 \cdots \rho_k$ is a reduced expression, then $\{\rho_1, \ldots, \rho_k\} \subset \langle B, \sigma\rangle = B \cup B\sigma B$, by Lemma B. Since ρ_i has order 2, it follows from Theorem 29.2(b) that $\rho_i = \sigma$, whence $\sigma \in S$ (and $k = 1$).

Finally, suppose a subset $S' \subset S$ generates W. It is obvious that (G, B, N, S') is again a Tits system, with Weyl group W, so the conclusion just reached is that S' may be characterized as the set of those $\sigma \in W$ for which $B \cup B\sigma B$ is a group. In particular, $S \subset S'$. $\square$

Lemma D. *Let* $\sigma \in W$, *and let* $I, J \subset S$. *If* $\sigma P_I \sigma^{-1} \subset P_J$, *then* $\sigma \in P_J$ *(i.e., all coset representatives of* σ *belong to* P_J).

Proof. By hypothesis, both B and $\sigma B \sigma^{-1}$ lie in P_J, so the coset representatives of σ do, by Lemma B. $\quad\square$

Call a subgroup of G **parabolic** if it includes a Borel subgroup (i.e., a conjugate of B). We can now prove that every parabolic subgroup is conjugate to some P_I.

Theorem. (a) *The only subgroups of G containing B are those of the form P_I, $I \subset S$.*

(b) *If P_I is conjugate to P_J, then $P_I = P_J$.*

(c) $N_G(P_I) = P_I$ *(in particular, $N_G(B) = B$).*

(d) *If $W_I \subset W_J$, then $I \subset J$.*

(e) *If $P_I \subset P_J$, then $I \subset J$.*

Proof. (a) By Theorem 29.2(a), any subgroup H of G including B is the union of certain double cosets $B\sigma B$. Let I be the set of all $\rho \in S$ appearing in reduced expressions for such $\sigma \in W$. Then clearly $H \subset P_I$. On the other hand, Lemma B implies that $P_I \subset H$.

(b) Since B and N (or W) generate G, two subgroups containing B are conjugate in G only if they are conjugate by some $\sigma \in W$. But then Lemma D applies.

(c) Apply Lemma D again.

(d) Each $\rho \in I$ is a product of certain elements of J, so Lemma C forces $I \subset J$.

(e) If $P_I \subset P_J$, $W_I \subset P_J$ and then $W_I \subset W_J$ (Theorem 29.2(b)), whence $I \subset J$ by part (d). $\quad\square$

The theorem shows that the lattice of subgroups of G containing B is isomorphic to the lattice of subsets of S (ordered by inclusion), or to the lattice of subgroups W_I of W.

29.4. Generators and Relations for W

When a Tits system has rank 1, its Weyl group is generated by a single element ρ, subject to the sole relation $\rho^2 = e$. In case the rank is 2, W has two generators ρ_1, ρ_2, and these satisfy at least the relations $\rho_i^2 = e\,(i = 1, 2)$. If $\rho_1\rho_2$ has finite order (say m), then W is a homomorphic image of the abstract group whose presentation is $\langle \hat{\rho}_1, \hat{\rho}_2 | \hat{\rho}_i^2 = e = (\hat{\rho}_1\hat{\rho}_2)^m \rangle$. It is well known (and easy to prove) that the latter group is isomorphic to the dihedral group D_m of order $2m$. But W also has order $2m$: Otherwise it would equal its *cyclic* subgroup $\langle \rho_1\rho_2 \rangle$ of order m, so there would be a unique subgroup $\langle \rho_1 \rangle = \langle \rho_2 \rangle$ of order 2, which is absurd. This shows that W is dihedral. In case $\rho_1\rho_2$ has infinite order, W is clearly isomorphic to the *infinite dihedral group* D_∞, with presentation $\langle \hat{\rho}_1, \hat{\rho}_2 | \hat{\rho}_i^2 = e \rangle$.

For notational ease, we shall assume that S has finite cardinality ℓ (though the following arguments go through for a generating set of arbitrary size). By definition, a **Coxeter group** of rank ℓ is a group with generators $\hat{\rho}_i(1 \leqslant i \leqslant \ell)$

and defining relations $(\hat{\rho}_i\hat{\rho}_j)^{m(i,\,j)} = e$, where $m(i, i) = 1$ and $2 \leqslant m(i, j) = m(j, i) \leqslant \infty$ for $i \neq j$. (When $m(i, j) = \infty$, we simply omit the corresponding relation.) The preceding discussion shows already that the Weyl group of a Tits system of rank $\leqslant 2$ is a Coxeter group. Our aim is to prove this in general. As a first step, let us observe that W satisfies an **exchange condition**.

Lemma. *Let $\sigma \in W$ have reduced expression $\rho_{i(1)} \cdots \rho_{i(t)}$. Suppose that $\ell(\rho_{i(0)}\sigma) \leqslant \ell(\sigma)$. Then there exists s, $1 \leqslant s \leqslant t$, such that $\rho_{i(0)}\rho_{i(1)} \cdots \rho_{i(s-1)} = \rho_{i(1)} \cdots \rho_{i(s)}$.*

Proof. By Lemma A of (29.3), $\rho_{i(0)}B \subset B\sigma B\sigma^{-1}B$, which (by repeated application of (T1)) lies in the union of all double cosets $B\sigma\rho_{i(n_1)} \cdots \rho_{i(n_r)}B$, where $r \leqslant t$ and $t \geqslant n_1 > n_2 > \cdots > n_r \geqslant 1$. In particular, thanks to Theorem 29.2, $\rho_{i(0)}$ is equal to one (and only one!) of the indicated elements of W. But $\ell(\sigma) = t$ and $\ell(\rho_{i(0)}) = 1$, so it is clear that $r = t - 1$, therefore that only one $\rho_{i(s)}$ has been omitted. It is this choice of s which verifies the lemma. $\square$

Theorem. *Let $m(i, j)\ (\leqslant \infty)$ be the order of $\rho_i\rho_j$ in $W(1 \leqslant i, j \leqslant \ell)$. Let $\pi : \hat{W} \to W$ be the canonical epimorphism, where $\hat{W}$ is the Coxeter group $\langle \hat{\rho}_i, 1 \leqslant i \leqslant \ell \,|\, (\hat{\rho}_i\hat{\rho}_j)^{m(i,\,j)} = e\rangle$. Then π is an isomorphism.*

Proof. (1) The first step is to show that if $\rho_{i(1)} \cdots \rho_{i(t)}$ and $\rho_{j(1)} \cdots \rho_{j(t)}$ are two *reduced* expressions for the same element σ of W, then $\hat{\rho}_{i(1)} \cdots \hat{\rho}_{i(t)} = \hat{\rho}_{j(1)} \cdots \hat{\rho}_{j(t)}$ in $\hat{W}$. For this we use induction on t, the case $t = 1$ being trivial. Assume the result proved for lengths $<t$. As an abbreviation, let us write $(i(1),\ldots,i(s)) \sim (j(1),\ldots,j(s))$ when $\rho_{i(1)} \cdots \rho_{i(s)} = \rho_{j(1)} \cdots \rho_{j(s)}$ and these expressions are reduced. So our assumption is that $(i(1),\ldots,i(t)) \sim (j(1),\ldots,j(t))$.

Since left multiplication by $j(1)$ reduces the length of σ, the exchange condition allows us to find s, $1 \leqslant s \leqslant t$, for which $(i(1),\ldots,i(s)) \sim (j(1), i(1),\ldots,i(s-1))$. Therefore, $(j(1),\ldots,j(t)) \sim (i(1),\ldots,i(t)) \sim (j(1), i(1),\ldots, i(s-1), i(s+1),\ldots, i(t))$, whence $(j(2),\ldots,j(t)) \sim (i(1),\ldots,i(s-1), i(s+1),\ldots, i(t))$. By induction, $\hat{\rho}_{j(2)} \cdots \hat{\rho}_{j(t)} = \hat{\rho}_{i(1)} \cdots \hat{\rho}_{i(s-1)}\hat{\rho}_{i(s+1)} \cdots \hat{\rho}_{i(t)}$, or $\hat{\rho}_{j(1)} \cdots \hat{\rho}_{j(t)} = \hat{\rho}_{j(1)}\hat{\rho}_{i(1)} \cdots \hat{\rho}_{i(s-1)}\hat{\rho}_{i(s+1)} \cdots \hat{\rho}_{i(t)}$. In case $s < t$, induction also yields $\hat{\rho}_{i(1)} \cdots \hat{\rho}_{i(s)} \cdots \hat{\rho}_{i(t)} = \hat{\rho}_{j(1)}\hat{\rho}_{i(1)} \cdots \hat{\rho}_{i(s-1)}\hat{\rho}_{i(s+1)} \cdots \hat{\rho}_{i(t)}$, which completes the proof.

This leaves the case $s = t$, where $\hat{\rho}_{j(1)} \cdots \hat{\rho}_{j(t)} = \hat{\rho}_{j(1)}\hat{\rho}_{i(1)} \cdots \hat{\rho}_{i(t-1)}$ (and $j(1) \neq i(1)$). The original problem has been replaced by one concerning $(i(1), i(2),\ldots, i(t)) \sim (j(1), i(1),\ldots, i(t-1))$. Repeat the above argument, this time using left multiplication by $i(1)$ to reduce length; either the induction step is completed or else we are left with $(i(1), j(1), i(1),\ldots, i(t-2)) \sim (j(1), i(1),\ldots, i(t-1))$. This leads finally to sequences of length t involving only $i(1), j(1)$ (in alternation). To say that the corresponding elements of W

are equal (and reduce) is just to say that t is $m = m(i(1), j(1))$, since the subgroup of W generated by $\rho_{i(1)}$ and $\rho_{j(1)}$ has been seen above to be *dihedral*. In particular, since $(\hat{\rho}_{i(1)}\hat{\rho}_{j(1)})^m = e$, while each $(\hat{\rho}_i)^2 = e$, we get the induction step.

(2) Step (1) allows us to define unambiguously a function $\varphi : W \to \hat{W}$ by letting $\varphi(\rho_{i(1)} \cdots \rho_{i(t)}) = \hat{\rho}_{i(1)} \cdots \hat{\rho}_{i(t)}$ whenever the first expression is reduced. Evidently $\pi \, \varphi = 1_W$, so φ is injective. It will suffice to prove φ surjective, or that Im φ (which contains all generators $\hat{\rho}_i$) is a *subgroup* of W. Since Im φ is obviously closed under inverses (the ρ_i and $\hat{\rho}_i$ being involutions), it is enough just to show that Im φ is closed under left multiplication by all $\hat{\rho}_{i(0)}$. Say $\rho_{i(0)}\rho_{i(1)} \cdots \rho_{i(t)}$ is reduced. Then $\hat{\rho}_{i(0)} \, \varphi(\rho_{i(1)} \cdots \rho_{i(t)}) = \varphi(\rho_{i(0)} \cdots \rho_{i(t)})$, so there is nothing to prove. Otherwise, the exchange condition allows us to write $\rho_{i(0)}\rho_{i(1)} \cdots \rho_{i(s-1)}\rho_{i(s+1)} \cdots \rho_{i(t)} = \rho_{i(1)} \cdots \rho_{i(t)}$. By step (1), the corresponding equation is true in $\hat{W}$. Multiply both sides by $\hat{\rho}_{i(0)}$ (noting that $\hat{\rho}_{i(0)}^2 = e$) to finish the proof. $\square$

A striking feature of a Coxeter group (such as W) is that its defining relations come from (Coxeter) subgroups of rank $\leqslant 2$. In §32 we shall see a similar phenomenon in reductive algebraic groups, which enables us to reduce the classification problem to the case of rank 2.

29.5. Normal Subgroups of G

Lemma. *Let H be a normal subgroup of G. Then there is a partition $S = I \cup J$ such that I, J commute elementwise and $HB = P_I$.*

Proof. Since H is normal, HB is a parabolic subgroup, therefore equal to some P_I (Theorem 29.3(a)). Let $J = S - I$. We have to show that I and J commute elementwise. Let $I' = \{\rho \in S | B\rho B \cap H \neq \emptyset\}$. We claim that $I = I'$. Since $HB = P_I$, the inclusion $I \subset I'$ is clear. Conversely, if $B\rho B$ meets H, then $B\rho B \subset HB$, so $\rho \in I$ by Theorem 29.2.

Now take $\rho \in I$, $\rho' \in J$. Clearly $\ell(\rho\rho') \geqslant 1 = \ell(\rho)$, so Lemma A of (29.3) says that $\rho B \rho' \subset B\rho\rho' B$, or $\rho' B\rho B\rho' \subset \rho' B\rho\rho' B \subset B\rho\rho' B \cup B\rho'\rho\rho' B$ (T1). By the preceding paragraph, H meets $B\rho B$; since H is normal, it also meets the conjugate $\rho' B\rho B\rho'$ ($\rho'^2 = e$), which lies in the indicated union of double cosets. Therefore H meets one of these cosets, and either $\rho\rho'$ or $\rho'\rho\rho'$ lies in W_I. The first is absurd, so $\rho'\rho\rho' \in W_I \cap W_{\{\rho, \rho'\}} = W_{I \cap \{\rho, \rho'\}} = W_{\{\rho\}} = \{e, \rho\}$ (here we use Theorem 29.3(d), cf. Exercise 5). Clearly $\rho'\rho\rho' \neq e$, so $\rho'\rho\rho' = \rho$, showing that ρ and ρ' commute. $\square$

Any partition of S into subsets I, J which commute elementwise yields a decomposition of W as a direct product $W_I \times W_J$. If no nontrivial decomposition of this sort is possible, we call W **irreducible**. In this case, a normal subgroup of G must (according to the lemma) be rather "large" or else be contained in B, hence in all conjugates of B. Let us write $Z = \bigcap_{x \in G} xBx^{-1}$. (When G is a reductive group, Z will be the center of G.) In order to obtain a useful simplicity criterion, we need to impose further conditions.

Theorem. *Let W be irreducible, and assume that G is generated by the conjugates of a normal solvable subgroup U of B, while $G = (G, G)$. Then G/Z is simple (or trivial).*

Proof. It just has to be shown that a normal subgroup H of G not included in Z is equal to G. Thanks to the lemma and the irreducibility of W, $G = HB$. The normality of U in B forces the subgroup HU already to contain all HB-conjugates of U (i.e., all G conjugates); the latter generate G, so in fact $G = HU$.

A standard isomorphism theorem shows that $U/U \cap H \cong HU/H \cong G/H$. The left side is solvable (since U is), while the right side is its own derived group (since this is true of G), forcing $G = H$. $\square$

Corollary. *Let G be a simple algebraic group, Z its (finite) center. Then G/Z is simple as an abstract group.*

Proof. For U take B_u (or even B itself!). Then use Theorem 27.5. $\square$

We remark that for a finite (or other) subgroup of an algebraic group, which need not equal its derived group, a less restrictive formulation of the theorem is needed. (Cf. PGL(2, k), k a subfield of K; its derived group, usually PSL(2, k), may be a proper subgroup.) As before, let W be irreducible. Require U to be a normal solvable subgroup of B for which $B = TU$, and require $G' = (G', G')$, where G' is the (normal) subgroup of G generated by all conjugates of U. Then one can show that each subgroup of G normalized by G' is included in Z or else includes G'; in particular, $G'/G' \cap Z$ is simple (or trivial).

Exercises

(G, B, N, S) is a Tits system, with Weyl group W.

1. For each subset $I \subset S$, (P_I, B, N_I, I) is a Tits system with Weyl group W_I; here N_I denotes the inverse image of W_I in N.

2. If (T2) is replaced by the requirement that $\rho B \rho^{-1} \not\subset B$ ($\rho \in S$), then the fact that S consists of involutions can be deduced from the other assumptions. [Note that $B\rho^{-1}B$ meets $B\rho B$, so $B \subset B\rho B\rho B$.]

3. Let $\sigma = \rho_1 \cdots \rho_k$, with $\ell(\sigma) = k$. Prove that $B\sigma B = (B\rho_1 B)(B\rho_2 B) \cdots (B\rho_k B)$. [Use Lemma 29.3A].

4. Deduce from Lemma 29.3B that the set of $\rho \in S$ occurring in a reduced expression for $\sigma \in W$ is uniquely determined by σ.

5. Prove that $W_{I \cap J} = W_I \cap W_J$ for $I, J \subset S$.

6. Let I be an index set, and let G' be a group which permutes I doubly transitively. Fix distinct $i, j \in I$, and set $B' = \{\pi \in G' | \pi(i) = i\}$, $N' = \{\pi \in G' | \pi\{i, j\} = \{i, j\}\}$. Show that $T' = B' \cap N'$ is normal in N', and that $W' = N'/T'$ has order 2. If $S' = \{\rho\}$, where ρ generates W', prove that (G', B', N', S') is a Tits system of rank 1.

7. Let (G, B, N, S) have rank 1, $S = \{\rho\}$, so $G = B \cup B\rho B$. G permutes G/B (via left multiplication), with $B = \{x \in G | x\, B = B\}$ and N the subgroup of G stabilizing the set $\{B, \rho B\}$. Prove that G acts doubly transitively on G/B (cf. Exercise 6).

8. A Tits system is said to be **saturated** if $\bigcap_{x \in N} xBx^{-1} = T$ (in general, this intersection includes T). Call the intersection T', and let $N' = \langle N, T' \rangle$. Prove that $W' = N'/T'$ is canonically isomorphic to $W = N/T$. If S' corresponds to S (under this isomorphism), prove that (G, B, N', S') is a saturated Tits system (with Weyl group W').

9. Prove that S can be partitioned uniquely into elementwise commuting sets $I_1, \ldots, I_t$ so that $W = W_{I_1} \times \cdots \times W_{I_t}$ and no further partition is possible.

10. In the proof of Theorem 29.5, show that the assumption that U is solvable may be weakened to require only that each nontrivial homomorphic image of U is distinct from its derived group. (S_n satisfies this requirement, for example).

Notes

For Tits systems and their applications to finite groups of Lie type, cf. Tits [3] [5] [6] [7] [12] (etc.), Bourbaki [1, Ch. 4], Carter [1], Curtis [3].

30. Parabolic Subgroups

We return to the situation of §28: G is a reductive group, T a maximal torus, B a Borel subgroup containing T, Δ the corresponding base of Φ.

Starting with the general description of parabolic subgroups in Tits systems (29.3), we shall investigate in more detail the internal structure of such subgroups of G. In particular, we shall show that a parabolic subgroup is the semidirect product of its unipotent radical and a reductive subgroup, the latter determined up to conjugacy (30.2). Then we shall use parabolic subgroups to get more information about the unipotent subgroups of G. Much of the material in this section is essential in the applications of algebraic groups, but only (30.1) is needed for the classification theory in Chapter XI.

30.1. Standard Parabolic Subgroups

In the context of Tits systems, Theorem 29.3 asserts that the conjugacy classes of parabolic subgroups of G correspond 1–1 with those groups which contain B. We call the latter the **standard parabolic subgroups** of G (relative to B). They correspond 1–1 to the 2^ℓ subsets of Δ ($\ell = \mathrm{rank}_{ss}(G)$), with $I \subset \Delta$ belonging to $P_I = BW_I B$. Here W_I is the subgroup of W generated by $\{\sigma_\alpha | \alpha \in I\}$. Of course, all of this could have been deduced directly from the Bruhat decomposition of G, without the formalism of §29. (Note that we use I to denote a subset of Δ rather than a subset of $\{\sigma_\alpha | \alpha \in \Delta\}$ as in §29.)

It is clear that $\mathscr{L}(P_I) = \mathfrak{t} \oplus \coprod \mathfrak{g}_\alpha$, where α ranges over some set of roots Θ including Φ^+. Our first task is to determine Θ. Since global and infinitesimal centralizers of tori correspond (18.4), we see that a *negative* root α lies in Θ if and only if $Z_\alpha \subset P_I$. By construction, if $\sigma_\alpha \in W_I$, then $Z_\alpha = B_\alpha \cup B_\alpha \sigma_\alpha B_\alpha \subset P_I$ (where $B_\alpha = B \cap Z_\alpha$). Now W_I is the Weyl group of the root system consisting of all $\mathbb{Z}$-linear combinations of I in Φ (A.6). It follows that all such roots lie in Θ. Conversely, suppose that a pair $\{\alpha, -\alpha\}$ is contained in Θ. Then $\sigma_\alpha \in W_I$, due to the uniqueness in the Bruhat decomposition. In turn, this shows that α is orthogonal to the orthogonal complement of the subspace spanned by I (in the euclidean space E where Φ lives); so α is an $\mathbb{R}$-linear combination of I and hence (being a root) a $\mathbb{Z}$-linear combination.

Let us summarize this discussion:

Theorem.   (a) *Each parabolic subgroup of G is conjugate to one and only one subgroup* $P_I = BW_IB$, *where* $I \subset \Delta$.

(b) *The roots of* P_I *relative to* T *are those in* Φ^+ *along with those roots in* Φ^- *which are* $\mathbb{Z}$-*linear combinations of* I. $\square$

30.2. Levi Decompositions

Consider the unipotent radical $V = R_u(P_I)$ of a standard parabolic subgroup. At the two extremes $I = \Delta$, $I = \emptyset$, we have no trouble describing V. In general, being a T-stable subgroup of $U = B_u$, V must be the product (in any order) of those U_α which it contains (28.1). We claim that these α are precisely the positive roots not in Ψ = subsytem of Φ spanned by I. Indeed, T is mapped isomorphically onto a maximal torus of the reductive group P_I/V (since $\mathfrak{t} \cap \mathfrak{v} = 0$), whose roots may therefore be identified with certain pairs of roots $\pm\alpha$ of P_I relative to T. In view of Theorem 30.1(b), all such pairs must lie in Ψ. On the other hand, no pair $\pm\alpha$ can occur in $\mathfrak{v}$, so our claim follows.

The Lie algebra of P_I decomposes neatly as $\mathfrak{l} + \mathfrak{v}$ (direct sum), where $\mathfrak{l} = \mathfrak{t} \oplus \coprod \mathfrak{g}_\alpha\ (\alpha \in \Psi)$. Since $[\mathfrak{g}_\alpha, \mathfrak{g}_\beta] \subset \mathfrak{g}_{\alpha+\beta}$, $\mathfrak{l}$ is even a subalgebra, while $\mathfrak{v}$ is an ideal (because $V \lhd P_I$). Is there a corresponding decomposition of P_I as a semidirect product LV? If so, we call it a **Levi decomposition**, and we call L a **Levi factor**. Notice that L must be reductive. (In general, such a decomposition (reductive times unipotent radical) exists in an *arbitrary* connected algebraic group when char $\mathsf{K} = 0$, but need not exist when char K is prime.)

In the case of P_I, there are several approaches to the construction of L. We could, for example, take the subgroup generated by T along with all $U_\alpha\ (\alpha \in \Psi)$. But then it would have to be shown that the group is not too big, i.e., does not contain other U_α as well. Alternatively, we could construct L as a set, since we know what its Bruhat decomposition must look like, and then verify that the set is a group. For this purpose we would take as our candidate for Borel subgroup of L the group $B' = B^- \cap P_I$ (B^- the Borel subgroup of G opposite B, relative to T), so that $L = B'W_IB'$.

A third approach is more satisfactory. We begin with the torus $Z = (\bigcap_{\alpha \in I} \operatorname{Ker} \alpha)^\circ$ and define $L = C_G(Z)$. Then L is automatically reductive, thanks to Corollary 26.2A. Those roots which are trivial on Z are just the roots in Ψ (since the characters of T trivial on Z are $\mathbb{R}$-linear combinations of I). In particular, $\mathfrak{l} = \mathfrak{t} \oplus \bigsqcup \mathfrak{g}_\alpha$ $(\alpha \in \Psi) \subset \mathscr{L}(P_I)$. Since global and infinitesimal centralizers of tori correspond, it follows that $L \subset P_I$ and then that $P_I = LV$ (since the Lie algebra of this subgroup is $\mathscr{L}(P_I)$). Since $\mathfrak{l} \cap \mathfrak{v} = 0$, $L \cap V$ is finite as well as T-stable, so $L \cap V = e$ (28.1). The semidirect product $P_I = L \ltimes V$ is not just group-theoretic: since the differential of the product map $\pi : L \times V \to P_I$ is an isomorphism of vector spaces $\mathfrak{l} \oplus \mathfrak{v} \to \mathscr{L}(P_I)$, it can be shown that π is an isomorphism of varieties. (But we shall not pursue this here.)

Notice that Z is nothing but the connected center of L, so that $ZV = R(P_I)$. If L were another Levi factor of P_I, it is clear that its connected center Z' would also be a maximal torus of $R(P_I)$, while $L' = C_G(Z')$. But then $xZ'x^{-1} = Z$ for some $x \in V$ (Theorem 19.3), so we would have $xL'x^{-1} = L$. This establishes:

Theorem. *Any parabolic subgroup P of G has a Levi decomposition $P = LV$ $(V = R_u(P))$, and any two Levi factors are conjugate by an element of V.* $\square$

When $P \neq G$, we have $V \neq e$. So the reductive group L has smaller semisimple rank than G. This provides an extremely useful inductive method for studying various properties of reductive groups (e.g., representations).

30.3. Parabolic Subgroups Associated to Certain Unipotent Groups

Let U be a closed, but not necessarily connected, unipotent subgroup of G. Define $N_1 = N_G(U)$, $U_1 = U \cdot R_u(N_1)$, and then (inductively) $N_i = N_G(U_{i-1})$, $U_i = U_{i-1} \cdot R_u(N_i)$. Since unipotent radicals are connected, it is clear that either $\dim U_{i+1} > \dim U_i$ or else $U_{i+1} = U_i$. In particular, these two sequences of closed subgroups of G must stabilize, say $U_k = U_{k+1} = \cdots$, $N_k = N_{k+1} = \cdots$. Set $\mathscr{P}(U) = N_k$, $V = U_k$. The first thing to observe is that any automorphism of U of the form Int x $(x \in G)$ stabilizes N_1, hence $R_u(N_1)$, hence $U_1, \ldots$, hence $\mathscr{P}(U)$. Therefore, $N_G(U)$ normalizes $\mathscr{P}(U)$.

Suppose next that that U lies in some Borel subgroup of G (this cannot be taken for granted if U is not connected!). Thanks to the following lemma, the same is true for each U_i.

Lemma. *Let H be an algebraic group, B a Borel subgroup and A a subset of B. If S is a connected solvable subgroup of H normalizing A, then $S \cdot A$ lies in some Borel subgroup of H.*

Proof. By assumption, the (closed) set X of fixed points of A in H/B is nonempty, and S stabilizes X. Since S is connected and solvable, it fixes a point of X (21.2), i.e., S lies in a Borel subgroup containing A. $\square$

Returning to our construction, $V = U_k$ includes the unipotent radical of $N_G(V)$, as well as U. *Assuming* that U (hence V) lies in some Borel subgroup, we shall prove that $\mathscr{P}(U)$ is *parabolic* and that V is *connected*. From this it will follow that $N_G(U) \subset \mathscr{P}(U)$, since parabolic subgroups are their own normalizers (Corollary 23.1B), and also that $U \subset R_u(\mathscr{P}(U)) = V$.

Proposition. *Let V be a closed unipotent subgroup of G, $N = N_G(V)$. Assume that V lies in some Borel subgroup of G, and that $V \supset R_u(N)$ (equivalently, $V^\circ = R_u(N)$). Then N is a parabolic subgroup of G, with $V = R_u(N)$.*

Proof. Let B be a Borel subgroup of G which includes V, and set $S = (B \cap N)^\circ$. Then S lies in a Borel subgroup B_1 of N, and we can choose an "opposite" $nB_1n^{-1}(n \in N)$, i.e., $R_u(B_1 \cap nB_1n^{-1}) = R_u(N)$ (apply Corollary 26.2C to $N^\circ/R_u(N)$). Evidently this choice implies that $R_u(N \cap B \cap nBn^{-1}) \subset R_u(N) \subset V$. On the other hand, n normalizes V, so V lies in the unipotent part of the solvable group $N \cap B \cap nBn^{-1}$. This forces equality to hold throughout; in particular, $V = R_u(N)$ is connected. Moreover, V lies in the intersection $B \cap B'$ $(B' = nBn^{-1})$. If V were smaller than $R_u(B \cap B')$, its normalizer in this group would have larger dimension than V (Proposition 17.4(b)), contradicting $R_u(N \cap B \cap B') = V$. Therefore, $V = R_u(B \cap B')$, whence $B \cap B' \subset N$. But $B \cap B'$ includes a maximal torus T of G (Corollary 28.4), so N has *maximal rank* in G.

Now we can use the root system of G to show that N is parabolic. Let Δ be the base of Φ determined by B, and let Ψ be the set of roots of N relative to T. If $U_\alpha \subset V(\alpha \in \Delta)$, then $\alpha \in \Psi$. Otherwise $U_\alpha \subset B$ but $U_\alpha \not\subset V$, forcing $U_\alpha \not\subset B'$, and in turn, $U_{-\alpha} \subset B'$. Let $B'' = \sigma_\alpha B \sigma_\alpha$; its roots are those of B, except that $-\alpha$ replaces α: the simple reflection σ_α permutes the positive roots other than α (A.5). Now $U_{-\alpha}$ normalizes both B' and $R_u(B \cap B'')$, hence also $R_u(B \cap B' \cap B'') = R_u(B \cap B') = V$. So $-\alpha \in \Psi$. Since $U_{-\alpha}$ lies in B' but not B, a similar argument (with B replaced by B' and B'' replaced by $\sigma_\alpha B' \sigma_\alpha$) shows that U_α normalizes V, so $\alpha \in \Psi$. It follows that $\Delta \subset \Psi$, and therefore $B \subset N$.

We conclude that N is parabolic; indeed, N is the standard parabolic subgroup (relative to B, T) determined by the subset $\{\alpha \in \Delta | U_\alpha \not\subset V\}$. $\square$

From the discussion preceding the proposition we draw an immediate inference:

Corollary A. *Let U be a closed unipotent subgroup of G contained in some Borel subgroup. Then the group $P = \mathscr{P}(U)$ constructed above is parabolic, with $N_G(U) \subset P$ and $U \subset R_u(P)$.* $\square$

Since each *connected* unipotent group does lie in a Borel subgroup, we also deduce:

Corollary B. *Let U be a connected unipotent subgroup of G. If $U = R_u(N_G(U))$, then $N_G(U)$ is parabolic.* $\square$

30.4. Maximal Subgroups and Maximal Unipotent Subgroups

The results of (30.3) lead to the following rather striking theorem.

Theorem. (a) *Let H be a maximal (proper) closed subgroup of G. Then either $H°$ is reductive or else H is parabolic.*

(b) *Each unipotent subgroup of G lies in some Borel subgroup; in particular, the maximal unipotent subgroups of G are the unipotent radicals of Borel subgroups.*

Proof. (a) If $H°$ is not reductive, then $U = R_u(H)$ is nontrivial. G being reductive, $N_G(U)$ is therefore a proper closed subgroup of G, including H. Maximality of H forces $H = N_G(U)$. Then Corollary B of (30.3) says that H is parabolic.

(b) We proceed by induction on dim G. Let U be a unipotent subgroup of G (not necessarily closed or connected), $U \neq e$. Since U is nilpotent (17.6), its center is nontrivial. Choose $e \neq u \in Z(U)$. The closure U_1 of the subgroup generated by u lies in the unipotent part of some Borel subgroup of G, since u does (by the density theorem (22.2)). So we obtain from Corollary A of (30.3) a parabolic subgroup $P = \mathscr{P}(U_1)$ such that $U \subset N_G(U_1) \subset P$ and $e \neq U_1 \subset R_u(P)$. In particular, $P \neq G$. Pass to the reductive group $P/R_u(P)$, of lower dimension than G. The image of U is a unipotent group (15.3), therefore contained (by induction) in some Borel subgroup of $P/R_u(P)$. It follows that U lies in a Borel subgroup of P, which in turn lies in (actually equals) a Borel subgroup of G. $\square$

We remark that part (b) extends at once to an arbitrary connected group G (not necessarily reductive). In characteristic 0, all unipotent groups are known to be connected (Exercise 15.6), so we have proved nothing new. In characteristic p, (b) asserts in effect that maximal "p-subgroups" of G exist and are conjugate (cf. the Sylow theorems for finite groups).

Exercises

1. Determine the standard parabolic subgroups of $GL(n, K)$ (relative to $T(n, K)$) and their dimensions. Describe their Levi factors.
2. Describe the parabolic subgroups of $Sp(4, K)$ and their Levi factors.
3. Let H be a closed subgroup of G containing T along with at least one of U_α, $U_{-\alpha}$ (for each $\alpha \in \Phi$). Prove that H is parabolic. (Example: $H = (P \cap P')R_u(P)$, where P and P' are parabolic.)
4. Does every element of G lie in a (proper) maximal closed subgroup?

Notes

The discussion in (30.1), (30.2) is based on Borel, Tits [1]; however, they also consider fields of definition. Similarly, (30.3) and (30.4) are drawn from Borel, Tits [2]. Part (b) of Theorem 30.4 is essentially due to Platonov [2].

Chapter XI

Representations and Classification of Semisimple Groups

In order to classify (up to isomorphism) the possible semisimple algebraic groups, we have to get a little information about their representations. For this purpose, (31.1)–(31.3), and the first few paragraphs of (31.4), will be essential. The reader is of course encouraged to read the rest of §31, even though this is not essential to the further development.

31. Representations

In this section we explore the rational representations of a semisimple group G. As in the Cartan-Weyl theory for semisimple Lie algebras over $\mathbb{C}$, the irreducible ones turn out to be parametrized by "highest weights". Let T be a maximal torus of G, $B = TU$ a Borel subgroup containing T, $B^- = TU^-$ the opposite Borel subgroup containing T, Δ the base of Φ determined by B.

31.1. Weights

If $\rho: G \to \mathrm{GL}(V)$ is a rational representation, the **weights** of ρ are the images in $\mathrm{X}(T)$ of the weights of $\rho(T)$ in V (16.4), via the canonical homomorphism $\mathrm{X}(\rho(T)) \to \mathrm{X}(T)$. Of course, ρ has only finitely many weights, since V is finite dimensional. It is often convenient to view V directly as a G-module, so that a **weight space** V_λ is described as $\{v \in V \mid t \cdot v = \lambda(t)v$ for all $t \in T\}$. We call dim V_λ the **multiplicity** of the weight λ. Notice that W permutes the weights of ρ. More precisely, if $n \in N_G(T)$ represents $\sigma \in W$, then $n \cdot V_\lambda = V_{\sigma(\lambda)}$; so all weights in a W-orbit have the same multiplicity.

It is always possible to find an isomorphism of G onto a closed subgroup of some $\mathrm{GL}(V)$ (8.6). In this case, the weights of the representation generate the group $\mathrm{X}(T)$. In general, Ker ρ will be a closed normal subgroup of G, for which there are a limited number of choices (27.5). For example, in case G is almost simple, Ker ρ is either all of G or else a finite normal (hence central) subgroup. In this latter situation, the weights of ρ generate a subgroup of finite index in $\mathrm{X}(T)$. When $\rho = \mathrm{Ad}$ the same is true for any semisimple group Corollary 26.2B(f). The weights of the adjoint representation are just the roots (each with multiplicity 1) and the trivial weight 0 (with multiplicity $\ell = \mathrm{rank}\ G$).

Let us see what can be said about the weights of an arbitrary rational representation $\rho: G \to \mathrm{GL}(V)$. By applying Proposition 27.2 to the group $\rho(G)$, which is either semisimple or trivial, and lifting back to G, we see

188

that for any $\alpha \in \Phi$, $\rho(U_\alpha)$ maps a weight space V_λ into $\sum V_{\lambda + k\alpha}$ $(k \in \mathbb{Z}^+)$.
It follows that $\rho(Z_\alpha)$ stabilizes $\sum V_{\lambda + k\alpha}$ $(k \in \mathbb{Z})$; in particular, $\sigma_\alpha(\lambda)$ is of
the form $\lambda + k\alpha$. But in the framework of abstract root systems (27.1),
σ_α acts in $\mathrm{E} = \mathbb{R} \otimes_{\mathbb{Z}} \mathrm{X}(T)$ as a reflection: $\sigma_\alpha(\lambda) = \lambda - \dfrac{2(\lambda, \alpha)}{(\alpha, \alpha)} \alpha$ (where the
inner product is W-invariant). The conclusion is that the number $\dfrac{2(\lambda, \alpha)}{(\alpha, \alpha)}$
is an *integer*. By definition, λ is therefore an **abstract weight** (A. 9).

The abstract weights in E form a lattice Λ (free abelian group of rank ℓ)
which contains the root lattice Λ_r as a subgroup of finite index. If
$\Delta = \{\alpha_1, \ldots, \alpha_\ell\}$, Λ has a corresponding basis $\{\lambda_1, \ldots, \lambda_\ell\}$ consisting of
fundamental dominant weights, where $2(\lambda_i, \alpha_j)/(\alpha_j, \alpha_j) = \delta_{ij}$. Recall further
that each abstract weight is conjugate under W to precisely one dominant
weight, where $\sum c_i \lambda_i$ $(c_i \in \mathbb{Z})$ is called **dominant** whenever all $c_i \geqslant 0$ (A. 10).

Combining these observations, we conclude that *all weights of rational
representations are abstract weights*; therefore, $\mathrm{X}(T)$ consists of abstract
weights and the index of the root lattice in $\mathrm{X}(T)$ is bounded by a constant
depending only on Φ. For example, let rank $G = 1$. Then $\lambda_1 = \alpha_1/2$, so
$[\Lambda : \Lambda_r] = 2$. This allows only two possibilities for the position of $\mathrm{X}(T)$. In
general, we call $\Lambda/\mathrm{X}(T)$ the **fundamental group** of G (because of a close analogy
with the topology of compact Lie groups), denoted $\pi(G)$. When this is trivial,
we call G **simply connected** (or of **universal type**). At the opposite extreme,
when $\mathrm{X}(T) = \Lambda_r$, we say that G is of **adjoint type** (e.g., Ad G for any semisimple
G). In this language, SL(2, K) is simply connected and PGL(2, K) is adjoint.
Evidently the fundamental group of G is an important invariant, which
must be taken into account if we wish to arrive at a classification of semisimple groups.

31.2. Maximal Vectors

Given a rational representation $\rho : G \to \mathrm{GL}(V)$, $V \neq 0$, the Lie-Kolchin
Theorem insures the existence of a 1-dimensional subspace of V stable
under $\rho(B)$. By definition, a **maximal vector** $v \in V$ is a vector which spans
such a subspace; equivalently, $0 \neq v$ lies in some weight space V_λ and is
fixed by all $\rho(U_\alpha)$, $\alpha > 0$. This notion of course depends on the choice of B.
In what follows we tacitly exclude the case $V = 0$ (but not the case in which
dim $V = 1$), so that maximal vectors always exist.

Proposition. *Let V be a (rational) G-module, v^+ a maximal vector of
weight λ, V' the G-submodule generated by v^+. Then the weights of V' are
of the form $\lambda - \sum c_\alpha \alpha$ $(\alpha \in \Phi^+, c_\alpha \in \mathbb{Z}^+)$, and λ itself has multiplicity 1. Moreover, V' has a unique maximal submodule.*

Proof. As remarked in (31.1), it follows from Proposition 27.2 that
for any α, U_α sends V_λ into $\sum V_{\lambda + k\alpha}$ $(k \in \mathbb{Z}^+)$. More precisely, if $x \in U_\alpha$,

$x \cdot v^+ = v^+ +$ (sum of vectors of weights $\lambda + k\alpha, k > 0$). Successive application of elements of U^- therefore yields sums of vectors of weights $\lambda - \sum c_\alpha \alpha$ ($\alpha \in \Phi^+$, $c_\alpha \in \mathbb{Z}^+$), with v^+ the only vector of weight λ occurring. Now the subspace of V spanned by $U^- \cdot v^+$ is also the subspace spanned by $U^- B \cdot v^+ = \Omega \cdot v^+$, where Ω is known to be *dense* in G (28.5). It follows that this is the same as the subspace V' spanned by $G \cdot v^+$. Finally, each proper submodule of V' is a sum of weight spaces for weights distinct from λ, so the sum of all these submodules is still proper. $\square$

The lattice Λ of abstract weights is partially ordered by: $\lambda > \mu$ if $\lambda - \mu$ is a sum (possibly 0) of positive roots (A.10). So the proposition shows that λ is the **highest weight** of V', all other weights being $< \lambda$. The reasoning which shows that each abstract weight is W-conjugate to precisely one dominant weight shows also that a dominant weight λ satisfies: $\lambda > \sigma(\lambda)$ for all $\sigma \in W$ (A.10). Since W permutes the weights of V', this establishes:

Corollary.   *In the notation of the proposition, λ is a dominant weight.* $\square$

31.3. Irreducible Representations

A G-module $V \neq 0$ is **irreducible** if it has no G-stable subspaces except 0 and itself. This is the case, for example, if $G = \mathrm{SL}(n, \mathsf{K})$ acts on $V = \mathsf{K}^n$ in the natural way, since G then permutes transitively the 1-dimensional subspaces of V. Any G-module V, being finite dimensional, has a composition series with irreducible composition factors, which makes it especially interesting to know what the irreducible modules look like.

We remark that in characteristic 0, Weyl's theorem on complete reducibility can be used to show that each G-module splits into a direct sum of irreducible submodules, cf. (14.3); but in prime characteristic this fails. There is, however, a weaker condition called **semireductivity** which all reductive groups satisfy: Given a G-module V with a maximal submodule W of codimension 1, such that G acts trivially on V/W, say $V = \mathsf{K}v + W$, there is a symmetric power of V which contains a G-invariant polynomial involving v. This condition was conjectured by Mumford [1] and recently proved by W. J. Haboush, cf. Seshadri [3]. It is adequate to insure finite generation of rings of invariants, as in (14.3); see Nagata [2].

Proposition 31.2 shows that in case V is irreducible, V must coincide with the submodule V' generated by a maximal vector. In turn, the weight λ is singled out for special attention.

Theorem.   *Let V be an irreducible (rational) G-module.*

(a) *There is a unique B-stable 1-dimensional subspace, spanned by a maximal vector of some dominant weight λ, whose multiplicity is 1. (λ is called the **highest weight** of V.)*

(b) *All other weights of V are of the form $\lambda - \sum c_\alpha \alpha$ ($\alpha \in \Phi^+$, $c_\alpha \in \mathbb{Z}^+$). They are permuted by W, with W-conjugate weights having the same multiplicity.*

(c) *If V' is another irreducible G-module, of highest weight λ', then V is isomorphic to V' (as G-module) if and only if $\lambda = \lambda'$.*

Proof. (a) (b) We have already observed, on the strength of Proposition 31.2, that V is generated by a maximal vector v^+ of (dominant) weight λ, that λ has multiplicity 1, and that all other weights are lower in the partial ordering. If V had a maximal vector nonproportional to v^+, its weight μ would have to be distinct from λ, with $\mu < \lambda$. But the same reasoning, applied to μ, would force $\lambda < \mu$, which is absurd. This proves all of (a) and (b), except for the last assertion of (b), which was already remarked in (31.1).

(c) If V is isomorphic to V', it follows at once from (a) that $\lambda = \lambda'$. For the converse, consider the G-module $V \oplus V'$ and its projections onto V, V' (which are G-module homomorphisms). Let $v \in V, v' \in V'$ be respective maximal vectors of weight $\lambda = \lambda'$. Then (v, v') is a maximal vector of weight λ in $V \oplus V'$, so the submodule V'' which it generates has λ as weight only with multiplicity 1 (31.2). Regarding V and V' as submodules of $V \oplus V'$, it follows in particular that $v \notin V', v' \notin V''$ (neither of these being proportional to (v, v')). So $V \cap V'', V' \cap V''$ are proper submodules of V, V', hence 0 (by irreducibility). This shows that the two projections map V'' isomorphically onto (nonzero) submodules of V, V', which must be all of V, V' (by irreducibility). Thus V is isomorphic to V'. $\square$

31.4. Construction of Irreducible Representations

The arguments in (31.3) closely parallel those for semisimple Lie algebras over $\mathbb{C}$. There remains the problem of constructing an irreducible G-module of highest weight λ, for each dominant $\lambda \in X(T)$. The solution here is less direct than in the case of Lie algebras.

First, we observe that it is enough to construct a G-module containing a maximal vector of weight λ. Indeed, Proposition 31.2 shows that the submodule generated by this vector has a maximal submodule excluding λ as a weight, so we need only pass to the (irreducible) quotient module. This observation is useful in the context of *tensor products*. Say V has a maximal vector v of weight λ and V' has a maximal vector v' of weight λ'. If $t \in T, t \cdot (v \otimes v') = t \cdot v \otimes t \cdot v' = \lambda(t)\lambda'(t) v \otimes v'$, while if $u \in U, u \cdot (v \otimes v') = v \otimes v'$; so $v \otimes v'$ is a maximal vector in $V \otimes V'$ of weight $\lambda + \lambda'$ (additive notation). This shows that the existence of an irreducible module of highest weight $\lambda + \lambda'$ follows from the existence of irreducible modules of highest weights λ, λ'. (Similarly for arbitrary finite sums.)

Still, an actual construction of some sort is needed in order to get things started. For this we recall Chevalley's Theorem 11.2. Each simple root $\alpha \in \Delta$ determines a maximal parabolic subgroup $P_{\Delta - \{\alpha\}}$, cf. (30.1). Say $\Delta = \{\alpha_1, \ldots, \alpha_\ell\}, P_i = P_{\Delta - \{\alpha_i\}}$. For each i, choose a rational representation $G \to \mathrm{GL}(V_i)$ containing a line Kv_i whose isotropy group in G is precisely P_i. Each $\sigma_{\alpha_k} (k \neq i)$ has a representative in P_i, which fixes v_i. By construction,

v_i is a maximal vector of some (dominant) weight $\mu_i = \sum d_j \lambda_j$. So for $k \neq i$, we have $\sigma_{\alpha_k}(\mu_i) = \mu_i$, forcing $d_k = 0$. On the other hand, G does not act trivially on v_i, so μ_i is a *positive* multiple $d_i \lambda_i$.

Combine the two preceding paragraphs: If $\lambda = \sum c_i \lambda_i$ ($c_i \in \mathbb{Z}^+$) belongs to $X(T)$, then $d\lambda$ is a positive $\mathbb{Z}$-linear combination of the $\mu_i = d_i \lambda_i$ (where $d = \prod d_i$), hence occurs as highest weight of an irreducible G-module. This is a first approximation to what we want.

To develop the right strategy, suppose for the moment that we already have an irreducible G-module V of highest weight λ, with a maximal vector v^+. Let V' be the subspace spanned by all weight vectors of weights other than λ, so $V = \mathsf{K}v^+ \oplus V'$ (as vector space). Let $r^+ \in V^*$ be the linear function which takes value 1 at v^+ and value 0 on V'. Then define a function c_λ on G by $c_\lambda(x) = r^+(x . v^+)$. It is clear that $c_\lambda \in \mathsf{K}[G]$, and that $x . v^+ \equiv c_\lambda(x)v^+$ (mod V') for all $x \in G$. (Think of c_λ as a "coordinate function".) We deduce that for all $x \in B^-$, $y \in G$, $z \in B$:

$$(*) \qquad\qquad c_\lambda(xyz) = \lambda(x)c_\lambda(y)\lambda(z),$$

where λ is viewed as a function on B^- or B by defining $\lambda(U^-) = \lambda(U) = 1$. Indeed, just use the above congruence along with the fact that V' is stable under B^- (cf. Proposition 27.2). As a special case of $(*)$ we get $c_\lambda(xy) = \lambda(y)$ for $x \in U^-$, $y \in B$. Since $\Omega = U^- B$ is dense in G (28.5), this implies that c_λ is completely determined by λ. Moreover, $(*)$ shows that c_λ is a maximal vector of weight λ in $\mathsf{K}[G]$, where G acts by right translations: $(\rho_x f)(y) = f(yx)$. Recall (8.6) that $\mathsf{K}[G]$ is the union of finite dimensional G-stable subspaces. It follows that $\mathsf{K}[G]$ itself contains a submodule of the type described in Proposition 31.2, so that passage to an irreducible quotient yields a copy of V (cf. (31.3)).

Now abandon the assumption that V exists. The preceding discussion tells us where to look in $\mathsf{K}[G]$ for a maximal vector of weight λ. But does c_λ exist? Note that is is possible to *define* a polynomial function on Ω by the rule: $c(xy) = \lambda(y)$ ($x \in U^-$, $y \in B$). Since Ω is open in G, $\mathsf{K}(\Omega) = \mathsf{K}(G)$ and c is at least a rational function on G. All we need show now is that c is everywhere defined (cf. Proposition 2.1). Thanks to Theorem 5.3B, it will even suffice to show that some (positive) *power* of c is everywhere defined. Take d as above, so the dominant weight $d\lambda$ does occur as highest weight of an irreducible G-module, and $c_{d\lambda}$ exists in $\mathsf{K}[G]$. On Ω, $c_{d\lambda}$ coincides with c^d. Ω being dense in G, it follows that c^d is everywhere defined, as required. We have proved:

Theorem. *Let $\lambda \in X(T)$ be dominant. Then there exists an irreducible G-module of highest weight λ.* $\square$

The role of the polynomial function c_λ introduced above can be made more precise. With $\lambda \in X(T)$ a dominant weight, viewed as a function on B or B^-, define $F_\lambda = \{f \in \mathsf{K}[G] \mid f(xy) = \lambda(x)f(y)$ for $x \in B^-$, $y \in G\}$. Evidently F_λ is a subspace of $\mathsf{K}[G]$, stable under right translations. Moreover, the equation (*) shows that $c_\lambda \in F_\lambda$. Suppose F is an irreducible G-submodule of F_λ (necessarily finite dimensional), with a maximal vector f^+ of weight μ. We claim that $\mu = \lambda$ and that f^+ is proportional to c_λ. Since f^+ is a maximal vector, $f^+(xy) = \mu(y)f^+(x)$ for $x \in G$, $y \in B$. In particular, for all $t \in T$, $f^+(e) = f^+(tet^{-1}) = \mu(t^{-1})f^+(te) = \mu(t^{-1})\lambda(t)f^+(e)$ (this last because $f^+ \in F_\lambda$). To deduce that $\lambda = \mu$, it just has to be observed that $f^+(e) \neq 0$. Since $\lambda(U^-) = e$, $f^+(xy) = f^+(y)$ for $x \in U^-$, $y \in G$; taking $y \in B$, we get $(**) f^+(xy) = f^+(e)\mu(y)$. If $f^+(e)$ were 0, f^+ would therefore vanish on $\Omega = U^- B$, hence on G, contrary to $f^+ \neq 0$. Now $\lambda = \mu$ follows. Comparison with $(**)$ shows that on Ω, $f^+ = f^+(e)c_\lambda$; by density, f^+ is proportional to c_λ, as claimed.

This discussion shows that an irreducible G-module of highest weight λ can be constructed canonically in $\mathsf{K}[G]$, as the submodule generated by c_λ. When char $\mathsf{K} = 0$, a complete reducibility argument shows further that this submodule is precisely F_λ. When char K is prime, this is usually not the case (but F_λ can still be proved to be finite dimensional).

31.5. Multiplicities and Minimal Highest Weights

Let $\lambda \in \mathsf{X}(T)$ be dominant. Theorems 31.3 and 31.4 show that there exists one and (up to isomorphism) only one irreducible G-module of highest weight λ, call it $V(\lambda)$. The weights of $V(\lambda)$ are certain dominant $\mu \in \mathsf{X}(T)$ satisfying $\mu < \lambda$, along with all $\sigma(\mu)$ $(\sigma \in W)$, W-conjugate weights having the same multiplicity. Moreover, λ and its W-conjugates occur with multiplicity 1. A satisfactory description of $V(\lambda)$ (e.g., a determination of its dimension) would certainly require detailed knowledge of these multiplicities. When char $\mathsf{K} = 0$, Lie algebra theory provides rather complete information (formulas of Weyl, Freudenthal, Kostant). Unfortunately, far less is known in prime characteristic.

One case in which the structure of $V(\lambda)$ is reasonably transparent occurs when λ is **minimal**, i.e., when no $\mu < \lambda$ distinct from λ is dominant. In this case, the weights are just the W-conjugates of λ, and dim $V(\lambda)$ is the cardinality of the W-orbit of λ. It is not too difficult to determine, for each irreducible root system Φ, exactly which $\lambda \in \Lambda$ are minimal. Of course, 0 is always minimal (and yields the trivial 1-dimensional representation of G). For $SL(\ell + 1, \mathsf{K})$, the other minimal weights are just the fundamental dominant weights $\lambda_1, \ldots, \lambda_\ell$ (e.g., λ_1 occurs as highest weight of the natural $\ell + 1$-dimensional representation). For root systems other than A_ℓ, there are fewer nonzero minimal weights, all of them again occurring among the fundamental weights.

31.6. Contragredients and Invariant Bilinear Forms

Let $V = V(\lambda)$, $\lambda \in \mathsf{X}(T)$ dominant. Recall that the dual space V^* also affords a rational representation of G, where G acts according to the rule $(x \cdot f)(v) = f(x^{-1} \cdot v)$ $(x \in G, f \in V^*, v \in V)$, cf. (8.2). It is clear that the

G-submodules of V^* are in 1–1 correspondence with the G-submodules of V, with inclusions reversed. Therefore V^* is also irreducible, hence isomorphic to $V(\mu)$ for some $\mu \in X(T)$.

It is not hard to find a maximal vector in V^*, hence to determine μ. As in (31.4), write $V = Kv^+ \oplus V'$ (v^+ a maximal vector), and require $r^+ \in V^*$ to take value 1 at v^+, 0 on V'. If $x \in B^-$, it follows that $(x \cdot r^+)(v^+) = r^+(x^{-1} \cdot v^+) = \lambda(x^{-1})$, while for $v \in V'$, $(x \cdot r^+)(v) = 0$, since B^- stabilizes V'. In other words, $x \cdot r^+ = \lambda(x^{-1})r^+$. This says that r^+ is a maximal vector of weight $-\lambda$ relative to the *opposite* Borel group B^-. Now W has a unique element σ_0 which interchanges B and B^-, so it follows that V^* has highest weight $-\sigma_0(\lambda)$ relative to B.

It is known that $\sigma_0 = -1$ for precisely the following irreducible root systems: A_1, B_ℓ, C_ℓ, D_ℓ (ℓ even), E_7, E_8, F_4, G_2. In these cases V^* is therefore always isomorphic (as G-module) to V. Even when $\sigma_0 \neq -1$, it can of course happen that $-\sigma_0(\lambda) = \lambda$ for certain λ.

Suppose that V is in fact isomorphic to V^*. This gives rise to a G-invariant bilinear form on V, as follows. Since V is irreducible, $\mathrm{Hom}_G(V, V)$ is easily seen to be 1-dimensional (Schur's Lemma). But under the standard identification (8.2) of $\mathrm{Hom}(V, V)$ with $V^* \otimes V$, this subspace goes over to the space of G-invariants. In turn, $V^* \otimes V \cong V^* \otimes V^* \cong (V \otimes V)^*$ may be viewed as the space of all bilinear forms on V. So we get a G-invariant bilinear form, which is unique (up to scalar multiples).

Suppose, for example, that G is a simply connected group of type C_2. From (31.5) one gets an irreducible representation $\rho: G \to GL(4, K)$, whose highest weight is minimal. The G-invariant bilinear form on $V = K^4$ can then be shown to be "alternating", from which it follows that the image of G lies in $Sp(4, K)$. Comparing dimensions, one sees that the image must be all of $Sp(4, K)$, while $\mathrm{Ker}\, \rho = e$. A little further argument shows that G is isomorphic as algebraic group to $Sp(4, K)$. This presupposes that G is simply connected: otherwise the highest weight in question would not lie in $X(T)$. But if we had introduced "projective" representations (as in Chevalley [8]), we could have used this method to classify all groups of type C_2.

Exercises

1. Let char $K = 2$. Prove that $SL(2, K)$ and $PGL(2, K)$ are isomorphic as abstract groups, but have distinct fundamental groups.

2. Show that Ad is an irreducible representation of $SL(3, K)$ unless char $K = 3$.

3. Let V be an irreducible G-module of highest weight λ, and define v^+, r^+, c_λ as in (31.4). For $v \in V$, $r \in V^*$, define $c_{r,v}(x) = r(x \cdot v)$ for all $x \in G$. Note that $\rho_x c_{r,v} = c_{r,x\cdot v}$ and that $c_\lambda = c_{r^+,v^+}$. Show that $v \mapsto c_{r^+,v}$ defines a G-module isomorphism of V onto the G-submodule of $K[G]$ generated by c_λ.

4. Keep the notation of Exercise 3. Set $I = \{\alpha \in \Delta \,|\, \sigma_\alpha(\lambda) = \lambda\}$ and let W_I be the corresponding subgroup of W, $P_I = BW_IB$. Verify that P_I is the isotropy group in G of $\mathsf{K}v^+$. Using the Bruhat decomposition, show that $c_\lambda\,(x\,n(\sigma)y) = \lambda(y)$ for $x \in U^-$, $y \in B$, $\sigma \in W_I$, while on other double cosets the value of c_λ is 0.

5. Let G be a simply connected, simple algebraic group of type A_2. Prove that G is isomorphic (at least as an abstract group) to $\mathrm{SL}(3, \mathsf{K})$. [Show that G has an irreducible 3-dimensional representation.]

Notes

We have followed Borel [6] [8], Steinberg [10, §12], Steinberg [13, 3.3–3.4]. Projective representations are studied in Chevalley [8, exposés 15, 16]. There are quite a few open questions in prime characteristic, cf. Carter, Lusztig [1], Humphreys [5], Jantzen [1] [2], Verma [1].

32. Isomorphism Theorem

In this section and the next we shall prove that a simple algebraic group G is determined up to isomorphism by its root system and its fundamental group. More generally, we consider the case when G is semisimple. Let T be a maximal torus of G, B a Borel subgroup containing T, Δ the corresponding base of Φ, $N = N(T)$, $W = N/T$. Recall that for each root α, $Z_\alpha = C_G(T_\alpha)$ (where $T_\alpha = (\mathrm{Ker}\ \alpha)^\circ$) is a reductive group of semisimple rank 1, with Borel subgroups TU_α and $TU_{-\alpha}$ containing T, and Weyl group $W_\alpha \subset W$ generated by the "reflection" σ_α.

Besides the groups Z_α we shall have to consider certain reductive subgroups of semisimple rank 2 in G. For each pair of distinct simple roots α, β, let $Z_{\alpha\beta} = C_G((T_\alpha \cap T_\beta)^\circ)$. As the centralizer of a torus, $Z_{\alpha\beta}$ is reductive, thanks to Corollary 26.2A. It is not difficult to see that $\mathscr{L}(Z_{\alpha\beta}) = \mathfrak{t} \oplus \coprod \mathfrak{g}_\gamma$, sum over all γ in the root subsystem $\Phi_{\alpha\beta}$ of rank 2 having $\{\alpha, \beta\}$ as base. (For example, use the fact that global and infinitesimal centralizers of tori correspond.) The reader who has studied (30.2) will recognize $Z_{\alpha\beta}$ as a Levi factor of the standard parabolic subgroup $P_{\{\alpha, \beta\}}$. As a matter of notation, let $Z_{\alpha\beta} = Z_\alpha$ when $\alpha = \beta$.

32.1. The Classification Problem

Assume first that G is simple, i.e., that its root system Φ is irreducible (27.5). Theorem 27.1 affirms that Φ is an abstract root system, in the sense of the Appendix; the classification of the latter being known, Φ is a suitable invariant to work with. Another invariant of G is the fundamental group $\pi(G) = \Lambda/\mathrm{X}(T)$, where Λ is the full lattice of abstract weights (31.1). Our efforts in §32 and §33 will be devoted to the proof of the following decisive result:

Theorem. *If G, G' are simple algebraic groups having isomorphic root systems and isomorphic fundamental groups, then G and G' are isomorphic (as algebraic groups).*

The reader who is acquainted with the isomorphism theorem for semi-simple Lie algebras over C will see that this is the best possible group-theoretic analogue. In fact, the proof we shall give resembles to some extent the Lie algebra proof; but it is much harder to construct a group isomorphism than to construct a Lie algebra isomorphism. We shall outline the method of proof in this subsection, then go on to fill in the details and to reduce the problem to rank 2, which will be treated separately in §33.

The theorem was first proved by Chevalley [8] in the more comprehensive setting of isogenies. By definition, an **isogeny** $\varphi: G \to G'$ is an epimorphism of algebraic groups whose kernel is finite (for example, $\mathrm{Ad}: G \to \mathrm{Ad}\, G$ when G is semisimple). Chevalley classified the possible isogenies, including iso-morphisms, which can exist between semisimple groups. These are almost all of the sort we might expect, e.g., from $SL(n, K)$ to the quotient by its center. But there are some unexpected isogenies between groups of type B_ℓ and C_ℓ in characteristic 2, or between a group of type F_4 (resp. G_2) and itself in characteristic 2 (resp. 3). We shall not enter into these constructions here.

How can we begin to exploit the hypothesis of the theorem, in order to arrive eventually at an isomorphism $\varphi: G \to G'$? Let T', B', Φ' (etc.) have the obvious significance for G'. Set $E = R \otimes X(T)$, $E' = R \otimes X(T')$. By assumption, there is a vector space isomorphism $f: E \to E'$ sending Φ onto Φ' and preserving the Cartan integers $\langle \alpha, \beta \rangle$ $(\alpha, \beta \in \Delta)$. If Λ and Λ' are the respective lattices of abstract weights, then f automatically maps Λ onto Λ'.

Now $\Lambda \supset X(T) \supset \Lambda_r$ (= subgroup generated by Φ), thanks to (31.1). Similarly, $\Lambda' \supset X(T') \supset \Lambda'_r$. The second part of the hypothesis says that $\Lambda/X(T)$ is isomorphic to $\Lambda'/X(T')$. In all cases except D_ℓ (ℓ even), Λ/Λ' is a *cyclic* group (A.9) and therefore has a unique subgroup of each possible order; it follows in such cases that f must map $X(T)$ onto $X(T')$. When Φ is of type D_ℓ (ℓ even), and $\pi(G)$ of order 2, this is not automatically the case. But an easy argument (Exercise 2) shows that the choice of f can be modified if necessary so as to send $X(T)$ onto $X(T')$. In effect, the subgroups of order 2 in the Klein 4-group are abstractly indistinguishable.

The study of tori and characters (16.2) shows that an isomorphism be-tween $X(T)$ and $X(T')$ induces uniquely an isomorphism between T and T' (as algebraic groups). Strictly speaking, this works contravariantly: $f^{-1}: X(T') \to X(T)$ belongs to a morphism $\varphi_T: T \to T'$. The remaining problem is to *extend* φ_T to an isomorphism of algebraic groups $\varphi: G \to G'$. Formulated thus, the problem can be stated when G is semisimple (not necessarily simple). Here the theorem to be proved is:

Theorem'. *Let G, G' be semisimple groups with respective maximal tori T, T', and root systems Φ, Φ'. If $\varphi_T: T \to T'$ is an isomorphism whose associated map $X(T') \to X(T)$ induces an isomorphism between Φ' and Φ, then φ_T can be extended to an isomorphism of G onto G'.*

The set-theoretic definition of φ will rest on the Bruhat normal form (28.4). To avoid a notational conflict here, and also to avoid a proliferation of primed objects, let us write H for G' and discard T', Φ', etc. Instead of giving explicit names to certain subgroups of H, we shall refer to them as the subgroups which "correspond" to T, B, U, U^-, U_α, Z_α, etc. The correspondence in question is of course based on the given isomorphism φ_T and the attached isomorphism of root systems. It will become clear to the reader that all the essential manipulations occur within G.

In (32.2) we shall show how to extend φ_T to $\varphi_N: N \to H$, *provided* the analogous extension exists for certain of the subgroups $Z_{\alpha\beta}$ of semisimple rank 1 or 2 in G. In (32.3) we shall show how to extend φ_T to an isomorphism of algebraic groups from Z_α ($\alpha \in \Delta$) onto the corresponding subgroup of H. Then in (32.4), (32.5), we shall show how to extend φ_T to $\varphi_B: B \to H$ in a way compatible with all $\varphi_{Z_\alpha}|U_\alpha$, again *provided* the analogous extension can be made for subgroups of semisimple rank 2. The rank 2 verifications on which these extensions are conditional will be deferred until §33, because they involve rather detailed calculations for the individual root systems of rank 2.

Assume for the moment that φ_N and φ_B have been defined. According to (28.4), each $x \in G$ has a unique expression of the form $x = u'n(\sigma)tu$ ($\sigma \in W$, $t \in T$, $u \in U$, $u' \in U'_\sigma$). This depends on a prior choice of coset representative $n(\sigma) \in N$ for each $\sigma \in W$; but no matter what choice is made, the N-component $n = n(\sigma)t$ is uniquely determined by x. Therefore we can define a *function* $\varphi: G \to H$ by the recipe $\varphi(x) = \varphi_B(u')\varphi_N(n)\varphi_B(u)$. It is by no means obvious that φ is a group homomorphism or that φ is a morphism of varieties. The former question is the harder one to settle. Subject to rank 2 conditions, whose proof is deferred to §33, we shall verify the multiplicativity of φ in (32.6). Once this is settled, we can then apply the following general criterion to conclude that φ is a *morphism*:

Lemma. *Let G be semisimple, B a Borel subgroup, H another algebraic group. If $\varphi: G \to H$ is a homomorphism of abstract groups whose restriction to B is a morphism, then φ is a morphism.*

Proof. Recall (28.5) the isomorphism (of varieties) $U^- \times B \to \Omega = U^-B$. Here $U^- = \sigma_0(U)$, where $\sigma_0 \in W$ interchanges positive and negative roots. Because φ is a group homomorphism and U^- is conjugate to U in G, the hypothesis implies that $\varphi|U^-$ is a morphism. But the following diagram commutes (the maps being the obvious ones):

$$
\begin{array}{ccc}
U^- \times B & \to & \varphi(U^-) \times \varphi(B) \\
\downarrow & & \downarrow \\
\Omega & \to & \varphi(\Omega)
\end{array}
$$

It follows that $\varphi|\Omega$ is a morphism (similarly for the translates of Ω). Since Ω is open in G (28.5), φ is given by rational functions on G, which are everywhere defined, i.e., belong to $\mathsf{K}[G]$ (Proposition 2.1). Therefore φ is a morphism. $\quad\square$

32.2. Extension of φ_T to $N(T)$

According to Theorem 29.4, W is generated by the elements σ_α ($\alpha \in \Delta$), subject only to the relations: $(\sigma_\alpha\sigma_\beta)^{m(\alpha,\,\beta)} = e$, where $m(\alpha, \beta)$ denotes the order of $\sigma_\alpha\sigma_\beta$ in W (in particular, $m(\alpha, \alpha) = 1$). Our task here is to transmute this presentation of W into a presentation of N, in which the relations will be those for T along with others coming from pairs of simple roots.

Consider the exact sequence:

$$e \to T \to N \to W \to e.$$

If this sequence were split (which it usually is not), our task would be simple. In general, we can select arbitrary representatives $n_\alpha \in N$ for the σ_α ($\alpha \in \Delta$) and assert that these elements along with T suffice to generate N. Next, set $t_{\alpha\beta} = (n_\alpha n_\beta)^{m(\alpha,\,\beta)} \in T$. We claim that these relations, along with relations defining T, suffice to determine N uniquely. In other words:

Proposition. *Let $\psi: T \to H$ be a homomorphism of abstract groups, and let $h_\alpha \in H$ ($\alpha \in \Delta$). Then ψ can be extended to a homomorphism $N \to H$, sending n_α to h_α ($\alpha \in \Delta$), if and only if $(h_\alpha h_\beta)^{m(\alpha,\,\beta)} = \psi(t_{\alpha\beta})$ for all $\alpha, \beta \in \Delta$.*

Proof. The "only if" part is clear. Suppose the condition is satisfied, and introduce a free group W^* on ℓ generators n_α^* ($\alpha \in \Delta$). Associate to n_α^* the automorphism $t \mapsto n_\alpha t n_\alpha^{-1}$ of T, thus obtaining canonically a homomorphism $W^* \to \operatorname{Aut} T$. Denote by N^* the resulting semidirect product $W^* \ltimes T$ (where T is identified with a subgroup of N^*). Next factor out the smallest normal subgroup of N^* containing the elements $t_{\alpha\beta}^{-1}(n_\alpha^* n_\beta^*)^{m(\alpha,\,\beta)}$ ($\alpha, \beta \in \Delta$), and call the resulting factor group $\tilde{N}$. Evidently N is a homomorphic image of $\tilde{N}$, with T (in N^*) being mapped isomorphically onto a subgroup of $\tilde{N}$ (also called T) and then onto T (in N). It is also clear that $\psi: T \to H$ extends (uniquely) to a homomorphism $\tilde{N} \to H$ sending $\tilde{n}_\alpha$ to h_α ($\alpha \in \Delta$), where $\tilde{n}_\alpha$ is the image of n_α^*.

All that remains is to show that the canonical map $\tilde{N} \to N$ is an isomorphism. For this it suffices to show that the epimorphism $\tilde{N}/T \to N/T = W$ is an isomorphism, which just requires that the two groups in question have the same (finite) cardinality. But $\tilde{N}/T$ is generated by elements (images of $\tilde{n}_\alpha$, $\alpha \in \Delta$) which satisfy the defining relations of W (Theorem 29.4), so this group is a homomorphic image of W. It therefore cannot be bigger than W. $\quad\square$

Return now to the situation of (32.1), where $\varphi_T: T \to H$ is given. In order to extend φ_T to $\varphi_N: N \to H$, we must select elements $n_\alpha \in N$ ($\alpha \in \Delta$), thus determining elements $t_{\alpha\beta} \in T$ $(\alpha, \beta \in \Delta)$, and then select images h_α in H for which $(h_\alpha h_\beta)^{m(\alpha, \beta)} = \varphi_T(t_{\alpha\beta})$. All of this depends only on what happens in the groups $Z_{\alpha\beta}$ of semisimple rank 1 or 2. The rank 1 case will be studied in (32.3) and the rank 2 cases in §33. Notice that since G, H have isomorphic Weyl groups, φ_N will automatically be a group isomorphism. It will also be a morphism of varieties, since φ_T is such and $[N:T] < \infty$.

32.3. Extension of φ_T to Z_α

Let us recall what we know about reductive groups of semisimple rank 1. To avoid notational problems, just call the group in question Z_α, $\pm\alpha$ being the roots relative to a maximal torus T (α positive relative to a Borel group $B_\alpha \supset T$). Let $U_{\pm\alpha}$ be the corresponding root groups in B_α and its opposite, and let σ_α be the nontrivial element of the Weyl group.

According to (25.3), Z_α/B_α is isomorphic to $\mathbf{P}^1$, and the natural action of Z_α on this homogeneous space induces an epimorphism of algebraic groups $\pi: Z_\alpha \to \mathrm{PGL}(2, \mathrm{K})$. The connected kernel here is $R(Z_\alpha) = Z(Z_\alpha)^\circ$. More precisely, we can say that Ker $\pi = Z(Z_\alpha) \subset T$ (Corollary 25.3).

The action of U_α on Z_α/B_α is straightforward to describe (28.2), if the coset representatives $U_\alpha\sigma_\alpha$ are taken to form a copy of $\mathbf{A}^1$ in $\mathbf{P}^1$ (the remaining point of $\mathbf{P}^1$ corresponds to the coset B_α). In effect, $\mathbf{G}_a$ just acts on $\mathbf{G}_a$ by translations, while fixing the point at infinity. This implies that the restriction of π to U_α is *separable*, as well as bijective and $\mathbf{G}_a$-equivariant ($\mathbf{G}_a$ acting in the evident way). So $\pi|U_\alpha$ is an isomorphism of algebraic groups (12.4). By symmetry, $\pi|U_{-\alpha}$ is also an isomorphism. For concreteness, we can even select admissible isomorphisms $\varepsilon_\alpha : \mathbf{G}_a \to U_\alpha$, $\varepsilon_{-\alpha} : \mathbf{G}_a \to U_{-\alpha}$, so that

$$\pi(\varepsilon_\alpha(x)) = \begin{bmatrix} 1 & x \\ 0 & 1 \end{bmatrix}, \quad \pi(\varepsilon_{-\alpha}(x)) = \begin{bmatrix} 1 & 0 \\ x & 1 \end{bmatrix} \text{ (notation of (6.4))}. \text{ With this choice,}$$

$\pi(T)$ is the "diagonal" subgroup of $\mathrm{PGL}(2, \mathrm{K})$. We can go a step further, by fixing a representative n_α of σ_α in $N_{Z_\alpha}(T)$: define $n_\alpha = \varepsilon_\alpha(1)\varepsilon_{-\alpha}(-1)\varepsilon_\alpha(1)$, so

$$\pi(n_\alpha) = \begin{bmatrix} 0 & 1 \\ -1 & 0 \end{bmatrix}. \text{ Then } t_\alpha = n_\alpha^2 \text{ is sent to } \begin{bmatrix} -1 & 0 \\ 0 & -1 \end{bmatrix} = \begin{bmatrix} 1 & 0 \\ 0 & 1 \end{bmatrix}.$$

These choices having been made for Z_α, let $Z'_{\alpha'}$ be another reductive group of semisimple rank 1, and use primes to denote corresponding objects.

Proposition. *With the above notation, let $\varphi_T : T \to T'$ be an isomorphism of tori whose associated isomorphism* $\mathrm{X}(T') \to \mathrm{X}(T)$ *sends α' to α. Then φ_T can be extended in one and only one way to an isomorphism of algebraic groups $\varphi : Z_\alpha \to Z'_{\alpha'}$ so that $\varphi(\varepsilon_\alpha(x)) = \varepsilon'_{\alpha'}(x)$ $(x \in \mathbf{G}_a)$ and $\varphi(n_\alpha) = n'_{\alpha'}$.*

Proof. The uniqueness of φ (if it exists) is clear, since T, n_α, U_α generate Z_α. It is also clear how to define an isomorphism of varieties between the corresponding big cells Ω_α, $\Omega'_{\alpha'} : \varphi(\varepsilon_{-\alpha}(x)t\varepsilon_\alpha(y)) = \varepsilon'_{-\alpha'}(x)\varphi_T(t)\varepsilon'_{\alpha'}(y)$. Next we exploit the epimorphisms π, π' onto PGL(2, K) to construct a common "covering" S as indicated in the following diagram:

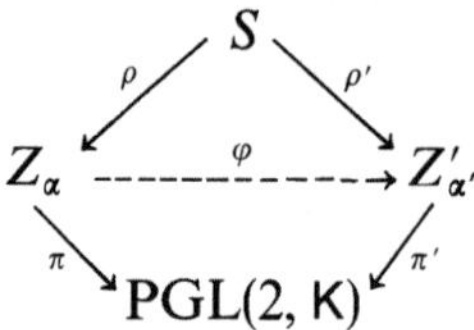

Start with the closed subgroup $R = \{(z, z') \in Z_\alpha \times Z'_{\alpha'} | \pi(z) = \pi'(z')\}$ (why is it closed?); then set $S = R^\circ$, and let ρ, ρ' be the respective projections. Evidently $\pi\rho = \pi'\rho'$. Since π, π' are surjective, so are the projections of R onto the two factors of the product. But S has finite index in R, so comparison of dimensions shows that ρ, ρ' are surjective. Note that, by construction, $\varphi\rho = \rho'$ whenever defined.

At this stage φ is given by rational functions on Z_α (Ω_α being an open subset). We claim that φ is everywhere defined and is then a group homomorphism. Say $x \in S$, $z = \rho(x)$, $z' = \rho'(x)$. Denote by λ_0, λ, λ' the respective left translations in S, Z_α, $Z'_{\alpha'}$ induced by these elements. The following equations hold whenever both sides are defined. First of all, $\varphi\rho = \rho'$ implies $\lambda'\varphi\rho = \lambda'\rho'$, which in turn equals $\rho'\lambda_0$ (ρ' being a homomorphism). But $\rho'\lambda_0 = \varphi\rho\lambda_0 = \varphi\lambda\rho$ (ρ being a homomorphism). Because ρ is surjective, it follows that (∗) $\lambda'\varphi = \varphi\lambda$. So if φ is defined at $y \in Z_\alpha$, it is also defined at $\lambda(y) = zy$, by the recipe $\varphi(zy) = \lambda'\varphi(y) = z'\varphi(y)$. Since z was arbitrary, φ is everywhere defined. From (∗) and $\varphi\rho = \rho'$ we then conclude that φ is a homomorphism.

The construction of φ forces $\varphi(n_\alpha) = n'_{\alpha'}$. To see that φ is an isomorphism, one just repeats the argument in the reverse direction to obtain its inverse. □

An immediate consequence of the proposition is the complete classification of semisimple groups of rank 1.

Corollary.   *A semisimple group of rank 1 is isomorphic to* SL(2, K) *or* PGL(2, K).

Proof.   As remarked in (31.1), a group whose root system has type A_1 must be of either simply connected or adjoint type. In these respective cases, the character group of a maximal torus is naturally isomorphic to the group $X(T) \cong X(G_m) \cong \mathbb{Z}$ for SL(2, K) or PGL(2, K), with the positive root α corresponding to 2 (resp. 1). Then the proposition takes over. □

32.4. Extension of φ_T to TU_α

In (32.3) we saw how to extend φ_T to an isomorphism of Z_α onto the corresponding subgroup of H, for any root α (say $\alpha \in \Phi^+$). The choice of $\varepsilon_\alpha : G_a \to U_\alpha$ was somewhat arbitrary there, giving us no reason to expect all of the maps φ_{Z_α} to be compatible, i.e., to admit some common extension to G. So let us begin by constructing extensions φ_{Z_α} only for *simple* roots α; here the choices are "independent". Denote by φ_α the restriction of φ_{Z_α} to U_α, and by n_α the distinguished representative of σ_α defined in (32.3): $n_\alpha = \varepsilon_\alpha(1)\varepsilon_{-\alpha}(-1)\varepsilon_\alpha(1)$, while $t_\alpha = n_\alpha^2 (= t_{\alpha\alpha})$. Set $h_\alpha = \varphi_{Z_\alpha}(n_\alpha)$.

According to (32.2), φ_T can be extended to a homomorphism $\varphi_N : N \to H$, sending $n_\alpha \mapsto h_\alpha$ ($\alpha \in \Delta$), provided the following condition is met:

(A) $\varphi_T(t_{\alpha\beta}) = (h_\alpha h_\beta)^{m(\alpha,\,\beta)}$ *for all* $\alpha, \beta \in \Delta$, *where* $t_{\alpha\beta} = (n_\alpha n_\beta)^{m(\alpha,\,\beta)}$, $m(\alpha,\beta) = $
order of $\sigma_\alpha\sigma_\beta$ *in* W.

This condition will be verified in §33, but in the meantime we shall assume it is true and utilize φ_N freely. By construction, φ_N and φ_{Z_α} agree on $N_{Z_\alpha}(T)$.

How can φ_T be extended to B? We insist, of course, that such an extension should agree with φ_α ($\alpha \in \Delta$), and should send every U_α ($\alpha > 0$) to the corresponding subgroup of H. In particular, if $U_{\alpha\beta}$ ($\alpha, \beta \in \Delta$) denotes $U \cap Z_{\alpha\beta}$, we require that the following condition be satisfied:

(B) *For each pair* $\alpha, \beta \in \Delta$, *there is an isomorphism* $\varphi_{\alpha\beta}$ (*of algebraic groups*) *from* $U_{\alpha\beta}$ *onto the corresponding subgroup of* H, *such that* $\varphi_{\alpha\beta}$ *extends both* φ_α *and* φ_β.

Like condition (A) above, this just involves rank 2 and will be verified in §33. For the remainder of §32 we assume its truth.

If φ_T is to be extended not only to B, but eventually to G, even more is required. Say β is a positive root, sent by $\sigma \in W$ to a simple root α, and let $n \in N$ represent σ. (For brevity, we write $\bar{n} = \sigma$.) Then $nU_\beta n^{-1} = U_\alpha$, or $U_\beta = n^{-1}U_\alpha n$. If there is to be a homomorphism $\varphi : G \to H$ extending both φ_N and φ_α, this equation tells us how φ *must* be defined on U_β. But there is a catch: β might be sent to α by another element of W, and β might also be sent to another simple root. We have to be certain that the choices made do not affect the definition of $\varphi_\beta : U_\beta \to H$. This problem arises already in rank 2, where the crucial condition to verify turns out to be the following ($\varphi_{\alpha\beta}$ as in (B)):

(C) *Let* $\alpha, \beta \in \Delta$. *If* $\gamma \in \Phi_{\alpha\beta}^+$ *with* $\gamma \neq \alpha$, *then* Int $h_\alpha \circ \varphi_{\alpha\beta}$ *agrees on* U_γ *with* $\varphi_{\alpha\beta} \circ$ Int n_α.

Like (A) and (B), this condition will be checked in §33; for now we accept it and proceed to show that φ_α can be defined consistently for all roots α (positive or negative).

Proposition. *Assume* (A), (B), (C) *above, so that* φ_N *and* φ_α, $\varphi_{\alpha\beta}$ ($\alpha, \beta \in \Delta$) *are defined. Then there exists for each* $\alpha \in \Phi$ *an isomorphism* φ_α *of* U_α *onto the corresponding subgroup of* H, *satisfying:*
(a) *If* $\alpha \in \Delta$, φ_α *agrees with the given morphism, and* $\varphi_{-\alpha} = \varphi_{Z_\alpha}|U_{-\alpha}$.
(b) *If* $\alpha, \beta \in \Delta$ *and* $\gamma \in \Phi_{\alpha\beta}^+$, *then* $\varphi_\gamma = \varphi_{\alpha\beta}|U_\gamma$.
(c) *If* $n \in N$, *and* $\bar{n}(\alpha) = \beta$ ($\alpha, \beta \in \Phi$), *then* Int $\varphi_N(n) \circ \varphi_\alpha$ *agrees with* $\varphi_\beta \circ$ Int n *on* U_α.

Proof. This proceeds in steps. Set $N_{\alpha\beta} = N_{Z_{\alpha\beta}}(T)$.

(1) *Let* α, $\beta \in \Delta$, $\alpha \neq \beta$. *If* $n \in N_{\alpha\beta}$ *and* $\bar{n}(\alpha) = \alpha$ *(resp.* $\bar{n}(\alpha) = \beta$*), then on* U_α, Int $\varphi_N(n) \circ \varphi_\alpha$ *agrees with* $\varphi_\alpha \circ$ Int n *(resp.* $\varphi_\beta \circ$ Int n*).*

Write $n = n_{\gamma_k} \cdots n_{\gamma_1} t$ for some $t \in T$ and $\gamma_i \in \{\alpha, \beta\}$. In case $n = t$ (so $\bar{n}(\alpha) = \alpha$), the assertion follows from the construction of φ_{Z_α}. So we can assume that $n = n_{\gamma_k} \cdots n_{\gamma_1}$ and use induction on k. When $k = 1$, $n = n_\alpha$ or n_β. In fact, since σ_α permutes the set $\Phi_{\alpha\beta}^+ - \{\alpha\}$ (A.5), it is clear that $n \neq n_\alpha$ (so we are in the case $\bar{n}(\alpha) = \alpha$). Condition (C) can then be applied (in view of condition (B)).

For the induction step, consider the sequence of roots $\alpha_i = \sigma_{\gamma_{i-1}} \cdots \sigma_{\gamma_1}(\alpha) \in \Phi_{\alpha\beta}$. If all $\alpha_i \in \Phi_{\alpha\beta}^+$, repeated application of condition (C) completes the proof, just as in the case $k = 1$. Otherwise find $i < k$ for which α_i is positive but α_{i+1} is negative. The only positive root which σ_{γ_i} sends to a negative root being γ_i, we get $\alpha_i = \alpha$ or β. Now write $n = n''n'$, where $n' = \sigma_{\gamma_{i-1}} \cdots \sigma_{\gamma_1}$, $n'' = \sigma_{\gamma_k} \cdots \sigma_{\gamma_i}$. Observe that each of n', n'' satisfies the original hypothesis on n (with the roles of α and β possibly interchanged for n''). Induction finishes the argument.

(2) *Let* α, $\beta \in \Delta$, *and suppose* $\bar{n}(\alpha) = \beta$, $n \in N$. *Then* Int $\varphi_N(n) \circ \varphi_\alpha$ *agrees with* $\varphi_\beta \circ$ Int n *on* U_α.

Note that this would follow at once from step (1) if $\alpha \neq \beta$ and $n \in N_{\alpha\beta}$, or even if $\alpha = \beta$ and $n \in N_{\alpha\gamma}$ for some other simple root γ. To reduce matters to one of these situations, we appeal to a lemma of Tits (A.12), which yields a sequence of simple roots $\gamma_0 = \alpha, \gamma_1, \ldots, \gamma_k = \beta$, and a sequence $\sigma_0, \ldots, \sigma_{k-1}$ from W, such that: $\bar{n} = \sigma_{k-1} \cdots \sigma_0$, $\sigma_i(\gamma_i) = \gamma_{i+1}$ $(0 \leqslant i \leqslant k - 1)$. The lemma also asserts that for each i there exists $\delta_i \in \Delta$ for which $\sigma_i \in W_{\gamma_i\delta_i}$ and $\gamma_{i+1} = \gamma_i$ or δ_i. So step (1) may be applied repeatedly.

(3) *Let* $\alpha \in \Phi$, n, $n' \in N$. *Suppose* $\bar{n}(\alpha) = \beta \in \Delta$ *and* $\bar{n}'(\alpha) = \beta' \in \Delta$. *Then* Int $\varphi_N(n)^{-1} \circ \varphi_\beta \circ$ Int n *agrees with* Int $\varphi_N(n')^{-1} \circ \varphi_{\beta'} \circ$ Int n' *on* U_α.

It just has to be shown that Int $\varphi_N(n'n^{-1}) \circ \varphi_\beta$ agrees with $\varphi_{\beta'} \circ$ Int$(n'n^{-1})$ on U_β. But this follows from step (2).

(4) Now we are in a position to define φ_α for all $\alpha \in \Phi$, in a way consistent with the cases already covered by (B). There exists a simple root β and an $n \in N$ for which $\bar{n}(\alpha) = \beta$ (A.4). For $x \in U_\alpha$, we then define $\varphi_\alpha(x)$ to be Int $\varphi_N(n)^{-1}\varphi_\beta$ Int $n(x)$. According to step (3), this recipe does not depend on which β and n we chose. In particular, parts (a) and (b) of the proposition follow. Moreover, φ_α is clearly an isomorphism of U_α onto the corresponding subgroup of H (since φ_β is an isomorphism). Finally, the construction, coupled with step (3), implies (c). $\quad \square$

32.5. Extension of φ_T to B

Assumptions (A), (B), (C) of (32.4) remain in force. The morphisms φ_α $(\alpha \in \Phi^+)$ constructed in Proposition 32.4, can now be combined with φ_T to yield an isomorphism (of varieties) from B onto the corresponding Borel subgroup of H. Indeed, B is isomorphic (as variety) to $T \times U_{\alpha_1} \times \cdots \times U_{\alpha_m}$, where the positive roots $\alpha_1, \ldots, \alpha_m$ are ordered in any way (28.5). So $\varphi_T \times \varphi_{\alpha_1} \times \cdots \times \varphi_{\alpha_m}$ induces the desired isomorphism φ_B. Of course, this much could have been accomplished by defining the φ_α arbitrarily (each of the 1-dimensional unipotent groups involved being isomorphic to $\mathbf{G}_a$). But we also want φ_B to be a *group homomorphism*. For this the main point to be settled involves the commutation in U. Proposition 28.1 says that the T-stable closed subgroup (U_α, U_β) (where $\alpha, \beta \in \Phi^+$) is the product of those U_γ it contains. Using the technique of Proposition 27.2 we can determine the possible γ which occur and at the same time obtain a fairly precise commutator formula.

Lemma. *Let* $\alpha, \beta \in \Phi^+$, *and let* Ψ *be the set of all roots of the form* $r\alpha + s\beta$ *(r, s positive integers). Fix any admissible isomorphisms* $\varepsilon_\delta : \mathbf{G}_a \to U_\delta$. *Then* $(\varepsilon_\alpha(x), \varepsilon_\beta(y)) = \prod \varepsilon_\gamma(c_{\alpha, \beta, \gamma} x^r y^s)$, *where the product is taken over all* $\gamma = r\alpha + s\beta \in \Psi$ *(in some fixed order) and where* $c_{\alpha, \beta, \gamma} \in \mathsf{K}$ *is independent of* x, y. *In particular,* $(U_\alpha, U_\beta) = e$ *if* $\Psi = \emptyset$ *(e.g., if* $\alpha + \beta \notin \Phi$ *(A.4))*.

Proof. Order the positive roots $\alpha_1, \ldots, \alpha_m$ in some way compatible with the chosen ordering of Ψ; as above, the product map $U_{\alpha_1} \times \cdots U_{\alpha_m} \to U$ is an isomorphism of varieties. Next define a morphism $\psi : \mathbf{G}_a \times \mathbf{G}_a \to U$ by $\psi(x, y) = (\varepsilon_\alpha(x), \varepsilon_\beta(y)) = \varepsilon_\alpha(x)\varepsilon_\beta(y)\varepsilon_\alpha(-x)\varepsilon_\beta(-y)$. The image has "normal form" $\varepsilon_{\alpha_1}(p_1(x, y)) \cdots \varepsilon_{\alpha_m}(p_m(x, y))$, where the p_i are polynomials in two variables:

$$p_i(\mathsf{X}, \mathsf{Y}) = \sum_{r, s \geqslant 0} a_{i, r, s} \mathsf{X}^r \mathsf{Y}^s.$$

By definition of ψ, $\psi(x, 0) = e = \psi(0, y)$, so that XY actually divides $p_i(\mathsf{X}, \mathsf{Y})$ (i.e., $r, s > 0$ whenever $a_{i, r, s} \neq 0$).

To pin down the image of ψ, we conjugate $\psi(x, y)$ by $t \in T$ and write the result in two different ways. On the one hand, $(\varepsilon_\alpha(\alpha(t)x), \varepsilon_\beta(\beta(t)y)) = \prod_i \varepsilon_{\alpha_i}(p_i(\alpha(t)x, \beta(t)y))$. But the right side equals $\prod_i \varepsilon_{\alpha_i}(\alpha_i(t)p_i(x, y))$. Comparison reveals that $a_{i, r, s} = 0$ unless $\alpha_i = r\alpha + s\beta$ ($r, s > 0$), in which case r and s are uniquely determined by i (it may be assumed that α and β are linearly independent). From this the lemma follows. $\square$

The lemma shows that the group law in U is determined by the group laws in certain "rank 2" subgroups. In case α, β are *simple* roots, the subgroup in question is $U_{\alpha\beta}$, and $\varphi_{\alpha\beta} = \varphi_B|U_{\alpha\beta}$ is a group isomorphism by assumption (B) of (32.4). In general, let α, β be positive roots, which may be assumed to be distinct (since φ_α is already a group isomorphism). According to (A.4), there exist $\sigma \in W$ and $\gamma, \delta \in \Delta$ so that $\sigma(\alpha) = \gamma$ while $\sigma(\beta) \in \Phi^+_{\gamma\delta}$. Say $\sigma = \bar{n} (n \in N)$. Then Proposition 32.4(c), combined with the fact the $\varphi_{\gamma\delta}$ is a group homomorphism, allows us to conclude that φ_B preserves the commutator formula for α, β developed in the lemma. From this it follows readily that $\varphi_B|U$ is an isomorphism of algebraic groups. In turn, φ_B respects the action of T (by conjugation) on each U_α, again by Proposition 32.4(c). So φ_B is a group isomorphism.

We remark that the verification of (B) to be made in §33 will yield very explicit descriptions of the constants which occur in the commutator formula above, for pairs of positive roots.

32.6. Multiplicativity of φ

Still working under the assumptions (A), (B), (C) of (32.4), we may now define as in (32.1) a set-theoretic bijection $\varphi : G \to H$ which extends both φ_N and φ_B, and whose restriction to the big cell Ω is an isomorphism of varieties. The proof of Theorem 32.1 will be completed (modulo (A), (B), (C)) if we can show that φ is a group homomorphism, since that will in turn force φ to be an isomorphism of varieties. The restrictions of φ to N, U, U^-, Z_α ($\alpha \in \varDelta$) are already known to be morphisms of algebraic groups.

To simplify matters, let us show that it suffices to prove that φ is "generically" multiplicative, i.e., that $\varphi(xy) = \varphi(x)\varphi(y)$ whenever x, y, xy all lie in Ω. Indeed, the following lemma implies that if $\varphi|\Omega$ has this property, then it admits an extension (unique, thanks to Proposition 2.5(b)) to a morphism $\varphi' : G \to H$ which is multiplicative. Since $\Omega \cap Z_\alpha$ is dense in Z_α ($\alpha \in \varDelta$), φ' must agree with φ on Z_α. But the Z_α generate G, so φ' must coincide with φ.

Lemma. *Let G_1, G_2 be connected algebraic groups, U an open (nonempty) subset of G_1, $\psi : U \to G_2$ a morphism such that $\psi(xy) = \psi(x)\psi(y)$ whenever x, y, $xy \in U$. Then ψ extends (uniquely) to a morphism of algebraic groups $\psi' : G_1 \to G_2$.*

Proof. Lemma 7.4 says that $G_1 = U \cdot U$. This suggests (in fact, dictates) a recipe for ψ'. If $x \in G_1$ can be written as $x = yz$ (y, $z \in U$), we should set $\psi'(x) = \psi(y)\psi(z)$. For this to define a function, however, it is essential that (*) $\psi(y)\psi(z) = \psi(y')\psi(z')$ whenever y', $z' \in U$ and $yz = y'z'$. Once this point is settled, ψ' will automatically be a *morphism*, since its restriction to each translate xU ($x \in U$) is a morphism. In turn, ψ' will automatically be *multiplicative*. To see this, observe first that the subset V of $U \times U$, consisting of pairs (x, y) for which $xy \in U$, is dense in $G_1 \times G_1$: V is the intersection of the open sets $U \times U$ and $\mu^{-1}(U)$, where $\mu : G_1 \times G_1 \to G_1$ is the product. Further, the morphism $G_1 \times G_1 \to G_2$ defined by $(x, y) \mapsto \psi'(xy)\psi'(y)^{-1}\psi'(x)^{-1}$ agrees on V with the constant morphism $(x, y) \mapsto e$, so it must agree everywhere (Proposition 2.5(b) again).

It remains to prove (*). Denote by D_1 (resp. D_2) the diagonal in $G_1 \times G_1$ (resp. $G_2 \times G_2$); so D_1 and D_2 are closed sets (2.5). Let X be the inverse image of D_1 under the morphism $(U \times U) \times (U \times U) \overset{\mu \times \mu}{\to} G_1 \times G_1$, and let Y be the inverse image of D_2 under the composite morphism $(U \times U) \times (U \times U) \overset{\psi \times \psi \times \psi \times \psi}{\longrightarrow} (G_2 \times G_2) \times (G_2 \times G_2) \overset{\mu \times \mu}{\to} G_2 \times G_2$. Then both X and Y are closed in $U \times U \times U \times U$. It is easy to see that X is *irreducible* (Exercise 7.12), since $(w, x, y, z) \mapsto (w, x, y)$ induces an isomorphism of varieties from the inverse image of D_1 in $(G_1 \times G_1) \times (G_1 \times G_1)$ onto $G_1 \times G_1 \times G_1$, and X is an open subset of the first of these.

If the open set V in $U \times U$ is defined as above, the original hypothesis on ψ implies that $X' = (V \times V) \cap X$ is contained in Y. But X' is open (therefore dense) in the irreducible closed set X, so the closedness of Y forces $X \subset Y$. $\square$

The remaining task of this section is to prove that $\varphi|\Omega$ is "generically" multiplicative.

Proposition. *Assume* (A), (B), (C) *of* (32.4), *and define* $\varphi : G \to H$ *as above. If* $x, y, xy \in \Omega$, *then* $\varphi(xy) = \varphi(x)\varphi(y)$.

Proof. A number of preliminary steps will be required, involving φ and not just $\varphi|\Omega$. We lean heavily on the fact that the restrictions of φ to Z_α, N, U, U^- are already known to be group homomorphisms.

(1) *Let* $n \in N$ *represent the element of* W *which interchanges positive and negative roots, so* $nUn^{-1} = U^-$, $nU^-n^{-1} = U$. *Then for all* $u \in U$ (*resp.* $u \in U^-$), $\varphi(nun^{-1}) = \varphi(n)\varphi(u)\varphi(n^{-1})$.

Since $\varphi|U$ and $\varphi|U^-$ are group homomorphisms, this is an immediate consequence of Proposition 32.4(c).

(2) *Let* $u \in B$, $v \in B^-$, $x \in \Omega$. *Then* $\varphi(vxu) = \varphi(v)\varphi(x)\varphi(u)$.

Write $v = v_1 t_1$, $x = v_2 t_2 u_2$, $u = t_3 u_3$ ($v_i \in U^-$, $t_i \in T$, $u_i \in U$). Then by definition, $\varphi(v)\varphi(x)\varphi(u) = \varphi(v_1)\varphi(t_1)\varphi(v_2)\varphi(t_2)\varphi(u_2)\varphi(t_3)\varphi(u_3)$. Similarly, $\varphi(vxu) = \varphi(v_1 t_1 v_2 t_1^{-1} t_1 t_2 t_3 t_3^{-1} u_2 t_3 u_3) = \varphi(v_1 t_1 v_2 t_1^{-1})\varphi(t_1 t_2 t_3)\varphi(t_3^{-1} u_2 t_3 u_3)$, this last because $t_1 v_2 t_1^{-1} \in U^-$ and $t_3^{-1} u_2 t_3 \in U$. Since $\varphi|U$ and $\varphi|U^-$ are homomorphisms, the left and right terms can be factored further. Comparison with the first equation leaves us with the following equations to verify:

$$\varphi(t_1 v_2 t_1^{-1}) = \varphi(t_1)\varphi(v_2)\varphi(t_1^{-1}),$$
$$\varphi(t_3^{-1} u_2 t_3) = \varphi(t_3^{-1})\varphi(u_2)\varphi(t_3).$$

But these follow readily from Proposition 32.4(c), thanks to the way φ is defined on U, U^-.

(3) *Let* $\alpha \in \Delta$, $x \in \Omega$. *If* $\mathrm{Int}\, n_\alpha(x)$ *lies in* Ω, *then* $\varphi(n_\alpha x n_\alpha^{-1}) = \varphi(n_\alpha)\varphi(x)\varphi(n_\alpha)^{-1}$.

Write $x = v x_{-\alpha} t x_\alpha u$, with $t \in T$, $x_{-\alpha} \in U_{-\alpha}$, $x_\alpha \in U_\alpha$, v in the product of all $U_{-\beta}$ and u in the product of all U_β ($\beta \in \Phi^+ - \{\alpha\}$). These latter products are normalized by $\mathrm{Int}\, n_\alpha$, since σ_α permutes $\Phi^+ - \{\alpha\}$ (A.5). When $x = u$ or v, the desired formula therefore follows from Proposition 32.4(c). Thanks to step (2), it is enough to verify the formula when $x = x_{-\alpha} t x_\alpha \in Z_\alpha$. But $\varphi|Z_\alpha$ is already known (32.3) to be a homomorphism, and $n_\alpha \in Z_\alpha$.

(4) *Let* $n \in N$, $x \in \Omega$. *If* $\mathrm{Int}\, n(x) \in \Omega$, *then* $\varphi(nxn^{-1}) = \varphi(n)\varphi(x)\varphi(n)^{-1}$.

This is true for $n \in T$, by Proposition 32.4(c) and the way φ is defined. It is also true when $n = n_\alpha$ ($\alpha \in \Delta$), by step (3). Any n can be built up from an element of T and a sequence of such elements n_α, but we cannot be sure that the successive images of x all lie in Ω, so the general case is not a trivial consequence of these special cases. Instead, notice that the two sides of the desired formula define morphisms of $\Omega \cap \mathrm{Int}\,(n^{-1})\Omega$ into H. For these morphisms to coincide, it is sufficient to verify that: (*) There exists an open subset V_n of Ω (depending on n) such that Int $n(V_n) \subset \Omega$ and such that $\varphi \circ$ Int n agrees with Int $\varphi(n) \circ \varphi$ on V_n. In turn, it is enough to prove that if (*) holds for $n' \in N$, then it holds for $n = n'n_\alpha$ ($\alpha \in \Delta$). Let $V_{n'}$ be given and define $V_n = \Omega \cap$ Int $(n_\alpha)^{-1}V_{n'}$, so that Int $n(V_n) \subset$ Int $n'(V_n') \subset \Omega$. If $x \in V_n$, then Int $n(x) =$ Int $n' \circ$ Int $n_\alpha(x)$. Since Int $n_\alpha(x) \in V_{n'}$, $\varphi($Int $n(x)) = \varphi($Int n' (Int $n_\alpha)(x)) =$ Int $\varphi(n')(\varphi($Int $n_\alpha(x)))$, by hypothesis. By step (3), $\varphi($Int $n_\alpha(x)) =$ Int $\varphi(n_\alpha)(\varphi(x))$, which yields (*).

(5) The proof of the proposition can now be completed. If $x, x' \in \Omega$, write $x = vtu$, $x' = v't'u'$ ($v, v' \in U^-$; $t, t' \in T$; $u, u' \in U$), so $xx' = vtuv't'u'$. Thanks to step (2) and the fact that $U^-\Omega U = \Omega$, it is enough to show that $uv' \in \Omega$ implies $\varphi(uv') = \varphi(u)\varphi(v')$. As in step (1), let $n \in N$ send U to U^-, so that

$$\varphi(u) = \varphi(n)^{-1}\varphi(nun^{-1})\varphi(n),$$
$$\varphi(v') = \varphi(n)^{-1}\varphi(nv'v^{-1})\varphi(n).$$

Then $\varphi(u)\varphi(v') = \varphi(n)^{-1}\varphi(nun^{-1})\varphi(nv'n^{-1})\varphi(n)$. Since $nun^{-1} \in U^-$, $nv'n^{-1} \in U$, the two inner terms may be combined (by definition of φ). The assumption that $uv' \in \Omega$ now allows us to complete the proof by appealing to step (4). $\square$

Exercises

1. Assume that all semisimple algebraic groups are known (up to isomorphism). To what extent can you classify all reductive groups?

2. Let G, G' be simple algebraic groups, with respective maximal tori T, T'. If G, G' have isomorphic root systems and isomorphic fundamental groups, prove that there exists an isomorphism of $X(T)$ onto $X(T')$ inducing isomorphisms of root systems and fundamental groups, as asserted in (32.1).

3. Deduce from Theorem 32.1 that if G is of simply connected or of adjoint type, every graph automorphism of Φ induces an automorphism of G (cf. Theorem 27.4). To what extent is this true for G of arbitrary type?

4. Deduce from Theorem 32.1 that all semisimple groups of type E_8 (resp. F_4, resp. G_2) are isomorphic.

5. Up to isomorphism, there exist precisely three distinct reductive groups over K having dimension 4 and semisimple rank 1. (Cf. Exercise 25.5.)

6. According to (32.3), (Z_α, Z_α) is isomorphic either to $\mathrm{SL}(2, \mathrm{K})$ or to $\mathrm{PGL}(2, \mathrm{K})$. Prove that the former is always the case if G is simply connected, while the latter is not always the case if G is adjoint. [Let λ_α be the fundamental dominant weight relative to α, $\alpha \in \Delta$. Its restriction to a maximal torus of (Z_α, Z_α) is a weight λ for which both $\sigma_\alpha\lambda = \lambda - \alpha$ and $\sigma_\alpha\lambda = -\lambda$.]

Notes

Theorem 32.1 (or a more general version, dealing with isogenies and not just isomorphisms) is due to Chevalley [8, exposés 23,24]. However, the method of proof followed here is closer to that of SGAD, exposé XXIII, except for Proposition 32.3, which is based on Chevalley [8, expos 18]. Lemma 32.6 comes from SGAD, exposé XVIII.

33. Root Systems of Rank 2

Retain the notation of §32.

33.1. Reformulation of (A), (B), (C)

The extension of $\varphi_T : T \to H$ to an isomorphism $\varphi : G \to H$ was carried out under several assumptions involving the canonical rank 1 or 2 subsystems $\Phi_{\alpha\beta}$ of Φ ($\alpha, \beta \in \Delta$). These assumptions were formulated as follows in (32.4):

(A) $\varphi_T(t_{\alpha\beta}) = (h_\alpha h_\beta)^{m(\alpha,\,\beta)}$ *for all* $\alpha, \beta \in \Delta$, *where* $t_{\alpha\beta} = (n_\alpha n_\beta)^{m(\alpha,\,\beta)}$, $m(\alpha, \beta) =$ *order of* $\sigma_\alpha \sigma_\beta$ *in* W.

(B) *For each pair* $\alpha, \beta \in \Delta$, *there is an isomorphism* $\varphi_{\alpha\beta}$ *(of algebraic groups) from* $U_{\alpha\beta}$ *onto the corresponding subgroup of* H, *such that* $\varphi_{\alpha\beta}$ *extends both* φ_α *and* φ_β.

(C) *Let* $\alpha, \beta \in \Delta$. *If* $\gamma \in \Phi_{\alpha\beta}^+$ *with* $\gamma \neq \alpha$, *then* Int $h_\alpha \circ \varphi_{\alpha\beta}$ *agrees on* U_γ *with* $\varphi_{\alpha\beta} \circ$ Int n_α.

Notice that when $\alpha = \beta$, (B) and (C) are vacuous, while (A) is satisfied by virtue of the classification of rank 1 groups in (32.3).

Ostensibly these three statements involve the pair of groups G, H. But to verify them it is actually sufficient to work inside G alone. All we have to show is that the choices made (ε_α, $\varepsilon_{-\alpha}$, hence n_α, for $\alpha \in \Delta$) completely determine the elements $t_{\alpha\beta}$, the commutation in $U_{\alpha\beta}$, and the action of Int α on certain root groups. Then the corresponding choices in H *must* lead to corresponding results. (This approach amounts to providing generators and relations for G.) With this in mind, let us reformulate (A), (B), (C).

To prove (A) it will be enough to express $t_{\alpha\beta}$ in terms of t_α and t_β in a way depending only on which root system of rank 2 is involved, since we already know that t_α and t_β are well determined by the given data.

For (B) we shall prescribe $\varepsilon_\gamma : \mathbf{G}_a \to U_\gamma$ for each $\gamma \in \Phi_{\alpha\beta}^+$ in a way depending only on the root system, then show that the constants occurring in the commutator formulas (Lemma 32.5) for all pairs of roots in $\Phi_{\alpha\beta}^+$ are completely determined. So the group-theoretic structure of $U_{\alpha\beta}$ will just depend on the root system. On the other hand, it is easy to construct an isomorphism of varieties from $U_{\alpha\beta}$ onto the corresponding subgroup of H (cf. (32.5)); the use of the ε_γ and corresponding $\varepsilon'_{\gamma'}$ will insure that this is a group homomorphism. As an example of how ε_γ is to be prescribed, consider type A_2, where the positive roots are α, β, $\alpha + \beta$. We are already given ε_α and n_β, so we just set $\varepsilon_{\alpha+\beta}(x) = $ Int $n_\beta(\varepsilon_\alpha(x))$. (Of course, it is then a problem to decide how this is related to the other natural choice, where α and β are interchanged.) In this case, we shall complete the description of the group law in $U_{\alpha\beta}$ by verifying that $(\varepsilon_\beta(y), \varepsilon_\alpha(x)) = \varepsilon_{\alpha+\beta}(xy)$, while $\varepsilon_{\alpha+\beta}(z)$ commutes with everything.

Next consider (C). The requirement that $\gamma \neq \alpha$ just insures that $\delta = \sigma_\alpha(\gamma)$ is again positive, so ε_γ and ε_δ are given (cf. the preceding discussion of (B)). Following ε_γ with Int n_α yields another admissible isomorphism $\mathbf{G}_a \to U_\delta$, which we shall relate to ε_δ in a way depending only on the root system. This will guarantee the same relationship in H, thereby proving (C). For example, in the discussion above of type A_2 we left open the question of how Int $n_\alpha \circ \varepsilon_\beta$ would be related to $\varepsilon_{\alpha+\beta}$; it will be shown that $n_\alpha\varepsilon_\beta(x)n_\alpha^{-1} = \varepsilon_{\alpha+\beta}(-x)(x \in \mathbf{G}_a)$. This sort of conclusion can be expressed a bit more efficiently if we use the Lie algebra. The choice of an admissible isomorphism $\varepsilon_\gamma: \mathbf{G}_a \to U_\gamma$ really amounts to a choice of basis vector $\mathbf{x}_\gamma = d\varepsilon_\gamma(1)$ for $\mathfrak{g}_\gamma$, cf. (26.3). This choice can only vary by a scalar multiple, i.e., by an automorphism of $\mathbf{G}_a$. The equation above is then equivalent to: Ad $n_\alpha(\mathbf{x}_\beta) = -\mathbf{x}_{\alpha+\beta}$, the choice $\mathbf{x}_{\alpha+\beta} =$ Ad $n_\beta(\mathbf{x}_\alpha)$ having been made earlier.

Whenever we define $\mathbf{x}_\gamma$ for a nonsimple positive root γ by a recipe of the form $\mathbf{x}_\gamma =$ Ad $n_\beta(\mathbf{x}_\alpha)$ $(\alpha, \beta \in \varDelta)$, we also decree that $\mathbf{x}_{-\gamma} =$ Ad $n_\beta(\mathbf{x}_{-\alpha})$. Having defined both ε_γ and $\varepsilon_{-\gamma}$, we can then set $n_\gamma = \varepsilon_\gamma(1)\varepsilon_{-\gamma}(-1)\varepsilon_\gamma(1)$. In this way, Int n_β transfers the canonical choices made in Z_α over to Z_γ. As a result, we can compute: $n_\beta n_\alpha n_\beta^{-1} = (n_\beta\varepsilon_\alpha(1)n_\beta^{-1})(n_\beta\varepsilon_{-\alpha}(-1)n_\beta^{-1})(n_\beta\varepsilon_\alpha(1)n_\beta^{-1}) = \varepsilon_\gamma(1)\varepsilon_{-\gamma}(-1)\varepsilon_\gamma(1) = n_\gamma$.

There are just four root systems of rank 2 (A.7): $\mathsf{A}_1 \times \mathsf{A}_1$, A_2, $\mathsf{B}_2 (= \mathsf{C}_2)$, G_2. Only G_2 will require lengthy calculations. It may be well to point out that G_2 cannot play the role of $\varPhi_{\alpha\beta}$ inside any irreducible root system other than itself; so the isomorphism theorem for types $\mathsf{A} - \mathsf{F}$ does not depend at all on these calculations.

It is also worth remarking that the fundamental group of G only influences the nature of the elements t_α (i.e., whether $t_\alpha = e$ or not). The root system alone determines the other relations involved in (A), (B), (C).

33.2. Some Preliminaries

Before tackling the root systems A_2, B_2, G_2, we have to make a few general observations.

The commutator formula (Lemma 32.5) was stated for pairs of positive roots, but of course it applies whenever two roots can be made positive by a suitable choice of base. We shall very frequently use the fact that $(U_\alpha, U_\beta) = e$ when $\alpha + \beta \notin \varPhi$, in order to rearrange formulas. In particular, when neither $\alpha + \beta$ nor $\alpha - \beta$ is a root, the definition of n_α shows that Ad $n_\alpha(\mathbf{x}_\beta) = \mathbf{x}_\beta$ and also that $n_\alpha n_\beta = n_\beta n_\alpha$. These considerations already settle the case $\mathsf{A}_1 \times \mathsf{A}_1$:

Proposition. *Let $\varPhi_{\alpha\beta}$ be of type $\mathsf{A}_1 \times \mathsf{A}_1$, with positive roots α, β. Then:*
(a) $t_{\alpha\beta} = (n_\alpha n_\beta)^2 = t_\alpha t_\beta = (n_\beta n_\alpha)^2 = t_{\beta\alpha}$.
(b) $(U_\alpha, U_\beta) = e$.
(c) Ad $n_\alpha(\mathbf{x}_\beta) = \mathbf{x}_\beta$, Ad $n_\beta(\mathbf{x}_\alpha) = \mathbf{x}_\alpha$. $\quad\square$

The other information we need concerns the elements t_α: What can be said about $\beta(t_\alpha)$ or about $n_\beta t_\alpha n_\beta^{-1}$ when $\beta \in \Delta$?

Recall (A.11) the notion of **dual root system** Φ^*. This is the set of vectors $\alpha^* = 2\alpha/(\alpha, \alpha)$ in the euclidean space E where Φ is defined. In the cases A$_2$, B$_2$, G$_2$, Φ^* is a root system of the same type; but a long root α gives rise to a short root α^* when there are two root lengths involved. The Weyl group of Φ^* is canonically isomorphic to W, with $\sigma_\alpha(\beta^*) = \beta^* - \langle \beta^*, \alpha^* \rangle \alpha^* = \beta^* - \langle \alpha, \beta \rangle \alpha^*$, where $\langle \alpha, \beta \rangle = 2(\alpha, \beta)/(\beta, \beta)$.

Our aim is to construct a copy of Φ^* directly in Y(T), so that in the natural pairing X(T) $\times$ Y(T) $\to \mathbb{Z}$, we shall have $\langle \beta, \alpha^* \rangle$ equal to the Cartan integer $\langle \beta, \alpha \rangle$. For each $\alpha \in \Delta$, (Z_α, Z_α) is isomorphic to SL(2, K) or to PGL(2, K) (32.3). In either case, the assignment $x \mapsto \begin{pmatrix} x & 0 \\ 0 & x^{-1} \end{pmatrix}$ yields a 1-psg $\gamma: \mathbf{G}_m \to T \cap (Z_\alpha, Z_\alpha)$ for which $t_\alpha = \gamma(-1)$ and for which $\langle \alpha, \gamma \rangle = 2$ (i.e. $\alpha(\gamma(x)) = x^2$). This is our candidate for α^*.

Now (Z_α, Z_α) lies in each maximal parabolic subgroup $P_{\Delta - \{\beta\}}$ ($\beta \neq \alpha$) and then fixes a vector (in some rational representation of G) whose weight is a positive multiple of the fundamental dominant weight λ_β attached to β (cf. first three paragraphs of (31.4)). It follows that for some positive c, $\lambda_\beta^c(\gamma(x)) = 1$ for all $x \in \mathbf{G}_m$, or $\langle c\lambda_\beta, \gamma \rangle = 0$. The vector spaces $\mathbb{R} \otimes$ Y(T) and E $= \mathbb{R} \otimes$ X(T) are in duality via the pairing, and γ therefore corresponds to the unique linear function on E which assigns 2 to α, 0 to all λ_β ($\beta \neq \alpha$). But in the natural pairing of E with itself via (,), α^* plays precisely this role (A.11). As a result, we obtain a base $\{\alpha^* | \alpha \in \Delta\}$ for a root system in $\mathbb{R} \otimes$ Y(T) isomorphic to Φ^*, with $\langle \beta, \alpha^* \rangle = \langle \beta, \alpha \rangle$ ($\alpha, \beta \in \Delta$), cf. (A.11). The action of W is then consistent with Exercise 24.8.

The following basic formulas are now evident:

(a) $\beta(t_\alpha) = \beta(\alpha^*(-1)) = (-1)^{\langle \beta, \alpha \rangle}$;

(b) $n_\beta t_\alpha n_\beta^{-1} = (\sigma_\beta \alpha^*)(-1) = (\alpha^* - \langle \beta, \alpha \rangle \beta^*)(-1)$
$= \alpha^*(-1)\beta^*(-1)^{-\langle \beta, \alpha \rangle} = t_\alpha t_\beta^{-\langle \beta, \alpha \rangle}$ ($\alpha, \beta \in \Delta$).

33.3. Type A$_2$

Proposition. *Let $\Phi_{\alpha\beta}$ be of type* A$_2$, *with positive roots* α, β, $\alpha + \beta$. *Set* $\mathsf{x}_{\alpha+\beta} = \mathrm{Ad}\, n_\beta(\mathsf{x}_\alpha)$. *Then:*

(a) $t_{\alpha\beta} = (n_\alpha n_\beta)^3 = e = (n_\beta n_\alpha)^3 = t_{\beta\alpha}$.

(b) *For all* $x, y \in \mathbf{G}_a$, $\varepsilon_\beta(y)\varepsilon_\alpha(x) = \varepsilon_\alpha(x)\varepsilon_\beta(y)\varepsilon_{\alpha+\beta}(xy)$.

(c) $\mathrm{Ad}\, n_\alpha(\mathsf{x}_\beta) = -\mathsf{x}_{\alpha+\beta}$, $\mathrm{Ad}\, n_\alpha(\mathsf{x}_{\alpha+\beta}) = \mathsf{x}_\beta$, $\mathrm{Ad}\, n_\beta(\mathsf{x}_{\alpha+\beta}) = -\mathsf{x}_\alpha$.

Proof. Since $\langle \alpha, \beta \rangle = \langle \beta, \alpha \rangle = -1$ (A.7), we have $\alpha(t_\beta) = \beta(t_\alpha) = -1$ (33.2)(a). From the definition of $\mathsf{x}_{\alpha+\beta}$ we deduce that $\mathrm{Ad}\, n_\beta(\mathsf{x}_{\alpha+\beta}) = \alpha(t_\beta)\mathsf{x}_\alpha = -\mathsf{x}_\alpha$, as asserted in (c). On the other hand, $\mathrm{Ad}\, n_\alpha(\mathsf{x}_\beta) = b\mathsf{x}_{\alpha+\beta}$ for some $b \in$ K, whence $\mathrm{Ad}\, n_\alpha(\mathsf{x}_{\alpha+\beta}) = \beta(t_\alpha)b^{-1}\mathsf{x}_\beta = -b^{-1}\mathsf{x}_\beta$. So the other assertions in (c) require b to be -1.

Lemma 32.5 says that for some $a \in \mathsf{K}$ independent of x, y, the following identity holds:

$$(1) \qquad \varepsilon_\beta(y)\varepsilon_\alpha(x) = \varepsilon_\alpha(x)\varepsilon_\beta(y)\varepsilon_{\alpha+\beta}(axy).$$

To prove (b) we have to show that $a = 1$. First apply Int n_β to both sides of (1), to get:

$$(2) \qquad \varepsilon_{-\beta}(-y)\varepsilon_{\alpha+\beta}(x) = \varepsilon_{\alpha+\beta}(x)\varepsilon_{-\beta}(-y)\varepsilon_\alpha(-axy).$$

By definition, $n_\beta\varepsilon_\alpha(x)n_\beta^{-1} = \varepsilon_{\alpha+\beta}(x)$, or

$$(3) \qquad \varepsilon_\beta(1)\varepsilon_{-\beta}(-1)\varepsilon_\beta(1)\varepsilon_\alpha(x)\varepsilon_\beta(-1)\varepsilon_{-\beta}(1)\varepsilon_\beta(-1) = \varepsilon_{\alpha+\beta}(x).$$

Since $\alpha + 2\beta \notin \Phi$, this can be rewritten:

$$(4) \qquad \varepsilon_\beta(1)\varepsilon_\alpha(x)\varepsilon_\beta(-1) = \varepsilon_{-\beta}(1)\varepsilon_{\alpha+\beta}(x)\varepsilon_{-\beta}(-1).$$

Applying (1) to the left side (with $y = 1$) and (2) to the right side (with $y = -1$), this becomes:

$$(5) \qquad \varepsilon_\alpha(x)\varepsilon_\beta(1)\varepsilon_{\alpha+\beta}(ax)\varepsilon_\beta(-1) = \varepsilon_{\alpha+\beta}(x)\varepsilon_{-\beta}(1)\varepsilon_\alpha(ax)\varepsilon_{-\beta}(-1).$$

Because $\alpha + 2\beta, \alpha - \beta, 2\alpha + \beta \notin \Phi$, (5) can be rewritten:

$$(6) \qquad \varepsilon_\alpha(x)\varepsilon_{\alpha+\beta}(ax) = \varepsilon_\alpha(ax)\varepsilon_{\alpha+\beta}(x).$$

It follows that $a = 1$, proving (b). Next apply Int n_α to both sides of (1):

$$(7) \qquad \varepsilon_{\alpha+\beta}(by)\varepsilon_{-\alpha}(-x) = \varepsilon_{-\alpha}(-x)\varepsilon_{\alpha+\beta}(by)\varepsilon_\beta(-b^{-1}xy).$$

From $n_\alpha\varepsilon_\beta(y)n_\alpha^{-1} = \varepsilon_{\alpha+\beta}(by)$ we derive (by the sort of argument that led to (4)):

$$(8) \qquad \varepsilon_\alpha(1)\varepsilon_\beta(y)\varepsilon_\alpha(-1) = \varepsilon_{-\alpha}(1)\varepsilon_{\alpha+\beta}(by)\varepsilon_{-\alpha}(-1).$$

Now apply (1) to the left side and (7) to the right side of (8):

$$(9) \qquad \varepsilon_\beta(y)\varepsilon_{\alpha+\beta}(-y) = \varepsilon_\beta(-b^{-1}y)\varepsilon_{\alpha+\beta}(by).$$

This forces $b = -1$, proving (c). It remains to prove (a). Start with the equation:

$$(10) \qquad n_\alpha n_\beta n_\alpha = n_\alpha n_\beta n_\alpha^{-1} n_\alpha^2 = n_{\alpha+\beta}^{-1} t_\alpha,$$

where $n_\alpha n_\beta n_\alpha^{-1}$ is evaluated using $n_\beta = \varepsilon_\beta(1)\varepsilon_{-\beta}(-1)\varepsilon_\beta(1)$, as described in (33.1). From (10) and (33.2) (formula (b)), we get:

$$(11) \qquad n_\beta n_\alpha n_\beta n_\alpha n_\beta = n_\beta n_{\alpha+\beta}^{-1} t_\alpha n_\beta = n_\beta n_{\alpha+\beta}^{-1} n_\beta^{-1} t_\alpha t_\beta^{-\langle\beta,\alpha\rangle} t_\beta = n_\alpha t_\alpha = n_\alpha^{-1}.$$

From this (a) follows. $\square$

33.4. Type B_2

Proposition. *Let $\Phi_{\alpha\beta}$ be of type B_2, with positive roots $\alpha, \beta, \alpha + \beta, 2\alpha + \beta$. Set $\mathsf{x}_{\alpha+\beta} = \mathrm{Ad}\, n_\beta(\mathsf{x}_\alpha), \mathsf{x}_{2\alpha+\beta} = \mathrm{Ad}\, n_\alpha(\mathsf{x}_\beta)$. Then:*
   (a) $t_{\alpha\beta} = (n_\alpha n_\beta)^4 = t_\alpha = (n_\beta n_\alpha)^4 = t_{\beta\alpha}.$

(b) *For* $x, y \in \mathbf{G}_a$, $\varepsilon_\beta(y)\varepsilon_\alpha(x) = \varepsilon_\alpha(x)\varepsilon_\beta(y)\varepsilon_{\alpha+\beta}(xy)\varepsilon_{2\alpha+\beta}(x^2y)$, *while* $\varepsilon_{\alpha+\beta}(y) \cdot$
$\varepsilon_\alpha(x) = \varepsilon_\alpha(x)\varepsilon_{\alpha+\beta}(y)\varepsilon_{2\alpha+\beta}(2xy)$.

(c) Ad $n_\alpha(\mathsf{x}_{\alpha+\beta}) = -\mathsf{x}_{\alpha+\beta}$, Ad $n_\alpha(\mathsf{x}_{2\alpha+\beta}) = \mathsf{x}_\beta$, Ad $n_\beta(\mathsf{x}_{\alpha+\beta}) = -\mathsf{x}_\alpha$,
Ad $n_\beta(\mathsf{x}_{2\alpha+\beta}) = \mathsf{x}_{2\alpha+\beta}$.

Proof. Since $\langle \alpha, \beta \rangle = -1, \langle \beta, \alpha \rangle = -2$, we have $\alpha(t_\beta) = -1, \beta(t_\alpha) = 1$
(33.2). From the definitions of $\mathsf{x}_{\alpha+\beta}$ and $\mathsf{x}_{2\alpha+\beta}$ we deduce that Ad $n_\beta(\mathsf{x}_{\alpha+\beta}) = \alpha(t_\beta)\mathsf{x}_\alpha = -\mathsf{x}_\alpha$ and that Ad $n_\alpha(\mathsf{x}_{2\alpha+\beta}) = \beta(t_\alpha)\mathsf{x}_\beta = \mathsf{x}_\beta$, as asserted in (c). Moreover, Ad $n_\beta(\mathsf{x}_{2\alpha+\beta}) = \mathsf{x}_{2\alpha+\beta}$ because $2\alpha + 2\beta, 2\alpha \notin \Phi$ (33.2). To complete the
proof of (c), we must show that $d = -1$, where Ad $n_\alpha(\mathsf{x}_{\alpha+\beta}) = d\mathsf{x}_{\alpha+\beta}$.

Lemma 32.5 yields scalars $a, b, c \in \mathsf{K}$ (independent of $x, y \in \mathbf{G}_a$) such that
the following formulas hold:

$$\text{(1)} \qquad \varepsilon_\beta(y)\varepsilon_\alpha(x) = \varepsilon_\alpha(x)\varepsilon_\beta(y)\varepsilon_{\alpha+\beta}(axy)\varepsilon_{2\alpha+\beta}(bx^2y),$$

$$\text{(2)} \qquad \varepsilon_{\alpha+\beta}(y)\varepsilon_\alpha(x) = \varepsilon_\alpha(x)\varepsilon_{\alpha+\beta}(y)\varepsilon_{2\alpha+\beta}(cxy).$$

To prove (b) we have to show that $a = b = 1, c = 2$. Apply Int n_α to both
sides of (1) (resp. (2)), to get:

$$\text{(3)} \qquad \varepsilon_{2\alpha+\beta}(y)\varepsilon_{-\alpha}(-x) = \varepsilon_{-\alpha}(-x)\varepsilon_{2\alpha+\beta}(y)\varepsilon_{\alpha+\beta}(adxy)\varepsilon_\beta(bx^2y),$$

$$\text{(4)} \qquad \varepsilon_{\alpha+\beta}(dy)\varepsilon_{-\alpha}(-x) = \varepsilon_{-\alpha}(-x)\varepsilon_{\alpha+\beta}(dy)\varepsilon_\beta(cxy).$$

Similarly, apply Int n_β to both sides of (1):

$$\text{(5)} \qquad \varepsilon_{-\beta}(-y)\varepsilon_{\alpha+\beta}(x) = \varepsilon_{\alpha+\beta}(x)\varepsilon_{-\beta}(-y)\varepsilon_\alpha(-axy)\varepsilon_{2\alpha+\beta}(bx^2y).$$

From $n_\beta\varepsilon_\alpha(x)n_\beta^{-1} = \varepsilon_{\alpha+\beta}(x)$ and the fact that $\alpha + 2\beta \notin \Phi$ we obtain:

$$\text{(6)} \qquad \varepsilon_\beta(1)\varepsilon_\alpha(x)\varepsilon_\beta(-1) = \varepsilon_{-\beta}(1)\varepsilon_{\alpha+\beta}(x)\varepsilon_{-\beta}(-1).$$

Applying (1) to the left side (with $y = 1$) and (5) to the right side (with $y = -1$),
this becomes:

$$\text{(7)} \qquad \varepsilon_\alpha(x)\varepsilon_\beta(1)\varepsilon_{\alpha+\beta}(ax)\varepsilon_{2\alpha+\beta}(bx^2)\varepsilon_\beta(-1)$$
$$= \varepsilon_{\alpha+\beta}(x)\varepsilon_{-\beta}(1)\varepsilon_\alpha(ax)\varepsilon_{2\alpha+\beta}(-bx^2)\varepsilon_{-\beta}(-1).$$

Because $\alpha + 2\beta, 2\alpha + 2\beta, \alpha - \beta, 2\alpha \notin \Phi$, this can be rearranged:

$$\text{(8)} \qquad \varepsilon_\alpha(x)\varepsilon_{\alpha+\beta}(ax)\varepsilon_{2\alpha+\beta}(bx^2) = \varepsilon_{\alpha+\beta}(x)\varepsilon_\alpha(ax)\varepsilon_{2\alpha+\beta}(-bx^2).$$

In turn, (2) can be applied to the right side to yield:

$$\text{(9)} \qquad \varepsilon_\alpha(x)\varepsilon_{\alpha+\beta}(ax)\varepsilon_{2\alpha+\beta}(bx^2) = \varepsilon_\alpha(ax)\varepsilon_{\alpha+\beta}(x)\varepsilon_{2\alpha+\beta}((ac - b)x^2),$$

whence $a = 1$ and $c = 2b$. From $n_\alpha\varepsilon_\beta(y)n_\alpha^{-1} = \varepsilon_{2\alpha+\beta}(y)$ and the fact that
$3\alpha + \beta \notin \Phi$ we next obtain:

$$\text{(10)} \qquad \varepsilon_\alpha(1)\varepsilon_\beta(y)\varepsilon_\alpha(-1) = \varepsilon_{-\alpha}(1)\varepsilon_{2\alpha+\beta}(y)\varepsilon_{-\alpha}(-1).$$

After (1) is applied to the left side, (3) to the right side, this becomes:

$$\text{(11)} \qquad \varepsilon_\beta(y)\varepsilon_{\alpha+\beta}(-ay)\varepsilon_{2\alpha+\beta}(by) = \varepsilon_{2\alpha+\beta}(y)\varepsilon_{\alpha+\beta}(ady)\varepsilon_\beta(by).$$

Since $U_{\alpha+\beta}, U_{2\alpha+\beta}, U_\beta$ all commute, we conclude at once that $b = 1, -a = ad,$

whence $a = b = 1, c = 2, d = -1$. This completes the proof of (b) and (c). It remains to prove (a). To begin with:

$$(12) \qquad n_\alpha n_\beta n_\alpha = n_\alpha n_\beta n_\alpha^{-1} t_\alpha = n_{2\alpha+\beta} t_\alpha.$$

In turn,

$$(13) \qquad n_\beta n_\alpha n_\beta n_\alpha n_\beta = n_\beta n_{2\alpha+\beta} n_\beta^{-1} t_\alpha t_\beta^{-\langle\beta,\alpha\rangle} t_\beta = n_{2\alpha+\beta} t_\alpha t_\beta,$$

because $\langle\beta,\alpha\rangle = -2$ and $t_\beta^2 = e$. Then:

$$(14) \qquad (n_\alpha n_\beta)^3 n_\alpha = (n_\alpha n_{2\alpha+\beta} n_\alpha^{-1})(n_\alpha t_\alpha t_\beta n_\alpha^{-1}) t_\alpha$$
$$= n_\beta t_\alpha t_\beta t_\alpha^{-\langle\alpha,\beta\rangle} t_\alpha = n_\beta t_\alpha t_\beta.$$

Finally, (14) implies that $(n_\beta n_\alpha)^4 = t_\alpha = (n_\alpha n_\beta)^4$. $\square$

33.5. Type G_2

Proposition.   *Let $\Phi_{\alpha\beta}$ be of type G_2, with positive roots $\alpha, \beta, \alpha + \beta, 2\alpha + \beta$, $3\alpha + \beta, 3\alpha + 2\beta$. Set $\mathsf{x}_{\alpha+\beta} = \mathrm{Ad}\, n_\beta(\mathsf{x}_\alpha), \mathsf{x}_{2\alpha+\beta} = \mathrm{Ad}\, n_\alpha(\mathsf{x}_{\alpha+\beta}), \mathsf{x}_{3\alpha+\beta} = -\mathrm{Ad}\, n_\alpha(\mathsf{x}_\beta)$, $\mathsf{x}_{3\alpha+2\beta} = \mathrm{Ad}\, n_\beta(\mathsf{x}_{3\alpha+\beta})$. Then:*
  (a) $t_{\alpha\beta} = (\sigma_\alpha\sigma_\beta)^6 = e = (\sigma_\beta\sigma_\alpha)^6 = t_{\beta\alpha}$.
  (b) *For all $x, y \in \mathbf{G}_a$,*
$$\varepsilon_\beta(y)\varepsilon_\alpha(x) = \varepsilon_\alpha(x)\varepsilon_\beta(y)\varepsilon_{\alpha+\beta}(xy)\varepsilon_{2\alpha+\beta}(x^2 y)\varepsilon_{3\alpha+\beta}(x^3 y)\varepsilon_{3\alpha+2\beta}(x^3 y^2);$$
$$\varepsilon_{\alpha+\beta}(y)\varepsilon_\alpha(x) = \varepsilon_\alpha(x)\varepsilon_{\alpha+\beta}(y)\varepsilon_{2\alpha+\beta}(2xy)\varepsilon_{3\alpha+\beta}(3x^2 y)\varepsilon_{3\alpha+2\beta}(3xy^2);$$
$$\varepsilon_{2\alpha+\beta}(y)\varepsilon_\alpha(x) = \varepsilon_\alpha(x)\varepsilon_{2\alpha+\beta}(y)\varepsilon_{3\alpha+\beta}(3xy);$$
$$\varepsilon_{3\alpha+\beta}(y)\varepsilon_\beta(x) = \varepsilon_\beta(x)\varepsilon_{3\alpha+\beta}(y)\varepsilon_{3\alpha+2\beta}(-xy);$$
$$\varepsilon_{2\alpha+\beta}(y)\varepsilon_{\alpha+\beta}(x) = \varepsilon_{\alpha+\beta}(x)\varepsilon_{2\alpha+\beta}(y)\varepsilon_{3\alpha+2\beta}(3xy).$$
  (c) $\mathrm{Ad}\, n_\alpha(\mathsf{x}_{2\alpha+\beta}) = -\mathsf{x}_{\alpha+\beta}$, $\mathrm{Ad}\, n_\alpha(\mathsf{x}_{3\alpha+\beta}) = \mathsf{x}_\beta$, $\mathrm{Ad}\, n_\alpha(\mathsf{x}_{3\alpha+2\beta}) = \mathsf{x}_{3\alpha+2\beta}$, $\mathrm{Ad}\, n_\beta(\mathsf{x}_{\alpha+\beta}) = -\mathsf{x}_\alpha$, $\mathrm{Ad}\, n_\beta(\mathsf{x}_{2\alpha+\beta}) = \mathsf{x}_{2\alpha+\beta}$, $\mathrm{Ad}\, n_\beta(\mathsf{x}_{3\alpha+2\beta}) = -\mathsf{x}_{3\alpha+\beta}$.

Proof.   Since $\langle\alpha, \beta\rangle = -1, \langle\beta, \alpha\rangle = -3$, we have $\beta(t_\alpha) = \alpha(t_\beta) = -1$. From the definitions we deduce that:

$$(1) \qquad \mathrm{Ad}\, n_\beta(\mathsf{x}_{\alpha+\beta}) = \alpha(t_\beta)\mathsf{x}_\alpha = -\mathsf{x}_\alpha,$$
$$(2) \qquad \mathrm{Ad}\, n_\alpha(\mathsf{x}_{2\alpha+\beta}) = \alpha(t_\alpha)\beta(t_\alpha)\mathsf{x}_{\alpha+\beta} = -\mathsf{x}_{\alpha+\beta},$$
$$(3) \qquad \mathrm{Ad}\, n_\alpha(\mathsf{x}_{3\alpha+\beta}) = -\beta(t_\alpha)\mathsf{x}_\beta = \mathsf{x}_\beta,$$
$$(4) \qquad \mathrm{Ad}\, n_\beta(\mathsf{x}_{3\alpha+2\beta}) = \alpha(t_\beta)^3\beta(t_\beta)\mathsf{x}_{3\alpha+\beta} = -\mathsf{x}_{3\alpha+\beta}.$$

Since $(3\alpha + 2\beta) \pm \alpha, (2\alpha + \beta) \pm \beta \notin \Phi$, we also get:

$$(5) \qquad \mathrm{Ad}\, n_\alpha(\mathsf{x}_{3\alpha+2\beta}) = \mathsf{x}_{3\alpha+2\beta},$$
$$(6) \qquad \mathrm{Ad}\, n_\beta(\mathsf{x}_{2\alpha+\beta}) = \mathsf{x}_{2\alpha+\beta}.$$

Statements (1)–(6) dispose of (c).

Turning next to (b), we use Lemma 32.5 to obtain scalars a, b, c, d, e, f, g, h, i (independent of x, y) such that the following formulas hold:

$$(7) \quad \varepsilon_\beta(y)\varepsilon_\alpha(x) = \varepsilon_\alpha(x)\varepsilon_\beta(y)\varepsilon_{\alpha+\beta}(axy)\varepsilon_{2\alpha+\beta}(bx^2 y)\varepsilon_{3\alpha+\beta}(cx^3 y)\varepsilon_{3\alpha+2\beta}(dx^3 y^2),$$

(8) $\qquad \varepsilon_{\alpha+\beta}(y)\varepsilon_{\alpha}(x) = \varepsilon_{\alpha}(x)\varepsilon_{\alpha+\beta}(y)\varepsilon_{2\alpha+\beta}(exy)\varepsilon_{3\alpha+\beta}(fx^2y)\varepsilon_{3\alpha+2\beta}(gxy^2),$

(9) $\qquad \varepsilon_{2\alpha+\beta}(y)\varepsilon_{\alpha}(x) = \varepsilon_{\alpha}(x)\varepsilon_{2\alpha+\beta}(y)\varepsilon_{3\alpha+\beta}(hxy),$

(10) $\qquad \varepsilon_{3\alpha+\beta}(y)\varepsilon_{\beta}(x) = \varepsilon_{\beta}(x)\varepsilon_{3\alpha+\beta}(y)\varepsilon_{3\alpha+2\beta}(ixy).$

Letting Int n_β act on both sides of (7), (9), (10), we obtain (respectively):

(11) $\varepsilon_{-\beta}(-y)\varepsilon_{\alpha+\beta}(x)$
$$= \varepsilon_{\alpha+\beta}(x)\varepsilon_{-\beta}(-y)\varepsilon_{\alpha}(-axy)\varepsilon_{2\alpha+\beta}(bx^2y)\varepsilon_{3\alpha+2\beta}(cx^3y)\varepsilon_{3\alpha+\beta}(-dx^3y^2),$$

(12) $\qquad \varepsilon_{2\alpha+\beta}(y)\varepsilon_{\alpha+\beta}(x) = \varepsilon_{\alpha+\beta}(x)\varepsilon_{2\alpha+\beta}(y)\varepsilon_{3\alpha+2\beta}(hxy),$

(13) $\qquad \varepsilon_{3\alpha+2\beta}(y)\varepsilon_{-\beta}(-x) = \varepsilon_{-\beta}(-x)\varepsilon_{3\alpha+2\beta}(y)\varepsilon_{3\alpha+\beta}(-ixy).$

Similarly, let Int n_α act on both sides of (7):

(14) $\varepsilon_{3\alpha+\beta}(-y)\varepsilon_{-\alpha}(-x)$
$$= \varepsilon_{-\alpha}(-x)\varepsilon_{3\alpha+\beta}(-y)\varepsilon_{2\alpha+\beta}(axy)\varepsilon_{\alpha+\beta}(-bx^2y)\varepsilon_{\beta}(cx^3y)\varepsilon_{3\alpha+\beta}(dx^3y^2).$$

Because $\alpha + 2\beta \notin \Phi$, the equation $n_\beta\varepsilon_\alpha(x)n_\beta^{-1} = \varepsilon_{\alpha+\beta}(x)$ simplifies to:

(15) $\qquad \varepsilon_{\beta}(1)\varepsilon_{\alpha}(x)\varepsilon_{\beta}(-1) = \varepsilon_{-\beta}(1)\varepsilon_{\alpha+\beta}(x)\varepsilon_{-\beta}(-1).$

Using first (7) and then (10), the left side of (15) can be written:

(16) $\varepsilon_{\beta}(1)\varepsilon_{\alpha}(x)\varepsilon_{\beta}(-1)$
$$= \varepsilon_{\alpha}(x)\varepsilon_{\beta}(1)\varepsilon_{\alpha+\beta}(ax)\varepsilon_{2\alpha+\beta}(bx^2)\varepsilon_{3\alpha+\beta}(cx^3)\varepsilon_{3\alpha+2\beta}(dx^3)\varepsilon_{\beta}(-1)$$
$$= \varepsilon_{\alpha}(x)\varepsilon_{\beta}(1)\varepsilon_{\alpha+\beta}(ax)\varepsilon_{2\alpha+\beta}(bx^2)\varepsilon_{\beta}(-1)\varepsilon_{3\alpha+\beta}(cx^3)\varepsilon_{3\alpha+2\beta}((d-ci)x^3)$$
$$= \varepsilon_{\alpha}(x)\varepsilon_{\alpha+\beta}(ax)\varepsilon_{2\alpha+\beta}(bx^2)\varepsilon_{3\alpha+\beta}(cx^3)\varepsilon_{3\alpha+2\beta}((d-ci)x^3).$$

Using first (11) and then (13), the right side of (15) can be written:

(17) $\varepsilon_{-\beta}(1)\varepsilon_{\alpha+\beta}(x)\varepsilon_{-\beta}(-1)$
$$= \varepsilon_{\alpha+\beta}(x)\varepsilon_{-\beta}(1)\varepsilon_{\alpha}(ax)\varepsilon_{2\alpha+\beta}(-bx^2)\varepsilon_{3\alpha+2\beta}(-cx^3)\varepsilon_{3\alpha+\beta}(-dx^3)\varepsilon_{-\beta}(-1)$$
$$= \varepsilon_{\alpha+\beta}(x)\varepsilon_{\alpha}(ax)\varepsilon_{2\alpha+\beta}(-bx^2)\varepsilon_{3\alpha+2\beta}(-cx^3)\varepsilon_{3\alpha+\beta}((ci-d)x^3).$$

In turn, (8) transforms the right side of (17) into:

(18) $\varepsilon_{\alpha}(ax)\varepsilon_{\alpha+\beta}(x)\varepsilon_{2\alpha+\beta}(aex^2)\varepsilon_{3\alpha+\beta}(a^2fx^3)\varepsilon_{3\alpha+2\beta}(agx^3)$
$$\varepsilon_{2\alpha+\beta}(-bx^2)\varepsilon_{3\alpha+2\beta}(-cx^3)\varepsilon_{3\alpha+\beta}((ci-d)x^3)$$
$$= \varepsilon_{\alpha}(ax)\varepsilon_{\alpha+\beta}(x)\varepsilon_{2\alpha+\beta}((ae-b)x^2)\varepsilon_{3\alpha+\beta}((a^2f+ci-d)x^3)$$
$$\varepsilon_{3\alpha+2\beta}((ag-c)x^3).$$

Comparison of (16) and (18) yields:

(19) $\qquad a = 1, \quad e = 2b, \quad c + d = f + ci, \quad f = g.$

Because $4\alpha + \beta \notin \Phi$, the equation $n_\alpha\varepsilon_\beta(y)n_\alpha^{-1} = \varepsilon_{3\alpha+\beta}(-y)$ can be simplified to:

(20) $\qquad \varepsilon_{\alpha}(1)\varepsilon_{\beta}(y)\varepsilon_{\alpha}(-1) = \varepsilon_{-\alpha}(1)\varepsilon_{3\alpha+\beta}(-y)\varepsilon_{-\alpha}(-1).$

Apply (7) to the left side to obtain:

(21) $\varepsilon_{\alpha}(1)\varepsilon_{\beta}(y)\varepsilon_{\alpha}(-1) = \varepsilon_{\beta}(y)\varepsilon_{\alpha+\beta}(-ay)\varepsilon_{2\alpha+\beta}(by)\varepsilon_{3\alpha+\beta}(-cy)\varepsilon_{3\alpha+2\beta}(-dy^2).$

Apply (14), (12), (10) successively to the right side of (20) to obtain:

$$(22) \quad \varepsilon_{-\alpha}(1)\varepsilon_{3\alpha+\beta}(-y)\varepsilon_{-\alpha}(-1)$$
$$= \varepsilon_{3\alpha+\beta}(-y)\varepsilon_{2\alpha+\beta}(ay)\varepsilon_{\alpha+\beta}(-by)\varepsilon_{\beta}(cy)\varepsilon_{3\alpha+2\beta}(dy^2)$$
$$= \varepsilon_{3\alpha+\beta}(-y)\varepsilon_{\alpha+\beta}(-by)\varepsilon_{2\alpha+\beta}(ay)\varepsilon_{3\alpha+2\beta}(-abhy^2)\varepsilon_{\beta}(cy)\varepsilon_{3\alpha+2\beta}(dy^2)$$
$$= \varepsilon_{\beta}(cy)\varepsilon_{3\alpha+\beta}(-y)\varepsilon_{\alpha+\beta}(-by)\varepsilon_{2\alpha+\beta}(ay)\varepsilon_{3\alpha+2\beta}((d - ci - abh)y^2)$$
$$= \varepsilon_{\beta}(cy)\varepsilon_{\alpha+\beta}(-by)\varepsilon_{2\alpha+\beta}(ay)\varepsilon_{3\alpha+\beta}(-y)\varepsilon_{3\alpha+2\beta}((d - ci - abh)y^2).$$

Comparison of (21) and (22) yields:

$$(23) \qquad c = 1, \quad a = b, \quad d - ci - abh = -d.$$

Together with (19), this implies:

$$(24) \quad a = b = c = 1, \quad e = 2, \quad f = g, \quad d + 1 = f + i, \quad 2d = h + i.$$

Next we use the equation $n_{\beta}\varepsilon_{3\alpha+\beta}(x)n_{\beta}^{-1} = \varepsilon_{3\alpha+2\beta}(x)$, which becomes:

$$(25) \qquad \varepsilon_{\beta}(1)\varepsilon_{3\alpha+\beta}(x)\varepsilon_{\beta}(-1) = \varepsilon_{-\beta}(1)\varepsilon_{3\alpha+2\beta}(x)\varepsilon_{-\beta}(-1).$$

Transform the left side by (10) and the right side by (13) to obtain:

$$(26) \qquad \varepsilon_{3\alpha+\beta}(x)\varepsilon_{3\alpha+2\beta}(-ix) = \varepsilon_{3\alpha+2\beta}(x)\varepsilon_{3\alpha+\beta}(-ix),$$

whence $i = -1$. Finally, use the equation $n_{\alpha}\varepsilon_{\alpha+\beta}(y)n_{\alpha}^{-1} = \varepsilon_{2\alpha+\beta}(y)$, in the form:

$$(27) \qquad \varepsilon_{\alpha}(1)\varepsilon_{\alpha+\beta}(y)\varepsilon_{\alpha}(-1) = \varepsilon_{-\alpha}(1)\varepsilon_{\alpha}(-1)\varepsilon_{2\alpha+\beta}(y)\varepsilon_{\alpha}(1)\varepsilon_{-\alpha}(-1).$$

Transform the left side by (8) and the right side by (9) to obtain:

$$(28) \qquad \varepsilon_{\alpha+\beta}(y)\varepsilon_{2\alpha+\beta}(-ey)\varepsilon_{3\alpha+\beta}(fy)\varepsilon_{3\alpha+2\beta}(-gy^2)$$
$$= \varepsilon_{-\alpha}(1)\varepsilon_{2\alpha+\beta}(y)\varepsilon_{3\alpha+\beta}(hy)\varepsilon_{-\alpha}(-1).$$

Moving $\varepsilon_{-\alpha}(-1)$ to the left, past $\varepsilon_{3\alpha+\beta}(hy)$ and then past $\varepsilon_{2\alpha+\beta}(y)$, introduces no further $\varepsilon_{3\alpha+\beta}$ term in the right side of (28). Accordingly, the right side of (28) has the form:

$$(29) \qquad \varepsilon_{\alpha+\beta}(\ )\varepsilon_{2\alpha+\beta}(\ )\varepsilon_{\beta}(\ )\varepsilon_{3\alpha+\beta}(hy)\varepsilon_{3\alpha+2\beta}(\ ),$$

the missing parameters being of no consequence. Comparison with the left side of (28) shows that $f = h$. Combined with (24) and the fact that $i = -1$, this shows that

$$(30) \qquad a = b = c = d = 1, \quad e = 2, \quad f = g = h = 3, \quad i = -1,$$

thus proving the first four formulas in (b). The fifth formula in (b) follows from (12), since $h = 3$.

It remains to prove (a). We proceed as before:

$$(31) \qquad n_{\alpha}n_{\beta}n_{\alpha} = n_{\alpha}n_{\beta}n_{\alpha}^{-1}t_{\alpha} = n_{3\alpha+\beta}^{-1}t_{\alpha}.$$
$$(32) \qquad n_{\beta}(n_{\alpha}n_{\beta})^2 = n_{\beta}n_{3\alpha+\beta}^{-1}n_{\beta}^{-1}(n_{\beta}t_{\alpha}n_{\beta}^{-1})t_{\beta} = n_{3\alpha+2\beta}^{-1}t_{\alpha}t_{\beta}t_{\beta} = n_{3\alpha+2\beta}^{-1}t_{\alpha}.$$
$$(33) \qquad n_{\alpha}(n_{\beta}n_{\alpha})^3 = n_{\alpha}n_{3\alpha+2\beta}^{-1}n_{\alpha}^{-1} = n_{3\alpha+2\beta}^{-1}.$$

(34) $n_\beta(n_\alpha n_\beta)^4 = n_\beta n_{3\alpha+2\beta}^{-1} n_\beta^{-1} t_\beta = n_{3\alpha+\beta} t_\beta.$

(35) $n_\alpha(n_\beta n_\alpha)^5 = n_\alpha n_{3\alpha+\beta} n_\alpha^{-1}(n_\alpha t_\beta n_\alpha^{-1}) t_\beta = n_\beta t_\beta t_\alpha t_\alpha = n_\beta^{-1}.$

Finally, $(n_\alpha n_\beta)^6 = (n_\beta n_\alpha)^6 = e.$ $\square$

33.6. The Existence Problem

Together with the discussion in (33.1), Propositions 33.2–33.5 dispose of assumptions (A), (B), (C), thereby completing the proof of Theorem 32.1.

It is natural to ask whether semisimple groups of all possible types exist. The answer is yes, but we shall not attempt to give a complete proof, only an outline of the methods involved. First consider the simple types A – G.

Some groups of type A_ℓ, B_ℓ, C_ℓ, D_ℓ are already known (cf. Exercise 27.1). It is not too hard to verify that $SL(\ell+1, K)$ and $Sp(2\ell, K)$ are simply connected, of respective types A_ℓ, C_ℓ. In turn, the image of $SL(\ell+1, K)$ (resp. $Sp(2\ell, K)$) in $PGL(\ell+1, K)$ (resp. $PGL(2\ell, K)$) is an adjoint group of the same type, denoted $PSL(\ell+1, K)$ (resp. $PSp(2\ell, K)$). (Since K is algebraically closed, $PSL(\ell+1, K)$ in fact equals $PGL(\ell+1, K)$, but over arbitrary fields these groups can differ.) For the root system C_ℓ, $[\Lambda:\Lambda_r] = 2$, so there are no further groups to be found. For the root system A_ℓ, $[\Lambda:\Lambda_r] = \ell+1$, and each divisor of $\ell+1$ corresponds to a possible intermediate group $X(T)$. In case K has $\ell+1$ distinct $\ell+1^{st}$ roots of unity (i.e., in case $p \nmid \ell+1$, $p = $ char K), the corresponding scalar matrices form the center of $SL(\ell+1, K)$, and one gets distinct groups of all desired types by factoring out the various subgroups of this center. This approach is not quite adequate if $p|\ell+1$, since then we get too few distinct groups. The problem is that isomorphic abstract groups need not be isomorphic as algebraic groups. To make a systematic construction of the groups of type A_ℓ, it is probably best to utilize exterior powers of the natural $\ell+1$ dimensional representation of $SL(\ell+1, K)$, cf. Chevalley [8], exposé 20.

Turning to types B_ℓ, D_ℓ, one can show that $SO(2\ell+1, K) = PSO(2\ell+1, K)$ is an adjoint group of type B_ℓ, while $SO(2\ell, K)$ is of type D_ℓ and has fundamental group of order 2, the corresponding adjoint group being $PSO(2\ell, K)$. The only missing groups are the simply connected ones, which turn out to be the "spin" groups associated with certain Clifford algebras.

A group of type G_2 (resp. F_4) can also be constructed explicitly, as the automorphism group of the 8-dimensional Cayley algebra (resp. the exceptional 27-dimensional Jordan algebra). In these cases there is only one distinct group of each type, since the roots span the full lattice of abstract weights. Groups of type E_6, E_7, E_8 can also be treated in similar (but more elaborate) ways.

Chevalley [7] gave a uniform construction of the groups of adjoint type (which also yields related groups over arbitrary fields). First he showed how to select an especially good basis for the semisimple Lie algebra over $\mathbb{C}$ having root system Φ; in particular, all structure constants are integral. (This of course presupposes the existence theorem for such Lie algebras.) Next he introduced a group of Lie algebra automorphisms, generated by those of the type "exp ad x_α", which leaves stable the $\mathbb{Z}$-span of the Chevalley basis. This group can be viewed as a matrix group "over $\mathbb{Z}$", whose entries may then be specialized to elements of an arbitrary field.

The construction of **Chevalley groups** is also described in Carter [1], Humphreys [6], Steinberg [10]. To show that the groups so constructed over K are simple algebraic groups of respective types A − G, one has to develop structural information leading to the Bruhat decomposition. In effect, one has to show that the groups are simple as abstract groups, cf. Theorem 29.5. The problem here is mainly group-theoretic, since the Chevalley groups over K are by construction connected algebraic groups, being generated by 1-dimensional unipotent subgroups.

It still remains to pass from the existence of *adjoint* groups to a general existence theorem. In Chevalley [8], exposé 23, there is a construction of a simply connected group of any given type, provided some group of that type already exists. The construction is based on the existence of "projective representations", which we have not discussed here. From this the existence of semisimple groups having all possible fundamental groups can be established by arguments like those stated above.

There is another method, based on the use of linear representations, which also leads to a construction of all possible types. By imitating Chevalley's original construction, with the adjoint representation of the Lie algebra replaced by other representations, one can construct matrix groups over K for which $X(T)$ occupies any desired position between Λ and Λ_r. This method is described in Humphreys [6], Steinberg [10]. It utilizes Kostant's $\mathbb{Z}$-form of the universal enveloping algebra.

Exercises

1. Assuming that simple algebraic groups of all possible types exist, discuss how to prove that semisimple groups of all types exist.
2. Formulate explicit generators and relations for a semisimple group G (viewed as an abstract group).

Notes

The calculations in (33.3)–(33.5) are reproduced from SGAD, exposé XXIII. The method of Chevalley [8] is entirely different: he gives an explicit classification of the groups of rank 2 via their projective and linear representations.

Chapter XII

Survey of Rationality Properties

This chapter summarizes, with examples but without proof, many of the known properties of algebraic groups relative to fields of definition. As before, K denotes an algebraically closed field. If k is an arbitrary subfield of K, we are interested in the closed subgroups of $GL(n, K)$ which are defined by polynomials with coefficients in k. Some general theory is developed in §34, to be followed by a discussion of special fields (finite, real, p-adic) in §35.

34. Fields of Definition

34.1. Foundations

A closed set X in $\mathbf{A}^n$ is said to be **k-closed** if X is the set of zeros of some collection of polynomials having coefficients in k. It is easy to check that the k-closed sets are the closed sets of a topology on $\mathbf{A}^n$, called the **Zariski k-topology**. So far, so good. But there is a subtle difficulty with this notion. To say that $X = \mathscr{V}(I)$ for *some* ideal I generated by its intersection with $k[T_1, \ldots, T_n]$ is not to say that $\mathscr{I}(X)$ has this property. (Recall from (1.1) that $\mathscr{I}(X)$ is the radical of I, which may be larger than I.) In case $\mathscr{I}(X)$ does happen to be generated by k-polynomials, we say that X is **defined over** k.

The property of being defined over k turns out to be more valuable than the weaker property of being k-closed. It is also more difficult (as a rule) to verify. For example, the matrices in $M(n, K)$ commuting with a given matrix in $M(n, k)$ obviously form a k-closed set, but the full ideal vanishing on this set is not readily described, so that its generation by k-polynomials remains uncertain. The problem here is brought into focus by the following result:

Lemma. *If X is k-closed in $\mathbf{A}^n$, then X is defined over a finite, purely inseparable extension of* k.

When k is *perfect* (e.g., of characteristic 0), the two notions therefore coincide. Quite a lot of the theory to be summarized below becomes appreciably simpler in the perfect case; then one can frequently verify that a group or other variety is defined over k by showing that it is stable under the galois group of K/k. However, most of the interesting results remain valid for arbitrary k, so it is rather unsatisfying to limit oneself to perfect fields. And imperfect fields do arise naturally in the arithmetic applications of algebraic groups.

The definitions given above are adequate so long as one deals only with closed subsets of affine spaces. To say what it means for an arbitrary variety to be defined over k, one has first to make this notion more intrinsic in the affine case, i.e., one has to look directly at the affine algebra $\mathsf{K}[X]$. When X is defined over k, $\mathscr{I}(X)$ is generated by $\mathscr{I}_k(X) = \mathscr{I}(X) \cap \mathsf{k}[\mathsf{T}_1,\ldots,\mathsf{T}_n]$, so we obtain a k-algebra $\mathsf{k}[X] = \mathsf{k}[\mathsf{T}]/\mathscr{I}_k(X)$. Then $\mathsf{K}[X] = \mathsf{K} \otimes_k \mathsf{k}[X]$. This k-structure on $\mathsf{K}[X]$ is the intrinsic feature to emphasize. With its aid one can then formulate the general idea of "variety defined over k", via affine open coverings. We shall not enter into the details, but shall assume that the notion is well founded.

If X is a closed subset of $\mathbf{A}^n$ defined over k, we let $X(\mathsf{k}) = X \cap \mathsf{k}^n$ be its set (possibly empty) of **k-rational points**. This subset of X is in fact intrinsically defined and can be introduced for any variety defined over k. In the affine case, when $X \subset \mathbf{A}^n$ and $Y \subset \mathbf{A}^m$ are defined over k, we say that a morphism $\varphi: X \to Y$ is **defined over** k (or is a **k-morphism**) if the coordinate functions all lie in $\mathsf{k}[\mathsf{T}_1,\ldots,\mathsf{T}_n]$. This notion too can be made intrinsic and then generalized to arbitrary varieties.

Once the foundations are well established, it is fairly easy to fill in the "relative" aspects of the earlier chapters (aspects relative to fields of definitions). However, it would have been somewhat distracting to do this from the outset, since the rationality proofs are usually unrelated to the group-theoretic arguments.

34.2. Review of Earlier Chapters

Here we shall indicate, section by section, some of the rationality assertions which can be established in the framework of Chapters II–IV, VI.

(7.1) Let G be an algebraic group. If G, along with $\mu: G \times G \to G$ and $\iota: G \to G$, are defined over k, we say that G is **defined over** k, or is a **k-group**. Examples: $\mathbf{G}_a$, $\mathbf{G}_m$, $\mathrm{GL}(n, \mathsf{K})$, $\mathrm{SL}(n, \mathsf{K})$, $\mathrm{T}(n, \mathsf{K})$ (etc.) are defined over the prime field of K. A direct product of k-groups is a k-group. If G is a k-group, the set $G(\mathsf{k})$ of its k-rational points is a subgroup of G. For example, $\mathrm{GL}(n, \mathsf{K})(\mathsf{k}) = \mathrm{GL}(n, \mathsf{k})$.

(7.3) If G is a k-group, so is G°.

(7.4) Let $\varphi: G \to G'$ be a k-morphism of k-groups. Then the image of φ is a k-subgroup of G' (but the kernel need not be defined over k).

(7.5) In the proposition, let G be a k-group, and let the f_i be k-morphisms from varieties X_i defined over k. Then $\mathscr{A}(M)$ is defined over k. In the corollary, G is therefore defined over k if all Y_i are.

(8.2) Given $\varphi: G \times X \to X$, say that G **acts k-morphically** on X if φ, G, X are all defined over k. In the proposition, let G act k-morphically on X. If $Y \subset X(\mathsf{k})$ and Z is k-closed, then $\mathrm{Tran}_G(Y, Z)$ is k-closed, as is $C_G(Y)$. Also, X^G is k-closed.

(8.3) When a k-group G acts morphically on X, its orbits are varieties defined over k, and the orbit maps are k-morphisms.

(8.6) In the proposition, let G be a k-group acting k-morphically on X. Given F, the subspace E in (a) can be chosen to be defined over k (relative to the k-structure on $K[X]$). In the theorem, if G is a k-group, then there exists a k-isomorphism of G onto a k-subgroup of $GL(n, K)$.

(9.1) If G is a k-group, its Lie algebra has a natural k-structure, with $\mathfrak{g} \cong K \otimes_k \mathfrak{g}(k)$.

(10.3) If G is a k-group, Ad is a k-morphism.

(11.2) In the theorem, let G and H be defined over k. Then V may be chosen with a k-structure so that L is defined over k and so that φ is a k-morphism.

(11.5) In the theorem, let G and N be defined over k. Then ψ may be taken to be a k-morphism, $GL(W)$ being given the structure of k-group.

(12.1) When G, H are k-groups, the variety Y and the morphism $\pi: G \to Y$ can be taken to be defined over k.

(15.1) In Lemma B, let $GL(V)$ be a k-group, where V and hence End V has been given a k-structure. If $x \in GL(V)(k)$ (or $GL(n, k)$ in matrix form), and k is perfect, then x_s, x_u are also k-rational; but in general a purely inseparable extension of k is required.

(15.3) If G is a k-group and $x \in G(k)$, then x_s, $x_u \in G(k)$ provided k is perfect. A similar statement holds for elements of $\mathfrak{g}(k)$. The set G_u is k-closed if G is a k-group.

34.3. Tori

A torus T defined over k is referred to simply as a **k-torus**. Let $X(T)_k$ denote the subgroup of $X(T)$ consisting of characters $T \to \mathbf{G}_m$ which are k-morphisms. We call T k-**split** in case $X(T)_k$ spans $k[T]$. Equivalently, T is k-isomorphic to $\mathbf{G}_m \times \cdots \times \mathbf{G}_m$ (d copies, $d = \dim T$); then $T(k) \cong k^* \times \cdots \times k^*$. At the opposite extreme, T is called k-**anisotropic** in case $X(T)_k = 0$.

Theorem. *Let T be a k-torus.*

(a) *There exists a finite galois extension of* k *over which T becomes split.*

(b) *There exist unique subtori T', T'' of T defined over* k, *such that:* $T = T'T''$, *T' is k-split, T'' is k-anisotropic. Here T' is the largest k-split subtorus of T, T'' the largest k-anisotropic subtorus. T'' is the identity component in the intersection of kernels of the characters of T defined over* k.

What do k-anisotropic tori look like? When $k = R$ the answer will be given in (35.3).

34.4. Some Basic Theorems

The main theme of this book has been the structure of reductive groups. In order to "relativize" the results on root systems, Bruhat decomposition, etc., we have to see which of the subgroups of a reductive group G defined over k are themselves defined over k. Let us call G k-**split** if G has a maximal torus T which is k-split (34.3) and if in addition the associated admissible isomorphisms $\varepsilon_\alpha: G_a \to U_\alpha$ can be taken to be k-isomorphisms (U_α being required to be defined over k). In this case the entire structure theory of G adapts quite satisfactorily to the group $G(k)$. For example, $GL(n, K)$ is split over any subfield of K, so that $GL(n, k)$ has a "Bruhat decomposition" of its own. Call G k-**anisotropic** if it has no k-split tori of positive dimension.

The first deep results of the relative theory are summarized in the following theorem.

Theorem.   *Let G be a connected k-group. Then:*

(a) G has a maximal torus defined over k.

(b) If G is reductive, G is k-split if and only if some maximal torus of G is k-split. In particular (cf. (34.3)), G splits over a finite galois extension of k.

(c) If G is reductive, S a k-torus in G, then $C_G(S)$ is reductive and defined over k. Moreover, S is contained in some maximal torus defined over k.

(d) If G is reductive and k is infinite, $G(k)$ is Zariski-dense in G. (This is true for any G if k is perfect, but can fail otherwise.)

These results, which take quite a bit of labor to prove, suggest already that tori and their centralizers will play a large role in the structure theory of a reductive group over k. By contrast, G need not possess any Borel subgroups defined over k (cf. (35.1)).

34.5. Borel-Tits Structure Theory

G denotes a reductive group defined over k. Choose a k-split torus S in G of largest possible dimension (which may be 0). Its dimension is an invariant of G (thanks to the following theorem), called the k-**rank**. The k-rank is 0 if and only if G is k-anisotropic. Define the k-**roots** $_k\Phi$ of G to be the nonzero weights of Ad S. (In case S is a maximal torus, these are just the usual roots.) Let $Z = C_G(S)$, $N = N_G(S)$; the finite group $_kW = N/Z$ is called the k-**Weyl group**.

Theorem A.   (a) *The maximal k-split tori of G are all conjugate under $G(k)$.*

(b) *Z is a reductive k-group and its derived group is k-anisotropic.*

(c) *$_k\Phi$ is a (possibly nonreduced) abstract root system in a suitable euclidean space, whose Weyl group is isomorphic to $_kW$.*

It has to be explained that by removing axiom (R2) in (A.1) we obtain a slightly more general notion of abstract root system in which $\alpha, 2\alpha$ can both be roots. When this happens, the root system is called **nonreduced**. But the only irreducible system of this type is the union of types B_ℓ and C_ℓ, long roots in B_ℓ being identified with short roots in C_ℓ. (This is called the system of type BC_ℓ.) So no "new" Weyl groups occur.

The proof of Theorem A is closely tied up with the study of parabolic subgroups of G defined over k, among which the Borel subgroups may or may not occur.

Theorem B. *Let P be minimal among the parabolic k-subgroups of G.*
(a) All other minimal k-parabolics are conjugate to P under G(k).
(b) P contains a maximal k-split torus S, and $Z = C_G(S)$ is a Levi factor (30.2) of P. Moreover, $R_u(P)$ is defined over k, and the k-roots occurring in $R_u(P)$ form a positive subsystem of $_k\Phi$.

This theorem shows that minimal k-parabolic subgroups play the role of Borel subgroups in the relative theory, while maximal k-split tori play the role of maximal tori.

Theorem C. *Let P be a minimal k-parabolic subgroup of G, $S \subset P$ a maximal k-split torus of G, $U = R_u(P)$, $N = N_G(S)$, $Z = C_G(S)$. Then:*
(a) $N = N$(k) $\cdot Z$, so G(k) contains a full set of representatives of $_kW$.
(b) G(k) is the disjoint union over $_kW$ of double cosets U(k)N(k)U(k) (**relative Bruhat decomposition**). *Indeed, G(k) has a Tits system (29.1) determined by P(k) and N(k) along with canonical generators of $_kW$.*

There is also a precise normal form in G(k), analogous to (28.4). But it should be pointed out that "root groups" are now no longer 1-dimensional (or even commutative) in general, so the internal structure of U(k) is not completely transparent. It should also be pointed out that the preceding theorems tell us nothing at all about the structure of G(k) when G is k-anisotropic. Examples show that this can vary considerably, depending on the nature of k. However, it is known when k is perfect that G(k) must consist of semisimple elements.

If G is not k-anisotropic, let G^+ be the subgroup of G(k) generated by all U(k), where U runs over the unipotent radicals of the minimal k-parabolic subgroups of G. When G is semisimple and simply connected, k arbitrary, Kneser and Tits have conjectured that G(k) $= G^+$. This is known to be true in many cases, e.g., when G is k-split.

34.6. An Example: Orthogonal Groups

Some of the preceding theorems can be illustrated effectively by means of an example. Let char $K \neq 2$, and let $G = SO(q)$, the special orthogonal group of a nondegenerate quadratic form q over k. If the form is defined on an n-dimensional vector space, and if its "Witt index" (= dimension of a maximal totally isotropic subspace) is r, the matrix of the associated bilinear form relative to a suitable basis becomes:

$$\begin{bmatrix} 0 & 0 & J \\ 0 & Q_0 & 0 \\ J & 0 & 0 \end{bmatrix}, \quad J = \begin{bmatrix} 0 & & .1 \\ & .1^{\cdot} & \\ 1^{\cdot} & & 0 \end{bmatrix}.$$

Here Q_0 is the matrix of an "anisotropic" quadratic form q_0 on a space of dimension $n - 2r$ (a form not taking the value 0 except at 0).

Then a maximal k-split torus S of G consists of block diagonal matrices diag (A, B, C), where $A = $ diag $(a_1, \ldots, a_r)$, $C = $ diag $(a_r^{-1}, \ldots, a_1^{-1})$, and B is the identity matrix. In turn, the group $Z = C_G(S)$ is the direct product of S and $\mathrm{SO}(q_0)$, the latter group being k-anisotropic. For a minimal k-parabolic subgroup P containing S we can take the group of matrices:

$$\begin{bmatrix} P_1 & P_2 & P_3 \\ 0 & P_4 & P_5 \\ 0 & 0 & P_6 \end{bmatrix},$$

where P_1, P_6 are $r \times r$ upper triangular matrices, $\mathrm{P_4} \in \mathrm{SO}(q_0)$, and det $P_1 \cdot $ det $P_6 = 1$. (P_6 depends on P_1 and P_3, while P_5 depends on P_1, P_2, P_4.) The unipotent radical U of P then consists of matrices (with U_1, U_6 unipotent):

$$\begin{bmatrix} U_1 & U_2 & U_3 \\ 0 & I & U_5 \\ 0 & 0 & U_6 \end{bmatrix},$$

subject to similar dependence requirements.

Closer study leads to an explicit description of the corresponding positive k-roots; e.g., send a typical matrix in S (as described above) to $a_i a_{i+1}^{-1}$ ($1 \leqslant i < r$). In case $n - 2r = 0$ or 1, G is k-split (of respective type D_r or B_r). In case $n - 2r = 2$, Z is actually a maximal torus (defined over k but not k-split), while P is a Borel subgroup. This is the so-called "quasisplit" case, to be discussed in (35.1).

Notes

General foundations for the theory of affine algebraic groups over arbitrary fields are laid in Rosenlicht [1] [2] [4] [6], Weil [2] [3] (cf. also Borel [1]), using an older language of algebraic geometry. The scheme approach is used in SGAD and in Demazure, Gabriel [1]. Borel [4, AG] attempts to provide a foundation adequate for the theory outlined in this section, without the full generality of schemes. For algebraic groups over perfect fields, cf. Godement [1], Satake [1] [5]. The theorems in (34.4), (34.5) are proved in Borel [4, Chapter V], Borel, Tits [1]; cf. also Borel, Tits [2] [3] [4], Borel, Springer [1] (and its predecessor in AGDS). Borel [3] treats the example in (34.6) in more detail.

35. Special Cases

In this section G denotes a semisimple group defined over k. We shall give some indications as to how the structure of $G(k)$ varies when special choices are made for k, then conclude by discussing the classification problem.

35.1. Split and Quasisplit Groups

According to (34.4), G is k-split if and only if some maximal torus T of G is k-split. Then the structure theory of (34.5) reduces to that of Chapter X, with $_k\Phi = \Phi$, $_kW = W$, etc. In turn, Chapter XI provides a classification of all such groups (modulo the existence theorem). The construction in Chevalley [7] may be viewed as providing for arbitrary k a concrete model of $G(k)$, when G is k-split of adjoint type. However, in Chevalley's framework the emphasis is placed not on $G(k)$ but on what is usually a proper subgroup, generated by unipotent elements, because the latter group is almost always simple. For example, in type A_ℓ, $G(k) = PGL(\ell + 1, k)$, while the "Chevalley group" of this type is $PSL(\ell + 1, k)$. Nowadays it is common to mean by **Chevalley group** any of the groups canonically associated with a k-split group G (not necessarily of adjoint type): the group $G(k)$, its derived group, the quotient of either by its center, etc.

Call G **k-quasisplit** if G has a Borel subgroup defined over k. The example in (34.6) indicates that this can happen even when G fails to be k-split. We can describe more precisely what is known; say G is of simple type and k-quasisplit but not k-split. Then a maximal k-split torus S has as centralizer a larger k-torus T, which is maximal in G and lies in a Borel k-subgroup B. Certain roots relative to T become "dependent" when restricted to S, the k-rank being smaller than the absolute rank. It is known that this can happen only when Φ admits a nontrivial graph automorphism (A.8), therefore only when Φ is of type A_ℓ ($\ell \geqslant 2$), D_ℓ, E_6.

Consider A_3, for example, with simple roots α_1, α_2, α_3 and remaining positive roots $\alpha_1 + \alpha_2, \alpha_2 + \alpha_3, \alpha_1 + \alpha_2 + \alpha_3$. On restriction to S, α_1 and α_3 become identical (call the common value β_1), while α_2 becomes a character β_2. It turns out that β_1 and β_2 form a base for $_k\Phi$, the other positive roots being $\beta_1 + \beta_2$ (occurring with multiplicity 2) and $2\beta_1 + \beta_2$. Thus $_k\Phi$ is of type C_2. In this case $G(k)$ can be described as a type of special unitary group.

Given a field k and a root system of type A_ℓ ($\ell \geqslant 2$), D_ℓ, or E_6, there does exist a k-quasisplit (but not k-split) group of this type. The relative root system $_k\Phi$ is of type C_r (resp. BC_r) when Φ is of type A_{2r-1} (resp. A_{2r}), of type F_4 when Φ is of type E_6, of type B_r when Φ is of type D_{r+1} (with yet another possibility when Φ is of type D_4, namely G_2). For example, the six simple roots of E_6 lead to the four simple roots of F_4 as indicated by loops around the nodes in the Dynkin diagram:

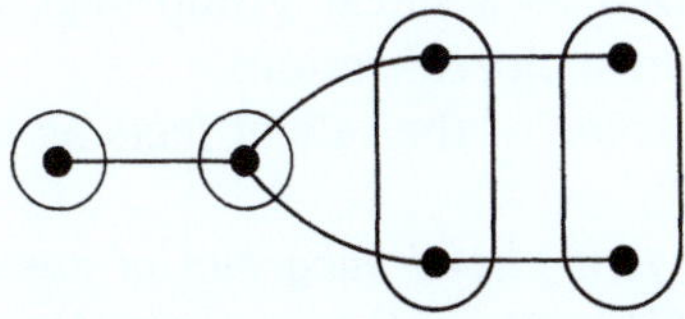

Over a finite field k these quasisplit groups yield new families of finite simple groups (often called "twisted" to distinguish them from the original

Chevalley groups). This was first appreciated by Hertzig, Steinberg, and Tits, who approached these groups independently from several directions. When char $k = 2, 3$, some further twisted groups were obtained by Ree and Suzuki from the root systems of types B_2, G_2, F_4, where graph automorphisms would exist if root lengths were ignored. Carter [1] and Steinberg [10] give rather thorough expositions of all these constructions.

35.2. Finite Fields

We have mentioned the importance of the k-split and k-quasisplit groups in the theory of finite simple groups. Indeed, the finite simple groups of "Lie type" along with the alternating groups account for all but a handful of the presently known simple groups. It is natural to ask whether other groups $G(k)$ can be exploited, apart from those already discussed. The answer is no.

Theorem. *If* k *is finite,* G *is automatically* k-*quasisplit.*

The proof rests on a theorem of Lang [1], which asserts that the morphism $G \to G$ defined by $x \mapsto x^{-1} x^{[q]}$ is *surjective*. Here $q = \text{Card } k$ and $[q]$ indicates, in case G is taken to be a matrix group, the operation of raising all matrix entries to the qth power. (This operation can be described more intrinsically. Note that it does map G into G when G is defined over k.) The criterion for a subgroup H of G to be defined over k is just that $H = H^{[q]}$. Using this, we deduce the theorem as follows. Let B be any Borel subgroup of G. Then $B^{[q]}$ is another such, so there exists $x \in G$ with $xB^{[q]}x^{-1} = B$. Using Lang's result, $x = y^{-1}y^{[q]}$ for some $y \in G$, or $yBy^{-1} = yxB^{[q]}x^{-1}y^{-1} = y^{[q]}B^{[q]}(y^{[q]})^{-1} = (yBy^{-1})^{[q]}$. So yBy^{-1} is defined over k.

This same argument yields part (a) of Theorem 34.4 when k is finite, and leads as well to some conjugacy results.

35.3. The Real Field

When $k = \mathbb{R}$, the theory of semisimple k-groups is very closely related to the theory of semisimple Lie groups. Indeed, $G(\mathbb{R})$ is a Lie group of this type, but is not necessarily connected in the real topology (or "usual topology"), by which we mean the euclidean topology on $\mathbb{R}^{n^2}$ relativized to subgroups of $GL(n, \mathbb{R})$. It is not true, however, that every semisimple Lie group can be realized as a linear group (e.g., the universal covering group of $SL(n, \mathbb{R})$ has no such realization).

We can summarize some of the salient facts as follows.

Theorem. (a) *If* G *is a closed subgroup of some* $GL(n, K)$, *then* $G(\mathbb{R})$ *is a closed subgroup of* $GL(n, \mathbb{R})$ *in the usual topology, and is therefore a Lie group. If* H *is any closed subgroup of* G, $H(\mathbb{R})$ *is similarly a Lie group.*

(b) *The connected component of the identity of* $G(\mathbb{R})$ *in the usual topology has finite index in* $G(\mathbb{R})$.

(c) *When G is of simply connected type, $G(\mathrm{R})$ is connected in the usual topology (but need not be simply connected in the topological sense).*

(d) *If G' is reductive and R-anisotropic in the sense of (34.4), then $G'(\mathrm{R})$ is connected and compact in the usual topology.*

(e) *If H is a compact subgroup of $\mathrm{GL}(n, \mathrm{R})$, then $H = G'(\mathrm{R})$ for some reductive R-subgroup G' of $\mathrm{GL}(n, \mathsf{K})$, i.e., all compact linear Lie groups are "algebraic".*

It is worth remarking also the behavior of tori defined over R. At one extreme, if T is an R-split torus of dimension d, then $T(\mathrm{R}) \cong \mathrm{R}^* \times \cdots \times \mathrm{R}^*$ (in particular, the identity component of $T(\mathrm{R})$ in the usual topology has index 2^d). At the other extreme, if T is R-anisotropic, then $T(\mathrm{R})$ is compact and connected, according to part (d) of the theorem. When dim $T = 1$, it can be shown that T is R-isomorphic to $\mathrm{SO}(2, \mathsf{K})$, hence that $T(\mathrm{R}) = \mathrm{SO}(2, \mathrm{R})$, which is topologically just a 1-torus. In higher dimensions, $T(\mathrm{R})$ is then a d-torus. This helps to explain the terminology introduced in (16.2).

35.4. Local Fields

When k is the completion of some field relative to a discrete nonarchimedean valuation, we may refer to it as a "local field". (Terminology varies quite a bit; for example, sometimes one also refers to R and C as local fields.) An elaborate theory of simple algebraic groups over local fields has been developed in recent years, beginning with Iwahori, Matsumoto [1] and culminating in the work of Bruhat, Tits [1]–[6]. Since this theory is rather formidable, we shall try to introduce the reader to it by means of the simplest example, SL_2.

Let v be the given valuation on k, R the corresponding ring of integers, $P = \pi R$ its unique maximal ideal, $\bar{\mathsf{k}} = R/P$ the residue class field. For example, k might be $\mathbb{Q}_p$, the p-adic completion of $\mathbb{Q}$ (p some rational prime); then R is $\mathbb{Z}_p$, the ring of p-adic integers, π may be taken to be p, and $\bar{\mathsf{k}}$ is the field of p elements.

Let $\mathscr{G} = \mathrm{SL}(2, \mathsf{k})$. As a simple algebraic group, split over the prime field of K, $\mathrm{SL}(2, \mathsf{K})$ has a Tits system (§29) which adapts readily to each subfield of K. In particular, G already has a Tits system of rank 1, with Weyl group of order 2, etc. But G also has another Tits system (with infinite Weyl group) which reflects more strikingly the special properties of the field k. Let us introduce the various constituents $\mathscr{B}, \mathscr{N}, \mathscr{S}, \mathscr{W}$ of this system.

For $\mathscr{N}$ we take the "usual" one, the group of all monomial matrices in $\mathscr{G}$. But for $\mathscr{B}$ we take the group of all matrices $\begin{pmatrix} a & b \\ c & d \end{pmatrix}$ for which $a, d \in R - P$, $b \in R, c \in P$. Notice that $\mathscr{B}$ is the inverse of the upper triangular group in $\mathrm{SL}(2, \bar{\mathsf{k}})$ under the natural homomorphism $\mathrm{SL}(2, R) \to \mathrm{SL}(2, \bar{\mathsf{k}})$. Now $\mathscr{T} = \mathscr{B} \cap \mathscr{N}$ just consists of the matrices diag (a, a^{-1}), $a \in R - P$, and is readily seen to be normal in $\mathscr{N}$. The quotient group $\mathscr{W}$ is generated by a set $\mathscr{S}$ of two elements s_0, s_1, each of order 2, where s_0 is the coset of $\begin{pmatrix} 0 & -1 \\ 1 & 0 \end{pmatrix}$ and s_1 is the coset of $\begin{pmatrix} 0 & \pi^{-1} \\ -\pi & 0 \end{pmatrix}$. It is not hard to verify that $s_0^2 = e = s_1^2$ gives a complete set of defining relations for $\mathscr{W}$, so $\mathscr{W}$ is the infinite dihedral group.

Theorem. $(\mathscr{G}, \mathscr{B}, \mathscr{N}, \mathscr{S})$ *is a Tits system in* $\mathscr{G}$.

Only axiom (T1) requires any labor. The idea of the proof is to exploit the usual Bruhat decomposition in $\mathrm{SL}(2, \bar{\mathsf{k}})$ by "lifting" it to $\mathrm{SL}(2, R)$. Some inductions on length in $\mathscr{W}$ are also involved.

Since $\mathscr{S}$ has cardinality 2, (29.3) implies that there must be four "parabolic" subgroups containing $\mathscr{B}$. Besides $\mathscr{B}$ and $\mathscr{G}$, there is one obvious candidate: $\mathscr{P}_0 = \mathrm{SL}(2, R)$. The other one is less obvious. Take $\mathscr{P}_1$ to be the group of all matrices $\begin{pmatrix} a & b \\ c & d \end{pmatrix}$ with $a, d \in R, b \in \pi^{-1}R, c \in \pi R = P$. Notice that $\mathscr{P}_0 \cap \mathscr{P}_1 = \mathscr{B}$.

When k is a locally compact field, $\mathscr{G}$ acquires from its embedding in k^4 the structure of locally compact topological group. Then $\mathscr{P}_0, \mathscr{P}_1$ are two types of maximal compact subgroup of $\mathscr{G}$, and one can proceed in this framework to do some harmonic analysis. One can also exploit the "building" (here a tree) attached to the Tits system to prove that any discrete subgroup of $\mathscr{G}$ having no elements $\neq e$ of finite order must be free.

35.5. Classification

Given an arbitrary field k, how can one hope to classify all possible semisimple k-groups? The group G is already "known" over K (or even over a finite galois extension of k, according to (34.4)). To compare the behavior over k and K, it is useful to "align" the root systems. As in (34.5), choose a maximal k-split torus S lying in a minimal k-parabolic subgroup P. Then find a maximal torus T of G containing S, contained in P, and defined over k (cf. (34.4)). If the choices are made carefully, one can arrange for the base $_\mathsf{k}\varDelta$ of $_\mathsf{k}\Phi$ determined by P to consist of all the nontrivial restrictions to S of roots relative to T in a base $\varDelta$ of Φ.

If G is k-anisotropic, one cannot hope to say much in this context (k arbitrary), so the classification proceeds modulo the determination of all anisotropic groups. (As mentioned earlier, the latter will depend heavily on the nature of k.) The structure theory of (34.5) yields in general a maximal reductive k-anisotropic subgroup M of G: take the product of the derived group of $Z = C_G(S)$ and the maximal anisotropic k-torus in the center of Z (the latter being complementary to S). We call M the **anisotropic kernel** of G; it is well determined by G (up to k-isomorphism).

The classification theorem of Tits says (roughly) that G is defined (up to k-isomorphism) by the giving of its anisotropic kernel (or just the derived group), its K-isomorphism class, and its "index". Without attempting a more precise statement, let us indicate what the index is. Certain roots $\alpha \in \varDelta$ have trivial restriction to S. If $\varDelta_0$ is the set of these, $\varDelta_0$ forms a base for the root system of the derived group of Z. The giving of Δ and Δ_0 (via the Dynkin diagram of Φ) constitutes part of the index. The remaining information involves a certain galois group action on Δ (or on the Dynkin diagram), whose orbits are shown by enclosing the corresponding nodes of the diagram in a common loop. Simple roots in the same orbit have the same restriction to S, determining therefore a single element of $_k\Delta$.

A couple of examples may be helpful. When G is k-quasisplit (35.1), $\Delta_0 = \emptyset$ and all nodes are circled; then G is k-split if and only if each galois orbit in Δ consists of a single root. (Cf. the picture of E_6 in (35.1).) For E_7, the diagram indicates that Δ_0 has three elements (not circled), while $_k\Delta$ has four elements; here $_k\Phi$ turns out to be of simple type F_4.

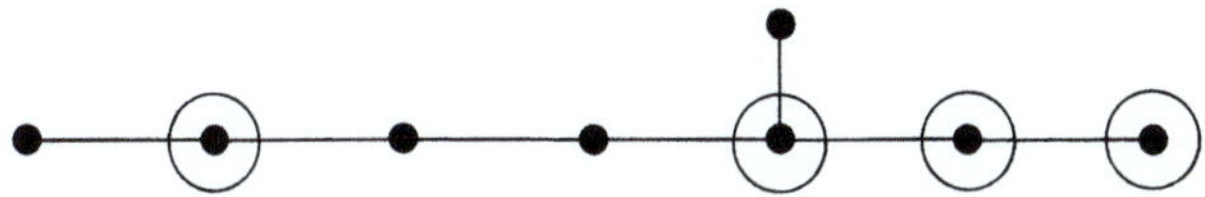

Tits has made a determination of all "admissible" diagrams of this sort, i.e., those which really do correspond to some group G over some field k. The diagram just discussed occurs, for example, when k = R or when k is a p-adic field, but of course not when k is finite (35.2).

Notes

Quasisplit groups over finite fields are treated in Hertzig [2], Steinberg [2] [6], Tits [2] [6]. Theorem 35.2, due to Lang [1], was generalized by Steinberg [11]. For finite groups of Lie type, cf. Carter [1], Curtis [3], Steinberg [10]. Results summarized in (35.3) are scattered throughout the literature, cf. Borel [3], Borel, Tits [1, §14] [3, §4], Chevalley [2] [4] [5, VI §5, no. 2], Matsumoto [1], and the papers of Hochschild, Mostow. For simple groups over local fields, see Iwahori, Matsumoto [1], Bruhat, Tits [1]–[6]. The example SL_2 is worked out in more detail in Humphreys [4, §15]; cf. Ihara [1]. The classification problem over a perfect field is dealt with by Satake [1] [5], and in full generality by Tits in an article in AGDS (cf. Tits [1]).

Appendix. Root Systems

Here we summarize, without proof, some basic properties of root systems. For fuller details consult, e.g., Bourbaki [1, Ch. VI], Humphreys [6, Ch. III], Serre [2, Ch. V], Steinberg [10, Appendix], SGAD, exposé XXI.

(A.1) Let E be a finite dimensional vector space over $\mathbb{R}$. Define a **reflection**, relative to a nonzero vector $\alpha \in E$, to be a linear transformation which sends α to $-\alpha$ and fixes pointwise a subspace of codimension 1. Such a transformation (which might more accurately be dubbed a "prereflection") is clearly its own inverse, but is *not* uniquely determined by α. Nevertheless, if Ψ is a finite set of nonzero vectors spanning E, and if a reflection τ relative to $\alpha \in \Psi$ maps Ψ into itself, then τ is uniquely determined by α.

(A.2) An (**abstract**) **root system** in the real vector space E is a subset Ψ of E satisfying:

(R1) Ψ is finite, spans E, and does not contain 0. (The elements of Ψ are called **roots**.)
(R2) If $\alpha \in \Psi$, the only multiples of α in Ψ are $\pm\alpha$.
(R3) If $\alpha \in \Psi$, there exists a reflection τ_α relative to α which leaves Ψ stable.
(R4) If $\alpha, \beta \in \Psi$, then $\tau_\alpha(\beta) - \beta$ is an integral multiple of α.

If Ψ' is an abstract root system in E', then Ψ' is said to be **isomorphic** to Ψ if there exists an isomorphism of vector spaces from E' onto E which maps Ψ' to Ψ and preserves the integers which occur in (R4).

Thanks to (A.1), τ_α in (R3) is uniquely determined by α, so (R4) is unambiguous. Call $\ell = \dim E$ the **rank** of Ψ.

(A.3) Let Ψ be a root system in E. Since Ψ spans E, and is finite, the group $W(\Psi) \subset GL(E)$ generated by the τ_α is also finite; it is called the **Weyl group** of Ψ and abbreviated as W. There is an inner product (α, β) on E relative to which W consists of orthogonal transformations. The formula for τ_α becomes: $\tau_\alpha(\beta) = \beta - <\beta, \alpha> \alpha$, where $<\beta, \alpha> = 2(\beta, \alpha)/(\alpha, \alpha)$.

(A.4) A subset Δ of Ψ is called a **base** if $\Delta = \{\alpha_1, \ldots, \alpha_\ell\}$ is a basis of E, relative to which each $\alpha \in \Psi$ has a (unique) expression $\alpha = \sum c_i \alpha_i$, where the c_i are integers of like sign. Bases exist; W permutes the collection of bases simply transitively, and every root lies in at least one base. Bases correspond 1–1 to **Weyl chambers** in E, which are the connected components in the complement of the union of hyperplanes orthogonal to roots. Elements of a base Δ are called **simple roots**. The roots which are nonnegative (resp. nonpositive) combinations of Δ comprise the set Ψ^+ (resp. Ψ^-) of **positive** (resp. **negative**) roots. If α, β are linearly independent, then there exist γ, $\delta \in \Delta$ and $\sigma \in W$ such that $\sigma(\alpha) = \gamma$, while $\sigma(\beta)$ is a $\mathbb{Z}^+$-linear combination of γ, δ. If $\alpha, \beta \in \Psi^+$ but $\alpha + \beta \notin \Psi$, then $r\alpha + s\beta \notin \Psi$ ($r, s \geq 1$).

(A.5) Let Δ be a base of Ψ. Then W is generated by $\{\tau_\alpha | \alpha \in \Delta\}$. The **length** $\ell(\tau)$ of $\tau \in W$ (relative to Δ) is defined to be the smallest t for which

$\tau = \tau_1 \cdots \tau_t$ (τ_i reflection relative to some simple root). Then $\ell(\tau)$ equals the number of positive roots α for which $\tau(\alpha)$ is negative. In particular, τ_α ($\alpha \in \Delta$) permutes $\Psi^+ - \{\alpha\}$.

(A.6) Let Δ be a base of Ψ, Δ' a subset of Δ. The roots which lie in the subspace E' of E spanned by Δ' form an abstract root system in E' having Δ' as a base; its Weyl group may be identified with the subgroup of W generated by all τ_α ($\alpha \in \Delta'$).

(A.7) Ψ is called **irreducible** if it (or equivalently, a base Δ) cannot be partitioned into the union of two mutually orthogonal proper subsets. Every root system is the disjoint union of (uniquely determined) irreducible root systems in suitable subspaces of E. Up to isomorphism, the irreducible root systems correspond 1–1 to the following **Dynkin diagrams**:

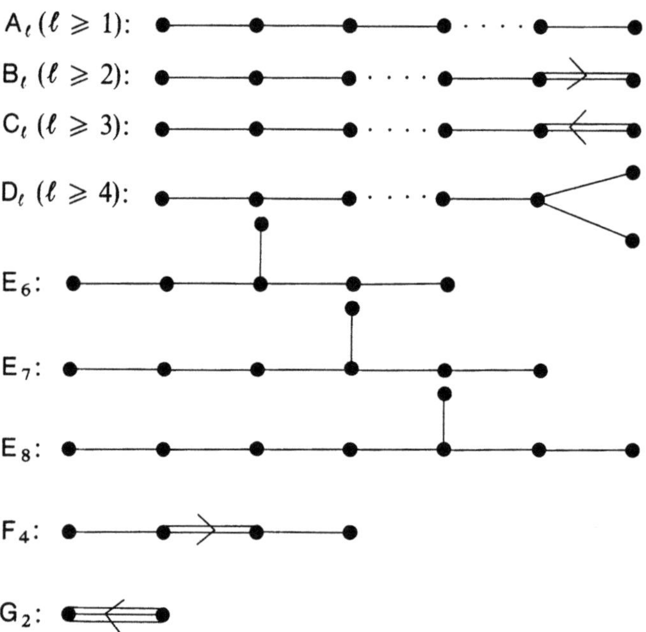

The nodes in the diagram correspond to the simple roots. Nodes belonging to α, β are joined by $\langle \alpha, \beta \rangle \langle \beta, \alpha \rangle$ bonds, with an arrow pointing to the shorter of the two roots if they are of unequal length. The order $m(\alpha, \beta)$ of $\tau_\alpha \tau_\beta$ in W is 2, 3, 4, or 6, according to whether α and β are joined by 0, 1, 2, or 3 bonds ($\alpha \neq \beta$). The giving of the Dynkin diagram is equivalent to the giving of the matrix of **Cartan integers** $\langle \alpha, \beta \rangle$ ($\alpha, \beta \in \Delta$).

In rank 2 there are just four distinct root systems, having the indicated Cartan integers:

$A_1 \times A_1$: $\begin{pmatrix} 2 & 0 \\ 0 & 2 \end{pmatrix}$ $\qquad\qquad$ B_2: $\begin{pmatrix} 2 & -2 \\ -1 & 2 \end{pmatrix}$ (α long, β short)

A_2: $\begin{pmatrix} 2 & -1 \\ -1 & 2 \end{pmatrix}$ $\qquad\qquad$ G_2: $\begin{pmatrix} 2 & -1 \\ -3 & 2 \end{pmatrix}$ (α short, β long).

(A.8) The automorphism group of Ψ is the semidirect product of W (which is normal) and the group Γ of **graph** (or **diagram**) **automorphisms**. When Ψ is irreducible, Γ is trivial except for types A_ℓ ($\ell \geqslant 2$), D_ℓ, E_6.

(A.9) A vector $\lambda \in E$ is called an **abstract weight** provided all $\langle \lambda, \alpha \rangle \in \mathbb{Z}$ ($\alpha \in \Psi$). These vectors form a lattice Λ, in which the lattice Λ_r spanned by Ψ is a subgroup of finite index. If $\Delta = \{\alpha_1, \ldots, \alpha_\ell\}$ is a base of Ψ, Λ has a corresponding basis of **fundamental dominant weights** $\{\lambda_1, \ldots, \lambda_\ell\}$ for which $\langle \lambda_i, \alpha_j \rangle = \delta_{ij}$ (Kronecker delta). The **fundamental group** Λ/Λ_r has the following structure for the irreducible types:

$$A_\ell: \mathbb{Z}/(\ell + 1)\mathbb{Z}$$
$$B_\ell, C_\ell, E_7: \mathbb{Z}/2\mathbb{Z}$$
$$D_\ell \ (\ell \text{ even}): \mathbb{Z}/2\mathbb{Z} \times \mathbb{Z}/2\mathbb{Z}$$
$$D_\ell \ (\ell \text{ odd}): \mathbb{Z}/4\mathbb{Z}$$
$$E_6: \mathbb{Z}/3\mathbb{Z}$$
$$E_8, F_4, G_2: 0$$

(A.10) Given $\Delta = \{\alpha_1, \ldots, \alpha_\ell\}$ and $\lambda_1, \ldots, \lambda_\ell$ as in (A.9), call $\lambda = \sum c_i \lambda_i$ **dominant** if all $c_i \in \mathbb{Z}^+$. Each $\lambda \in \Lambda$ is W-conjugate to one and only one dominant weight. E is partially ordered by: $\lambda > \mu$ if $\lambda - \mu$ is a sum (possibly 0) of positive roots. (This ordering depends on Δ.) If $\lambda \in \Lambda$ is dominant, then $\lambda > \sigma(\lambda)$ for all $\sigma \in W$.

(A.11) The vectors $\alpha^* = 2\alpha/(\alpha, \alpha)$ ($\alpha \in \Psi$) form a root system in E, called the **dual** of Ψ. The Cartan integer $\langle \alpha^*, \beta^* \rangle$ is equal to $\langle \beta, \alpha \rangle$. The Weyl group of Ψ^* is canonically isomorphic to W (via $\sigma_{\alpha^*} \mapsto \sigma_\alpha$). If E is identified with its dual space by means of the inner product, α^* ($\alpha \in \Delta$) becomes identified with the (unique) linear function on E taking value 2 at α and value 0 at all fundamental dominant weights λ_β (A.9) corresponding to simple roots $\beta \neq \alpha$. (The uniqueness of this linear function follows from the fact that α along with these λ_β form a basis of E.)

(A.12) The following lemma due to Tits is needed in (32.4). Since it is not so well known as the preceding results, we reproduce here the proof given in SGAD, exposé XXI, 5.6.

Lemma. *Let α, $\beta \in \Delta$ and suppose $\sigma(\alpha) = \beta$ ($\sigma \in W$). Then there exist $\gamma_0 = \alpha, \gamma_1, \ldots, \gamma_k = \beta$ in Δ and elements $\sigma_0, \sigma_1, \ldots, \sigma_{k-1} \in W$ satisfying the following conditions:*
(a) $\sigma = \sigma_{k-1} \cdots \sigma_0$.
(b) $\sigma_i(\gamma_i) = \gamma_{i+1}(0 \leqslant i \leqslant k - 1)$.
(c) Let $0 \leqslant i \leqslant k - 1$. If $\gamma_i \neq \gamma_{i+1}$, $\sigma_i \in W_{\gamma_i \gamma_{i+1}}$, while if $\gamma_i = \gamma_{i+1}$, there exists $\delta_i \in \Delta$ for which $\sigma_i \in W_{\gamma_i \delta_i}$.

Proof. Let $n(\sigma) = \text{Card}(\Psi^+ \cap -\sigma^{-1}(\Psi^+))$. According to (A.5), $n(\sigma) =$ number of positive roots γ for which $\sigma^{-1}(\gamma)$ is negative $= \ell(\sigma^{-1}) = \ell(\sigma)$. In particular, $n(\sigma) = 0$ implies $\sigma = e$, in which case there is nothing to prove. Proceed by induction on $n(\sigma)$.

If $n(\sigma) > 0$, there exists $\gamma \in \Delta$ with $\sigma(\gamma) < 0$ (otherwise $\sigma = e$). Set $\gamma_0 = \alpha$ and consider the subsystem $\Psi_{\gamma_0\gamma}$, one of whose possible sets of positive roots is $\sigma^{-1}(\Psi^+) \cap \Psi_{\gamma_0\gamma}$, call it Θ. Thanks to (A.4), (A.6), there exists $\sigma_0 \in W_{\gamma_0\gamma}$ such that $\sigma_0(\Theta) = \Psi^+_{\gamma_0\gamma}$. Set $\tau = \sigma\sigma_0^{-1}$. Now each of σ_{γ_0}, σ_γ stabilizes $\Theta' = \Psi^+ - \Psi^+_{\gamma_0\gamma}$ (A.5), so $\sigma_0 \in W_{\gamma_0\gamma}$ does likewise. It follows that $\Theta' \cap -\sigma^{-1}(\Psi^+) = \Theta' \cap -\tau^{-1}(\Psi^+)$. On the other hand, $\Psi^+_{\gamma_0\gamma} \cap -\tau^{-1}(\Psi^+) = \Psi^+_{\gamma_0\gamma} \cap -\sigma_0(\Theta) = \Psi^+_{\gamma_0\gamma} \cap -\Psi^+_{\gamma_0\gamma} = \emptyset$, while $\gamma \in \Psi^+_{\gamma_0\gamma} \cap -\sigma^{-1}(\Psi^+)$. This shows that $n(\tau) < n(\sigma)$.

Define $\gamma_1 = \sigma_0(\gamma_0)\ (= \sigma_0(\alpha))$. By assumption, $\gamma_0 \in \sigma^{-1}(\Delta)$, so γ_0 belongs to the base $\sigma^{-1}(\Delta) \cap \Psi_{\gamma_0\gamma}$ of $\Psi_{\gamma_0\gamma}$, which lies in $\Theta = \sigma_0^{-1}(\Psi^+_{\gamma_0\gamma})$. Therefore $\gamma_0 \in \sigma_0^{-1}(\Delta \cap \Psi^+_{\gamma_0\gamma}) \subset \sigma_0^{-1}(\Delta)$. So γ_1 is *simple* and lies in $\Psi^+_{\gamma_0\gamma}$. If $\gamma_1 \neq \gamma_0$, this forces $\gamma_1 = \gamma$, so $\beta = \tau(\gamma_1)$. The proof of the lemma may now be completed by induction. $\square$

Bibliography

The following list of articles and books includes the essential source material on linear algebraic groups as well as a broad sampling of related work. This sampling is somewhat arbitrary and tends to ignore those articles whose primary emphasis is on Lie groups, arithmetic groups, finite groups, etc. In spite of this limitation, the listing may help to orient the reader to the rather extensive literature of the last two decades.

The articles in the volume Algebraic Groups and Discontinuous Subgroups (abbreviated AGDS), edited by A. Borel and G. D. Mostow, are not listed individually here. They give a good indication of the state of the subject in 1965, although some are now superseded by newer developments.

One further note: If an article in Russian is not available in translation, a reference to Mathematical Reviews (MR) is added.

E. Abe, T. Kanno, [1] Some remarks on algebraic groups. *Tôhoku Math. J.* **11**, 376–384 (1959).

AGDS = A. Borel, G. D. Mostow [1]

L. Bai, C. Musili, C. S. Seshadri, [1] Cohomology of line bundles on G/B. *Ann. Sci. École Norm. Sup.* (4) 7, 89–138 (1974).

P. Bala, R. W. Carter, [1] The classification of unipotent and nilpotent elements. *Indag. Math.* **36**, 94–97 (1974).

H. Behr, [1] Zur starken Approximation in algebraischen Gruppen über globalen Körpern. *J. Reine Angew. Math.* **229**, 107–116 (1968). [2] Endliche Erzeugbarkeit arithmetischer Gruppen über Funktionenkörpern. *Invent. Math.* 7, 1–32 (1969). [3] Explizite Präsentation von Chevalleygruppen über Z. to appear.

A. Bialynicki-Birula, [1] On homogeneous affine spaces of linear algebraic groups. *Amer. J. Math.* **85**, 577–582 (1963). [2] Some theorems on actions of algebraic groups. *Ann. of Math.* **98**, 480–497 (1973).

D. Birkes, [1] Orbits of linear algebraic groups. *Ann. of Math.* **93**, 459–475 (1971).

A. Borel, [1] Groupes linéaires algébriques. *Ann. of Math.* **64**, 20–80 (1956). [2] Some finiteness properties of adele groups over number fields. *Inst. Hautes Études Sci. Publ. Math.* **16**, 101–126 (1963). [3] *Introduction aux groupes arithmétiques*. Paris: Hermann 1969. [4] *Linear Algebraic Groups*, notes by H. Bass. New York: W. A. Benjamin 1969. [5] On the automorphisms of certain subgroups of semi-simple Lie groups. In: *Algebraic Geometry*, ed. S. Abhyankar, pp. 43–74. London: Oxford Univ. Press 1969. [6] Properties and linear representations of Chevalley groups. In: *Seminar on Algebraic Groups and Related Finite Groups*, *Lect. Notes in Math.* **131**, pp. 1–55. Berlin–Heidelberg–New York: Springer 1970. [7] Cohomologie réelle stable de groupes S-arithmétiques classiques. *C. R. Acad. Sci. Paris* **274**, 1700–1702 (1972). [8] Linear representations of semi-simple algebraic groups. Proc. A. M. S. Summer Inst. (Arcata), to appear.

A. Borel, Harish-Chandra, [1] Arithmetic subgroups of algebraic groups. *Ann. of Math.* **75**, 485–535 (1962).

A. Borel, G. D. Mostow, ed., [1] *Algebraic Groups and Discontinuous Subgroups. Proc. Symp. Pure Math. IX.* Providence: Amer. Math. Soc. 1966.

A. Borel, J. -P. Serre, [1] Théorèmes de finitude en cohomologie galoisienne. *Comm. Math. Helv.* **39**, 111–164 (1964). [2] Adjonction de coins aux espaces symétriques. Applications à la cohomologie des groupes arithmétiques. *C. R. Acad. Sci. Paris* **271**, 1156–1158 (1970). [3] Cohomologie à supports compacts des immeubles de Bruhat-Tits. Applications à la cohomologie des groupes S-arithmétiques. *C. R. Acad. Sci. Paris* **272**, 110–113 (1971). [4] Corners and arithmetic groups. *Comm. Math. Helv.* **48**, 436–491 (1973).

A. Borel, T. A. Springer, [1] Rationality properties of linear algebraic groups II. *Tôhoku Math. J.* **20**, 443–497 (1968).

A. Borel, J. Tits, [1] Groupes réductifs. *Inst. Hautes Etudes Sci. Publ. Math.* **27**, 55–150 (1965). [2] Eléments unipotents et sous-groupes paraboliques de groupes réductifs I. *Invent. Math.* **12**, 95–104 (1971). [3] Compléments à l'article "Groupes réductifs". *Inst. Hautes Etudes Sci. Publ. Math.* **41**, 253–276 (1972). [4] Homomorphismes "abstraits" de groupes algébriques simples, *Ann. of Math.* **97**, 499–571 (1973).

N. Bourbaki, [1] *Groupes et algèbres de Lie*. Paris: Hermann. Ch. 1 (2nd ed.), 1971; Ch. 2–3, 1972; Ch. 4–6, 1968; Ch. 7–8, 1974.

E. Brieskorn, [1] Singular elements of semi-simple algebraic groups. In: *Actes, Congrès Intern. Math. 1970*, t. 2, pp. 279–284. Paris: Gauthier-Villars 1971. [2] Die Fundamentalgruppe des Raumes der regulären Orbits einer endlichen komplexen Spiegelungsgruppe. *Invent. Math.* **12**, 57–61 (1971).

F. Bruhat, J. Tits, [1] Groupes algébriques simples sur un corps local. In: *Proceedings of a Conference on Local Fields*, pp. 23–36. Berlin–Heidelberg–New York: Springer 1967. [2] BN-paires de type affine et données radicielles. *C. R. Acad. Sci. Paris* **263**, 598–601 (1966). [3] Groupes simples résiduellement déployés sur un corps local. *C. R. Acad. Sci. Paris* **263**, 766–768 (1966). [4] Groupes algébriques simples sur un corps local. *C. R. Acad. Sci. Paris* **263**, 822–825 (1966). [5] Groupes algébriques simples sur un corps local; cohomologie galoisienne, décompositions d'Iwasawa et de Cartan. *C. R. Acad. Sci. Paris* **263**, 867–869 (1966). [6] Groupes réductifs sur un corps local. *Inst. Hautes Études Sci. Publ. Math.* **41**, 5–252 (1972).

R. W. Carter, [1] *Simple Groups of Lie Type*. London–New York: Wiley 1972.

R. W. Carter, G. B. Elkington, [1] A note on the parametrization of conjugacy classes. *J. Algebra* **20**, 350–354 (1972).

R. W. Carter, G. Lusztig, [1] On the modular representations of the general linear and symmetric groups. *Math. Z.* **136**, 193–242 (1974).

P. Cartier, [1] Groupes algébriques et groupes formels. In: *Colloq. Théorie des Groupes Algébriques (Bruxelles, 1962)*, pp. 87–111. Paris: Gauthier-Villars 1962.

P. J. Cassidy, [1] Differential algebraic groups. *Amer. J. Math.* **94**, 891–954 (1972).

C. Chevalley, [1] A new relationship between matrices. *Amer. J. Math.* **65**, 521–531 (1943). [2] *Theory of Lie Groups*. Princeton: Princeton Univ. Press 1946. [3] Algebraic Lie algebras. *Ann. of Math.* **48**, 91–100 (1947). [4] *Théorie des groupes de Lie. Tome II, Groupes algébriques*. Paris: Hermann 1951. [5] *Théorie des groupes de Lie. Tome III, Théorèmes généraux sur les algèbres de Lie*. Paris: Hermann 1955. [6] On algebraic group varieties. *J. Math. Soc. Japan* **6**, 303–324 (1954). [7] Sur certains groupes simples. *Tôhoku Math. J.* **7**, 14–66 (1955). [8] *Séminaire sur la classification des groupes de Lie algébriques*. Paris: Ecole Norm. Sup. 1956–1958. [9] La théorie des groupes algébriques. In: *Proc. Int. Congress Mathematicians (Edinburgh 1958)*, pp. 53–68. Cambridge Univ. Press 1960. [10] *Fondements de la géométrie algébrique*. Paris: Secrétariat Mathématique 1958. [11] *Certains schémas de groupes semi-simples. Sém. Bourbaki (1960–61), Exp. 219*. New York: W. A. Benjamin 1966.

C. W. Curtis, [1] Representations of Lie algebras of classical type with applications to linear groups. *J. Math. Mech.* **9**, 307–326 (1960). [2] The classical groups as a source of algebraic problems. *Amer. Math. Monthly* **74**, Part II of No. 1, 80–91 (1967). [3] Chevalley groups and related topics. In: *Finite Simple Groups*, ed. M. B. Powell, G. Higman, pp. 135–189. London–New York: Academic Press 1971.

M. Demazure, [1] Schémas en groupes réductifs. *Bull. Soc. Math. France* **93**, 369–413 (1965). [2] Sous-groupes algébriques de rang maximum du groupe de Cremona. *Ann. Sci. Ecole Norm. Sup.* (4) **3**, 507–588 (1971). [3] Désingularisation des variétés de Schubert généralisées. *Ann. Sci. École Norm. Sup.* (4) **7**, 53–88 (1974).

M. Demazure, P. Gabriel, [1] *Groupes Algébriques. Tome I: Géométrie algébrique, généralités, groupes commutatifs*. Paris: Masson, Amsterdam: North-Holland 1970.

M. Demazure, A. Grothendieck, [1] Schémas en Groupes. *Lect. Notes in Math.* **151, 152, 153**. Berlin–Heidelberg–New York: Springer 1970.

V. V. Deodhar, [1] On central extensions of rational points of algebraic groups. To appear.

J. Dieudonné, [1] Sur les groupes de Lie algébriques sur un corps de caractéristique $p > 0$. *Rend. Circ. Mat. Palermo* **1**, 380–402 (1952). [2] Sur quelques groupes de Lie abéliens sur un corps de caractéristique $p > 0$. *Arch. Math.* **5**, 274–281 (1954); correction, ibid. **6**, 88 (1955). [3] Groupes de Lie et hyperalgèbres de Lie sur un corps de caractéristique $p > 0$. *Comm. Math. Helv.* **28**, 87–118 (1954). [4] Lie groups and Lie hyperalgebras over a field of characteristic $p > 0$. II. *Amer. J. Math.* **77**, 218–244 (1955). [5] Groupes de Lie et hyperalgèbres de Lie sur un corps de caractéristique $p > 0$. III. *Math. Z.* **63**, 53–75 (1955). [6] Lie groups and Lie hyperalgebras over a field of characteristic $p > 0$. IV. *Amer. J. Math.* **77**, 429–452 (1955). [7] Groupes de Lie et hyperalgèbres de Lie sur un corps de caractéristique $p > 0$. V. *Bull. Soc. Math. France* **84**, 207–239 (1956). [8] Lie groups and Lie hyperalgebras over a field of characteristic $p > 0$. VI. *Amer. J. Math.* **79**, 331–388 (1957). [9] Groupes de Lie et hyperalgèbres de Lie sur un corps de caractéristique $p > 0$. VII. *Math. Ann.* **134**, 114–133 (1957). [10] Les algèbres de Lie simples associées aux groupes simples algébriques sur un corps de caractéristique $p > 0$. *Rend. Circ. Mat. Palermo* **6**, 198–204 (1957). [11] Lie groups and Lie hyperalgebras over a field of characteristic $p > 0$. VIII. *Amer. J. Math.* **80**, 740–772 (1958). [12] Remarques sur la réduction mod. p des groupes linéaires algébriques. *Osaka J. Math.* **10**, 75–82 (1958). [13] *La géométrie des groupes classiques*. 3rd ed. Berlin–Heidelberg–New York: Springer 1971.

J. Dieudonné, J. B. Carrell, [1] Invariant theory, old and new. *Advances in Math.* **4**, 1–80 (1970).

G. B. Elkington, [1] Centralizers of unipotent elements in semisimple algebraic groups. *J. Algebra* **23**, 137–163 (1972).

J. Fogarty, [1] *Invariant Theory*. New York: W. A. Benjamin 1969.

H. Freudenthal, H. de Vries, [1] *Linear Lie Groups*. New York–London: Academic Press 1969.

H. Garland, [1] p-adic curvature and a conjecture of Serre. *Bull. Amer. Math. Soc.* **78**, 259–261 (1972). [2] p-adic curvature and the cohomology of discrete subgroups of p-adic groups. *Ann. of Math.* **97**, 375–423 (1973).

R. Godement, [1] *Groupes linéaires algébriques sur un corps parfait. Sém. Bourbaki (1960–61)*, Exp. 206. New York: W. A. Benjamin 1966.

F. Grosshans, [1] Orthogonal representations of algebraic groups. *Trans. Amer. Math. Soc.* **137**, 519–531 (1969). [2] Representations of algebraic groups preserving quaternion skew-hermitian forms. *Proc. Amer. Math. Soc.* **24**, 497–501 (1970). [3] Semi-simple algebraic groups defined over a real closed field. *Amer. J. Math.* **94**, 473–485 (1972). [4] Observable groups and Hilbert's fourteenth problem. *Amer. J. Math.* **95**, 229–253 (1973). [5] Open sets of points with good stabilizers. *Bull. Amer. Math. Soc.* **80**, 518–521 (1974).

W. J. Haboush, [1] Deformation theoretic methods in the theory of algebraic transformation spaces. *J. Math. Kyoto Univ.* **14**, 341–370 (1974).

G. Harder, [1] Uber einen Satz von E. Cartan. *Abh. Math. Sem. Univ. Hamburg* **28**, 208–214 (1965). [2] Uber die Galoiskohomologie halbeinfacher Matrizengruppen. I. *Math. Z.* **90**, 404–428 (1965). II. ibid. **92**, 396–415 (1966). [3] Halbeinfache Gruppenschemata über Dedekindringen. *Invent. Math.* **4**, 165–191 (1967). [4] Halbeinfache Gruppenschemata über vollständigen Kurven. *Invent. Math.* **6**, 107–149 (1968). [5] Bericht über neuere Resultate der Galoiskohomologie halbeinfacher Gruppen. *Jber. Deutsch. Math.-Verein* **70**, 182–216 (1968). [6] Eine Bemerkung zum schwachen Approximationssatz. *Arch. Math.* **19**, 465–471 (1968). [7] Minkowskische Reduktionstheorie über Funktionenkörpern. *Invent. Math.* **7**, 33–54 (1969). [8] A Gauss-Bonnet formula for discrete arithmetically defined groups. *Ann. Sci. École Norm. Sup.* (4) **4**, 409–455 (1971). [9] Chevalley groups over function fields and automorphic forms. *Ann. of Math.* **100**, 249–306 (1974).

S. J. Haris, [1] Some irreducible representations of exceptional algebraic groups. *Amer. J. Math.* **93**, 75–106 (1971).

D. Hertzig, [1] The structure of Frobenius algebraic groups, *Amer. J. Math.* **83**, 421–431 (1961). [2] Forms of algebraic groups. *Proc. Amer. Math. Soc.* **12**, 657–660 (1961). [3] Cohomology of algebraic groups. *J. Algebra* **6**, 317–334 (1967). [4] Fixed-point-free automorphisms of algebraic tori. *Amer. J. Math.* **90**, 1041–1047 (1968). [5] Cohomology of certain Steinberg groups. *Bull. Amer. Math. Soc.* **75**, 35–36 (1969).

H. Hijikata, [1] A note on the groups of type G_2 and F_4. *J. Math. Soc. Japan* **15**, 159–164 (1963).

G. Hochschild, [1] Algebraic Lie algebras and representative functions. *Illinois J. Math.* **3**, 499–523 (1959); supplement, ibid. **4**, 609–618 (1960). [2] On the algebraic hull of a Lie algebra. *Proc. Amer. Math. Soc.* **11**, 195–199 (1960). [3] Cohomology of algebraic linear groups. *Illinois J. Math.* **5**, 492–519 (1961). [4] Cohomology of affine algebraic homogeneous spaces. *Illinois J. Math.* **11**, 635–643 (1967). [5] Coverings of pro-affine algebraic groups. *Pacific J. Math.* **35**, 399–415 (1970). [6] Algebraic groups and Hopf algebras. *Illinois J. Math.* **14**, 52–65 (1970). [7] Note on algebraic Lie algebras. *Proc. Amer. Math. Soc.* **29**, 10–16 (1971). [8] *Introduction to Affine Algebraic Groups*. San Francisco: Holden-Day 1971. [9] Automorphism towers of affine algebraic groups. *J. Algebra* **22**, 365–373 (1972). [10] Lie algebra cohomology and affine algebraic groups. *Illinois J. Math.* **18**, 170–176 (1974).

G. Hochschild, G. D. Mostow, [1] Pro-affine algebraic groups. *Amer. J. Math.* **91**, 1127–1140 (1969). [2] Automorphisms of affine algebraic groups. *J. Algebra* **13**, 535–543 (1969). [3] Analytic and rational automorphisms of complex algebraic groups. *J. Algebra* **25**, 146–151 (1973). [4] Unipotent groups in invariant theory. *Proc. Nat. Acad. Sci. U.S.A.* **70**, 646–648 (1973).

R. Hooke, [1] Linear p-adic groups and their Lie algebras. *Ann. of Math.* **43**, 641–655 (1942).

J. E. Humphreys, [1] Algebraic groups and modular Lie algebras. *Mem Amer. Math. Soc.* **71** (1967). [2] Existence of Levi factors in certain algebraic groups. *Pacific J. Math.* **23**, 543–546 (1967). [3] On the automorphisms of infinite Chevalley groups. *Canad. J. Math.* **21**, 908–911 (1969). [4] *Arithmetic Groups, mimeographed lecture notes.* New York: Courant Institute of Mathematical Sciences 1971. [5] Modular representations of classical Lie algebras and semi-simple groups. *J. Algebra* **19**, 51–79 (1971). [6] *Introduction to Lie Algebras and Representation Theory*. Berlin–Heidelberg–New York: Springer 1972.

J. E. Humphreys, D. -N. Verma, [1] Projective modules for finite Chevalley groups. *Bull. Amer. Math. Soc.* **79**, 467–468 (1973).

J. Igusa, [1] On certain representations of semi-simple algebraic groups and the arithmetic of the corresponding invariants. I. *Invent. Math.* **12**, 62–94 (1971).

Y. Ihara, [1] On discrete subgroups of the two by two projective linear group over p-adic fields. *J. Math. Soc. Japan* **18**, 219–235 (1966).

B. Iversen, [1] A fixed point formula for action of tori on algebraic varieties. *Invent. Math.* **16**, 229–236 (1972).

N. Iwahori, [1] On the structure of a Hecke ring of a Chevalley group over a finite field. *J. Fac. Sci. Univ. Tokyo* **10**, 215–235 (1964).

N. Iwahori, H. Matsumoto, [1] On some Bruhat decomposition and the structure of the Hecke rings of p-adic Chevalley groups. *Inst. Hautes Études Sci. Publ. Math.* **25**, 5–48 (1965).

K. Iyanaga, [1] On certain double coset spaces of algebraic groups. *J. Math. Soc. Japan* **23**, 103–122 (1971).

N. Jacobson, [1] *Lie Algebras*. New York: Interscience 1962.

J. C. Jantzen, [1] Darstellungen halbeinfacher algebraischer Gruppen und zugeordnete kontravariante Formen. *Bonner Math. Schriften* **67**. Bonn: Math. Inst. der Universität 1973. [2] Zur Charakterformel gewisser Darstellungen halbeinfacher Gruppen und Lie-Algebren. *Math. Z.*, **140**, 127–149 (1974).

W. L. J. van der Kallen, [1] Infinitesimally Central Extensions of Chevalley Groups. *Lect. Notes in Math.* **356**. Berlin–Heidelberg–New York: Springer 1973.

T. Kambayashi, [1] Projective representations of algebraic linear groups of transformations. *Amer. J. Math.* **88**, 199–205 (1966).

I. Kaplansky, [1] *An Introduction to Differential Algebra*. Paris: Hermann 1957.

H. Kimura, [1] Relative cohomology of algebraic linear groups. II. *Nagoya Math. J.* **26**, 89–99 (1966).

M. Kneser, [1] Starke Approximation in algebraischen Gruppen. I. *J. Reine Angew. Math.* **218**, 190–203 (1965). [2] Galois-Kohomologie halbeinfacher algebraischer Gruppen über p-adischen Körpern. I. *Math. Z.* **88**, 40–47 (1965); II. ibid. **89**, 250–272 (1965). [3] Semisimple algebraic groups. In: *Algebraic Number Theory*, ed. J. W. S. Cassels, A. Fröhlich, pp. 250–265. Washington, D.C.: Thompson 1967. [4] Normal subgroups of integral orthogonal groups. In: *Algebraic K-Theory and its Geometric Applications. Lect. Notes in Math.* **108**. Berlin–Heidelberg–New York: Springer 1969. [5] *Lectures on Galois Cohomology of Classical Groups*. Bombay: Tata Inst. of Fundamental Research 1969.

E. R. Kolchin, [1] Algebraic matric groups and the Picard–Vessiot theory of homogeneous linear ordinary differential equations. *Ann. of Math.* **49**, 1–42 (1948). [2] On certain concepts in the theory of algebraic matric groups. *Ann. of Math.* **49**, 774–789 (1948). [3] *Differential Algebra and Algebraic Groups*. New York–London: Academic Press 1973.

J. Kovacic, [1] Pro-algebraic groups and the Galois theory of differential fields. *Amer. J. Math.* **95**, 507–536 (1973).

S. Lang, [1] Algebraic groups over finite fields. *Amer. J. Math.* **78**, 555–563 (1956). [2] *Abelian Varieties*. New York: Interscience 1959.

R. P. Langlands, [1] *Euler Products. Yale Math. Monographs* **1**. New Haven–London: Yale Univ. Press 1971.

M. Lazard, [1] Sur le nilpotence de certains groupes algébriques. *C. R. Acad. Sci. Paris* **241**, 1687–1689 (1955). [2] Groupes analytiques p-adiques. *Inst. Hautes Etudes Sci. Publ. Math.* **26**, 1–219 (1965).

B. Lou, [1] The centralizer of a regular unipotent element in a semisimple algebraic group. *Bull. Amer. Math. Soc.* **74**, 1144–1147 (1968).

D. Luna, [1] Sur les orbites fermées des groupes algébriques réductifs. *Invent. Math.* **16**, 1–5 (1972).

D. Luna, T. Vust, [1] Un théorème sur les orbites affines des groupes algébriques semi-simples. *Ann. Scuola Norm. Sup. Pisa* (3) **27**, 527–535 (1973).

I. G. Macdonald, [1] Spherical functions on a p-adic Chevalley group. *Bull. Amer. Math. Soc.* **74**, 520–525 (1968). [2] *Spherical Functions on a Group of p-adic Type*. Madras: Ramanujan Inst. 1971.

L. Markus, [1] Exponentials in algebraic matrix groups. *Advances in Math.* **11**, 351–367 (1973).

R. Marlin, [1] Anneaux de Chow des groupes algébriques SU(n), Sp(n), SO(n), Spin(n), G_2, F_4. *C. R. Acad. Sci. Paris* **279**, 119–122 (1974).

J. G. M. Mars, [1] Les nombres de Tamagawa de certains groupes exceptionnels. *Bull. Soc. Math. France* **94**, 97–140 (1966). [2] Solution d'un problème posé par A. Weil. *C. R. Acad. Sci. Paris* **266**, 484–486 (1968). [3] The Tamagawa number of ^2A_n. *Ann. of Math.* **89**, 557–574 (1969). [4] *Les nombres de Tamagawa de groupes semi-simples. Sém. Bourbaki (1968–69)*, Exp. 351. *Lect. Notes in Math.* **179**. Berlin–Heidelberg–New York: Springer 1971.

H. Matsumoto, [1] Quelques remarques sur les groupes de Lie algébriques réels. *J. Math. Soc. Japan* **16**, 419–446 (1964). [2] Générateurs et relations des groupes de Weyl généralisés. *C. R. Acad. Sci. Paris* **258**, 3419–3422 (1964). [3] Un théorème de Sylow pour les groupes semi-simples p-adiques. *C. R. Acad. Sci. Paris* **262**, 425–427 (1966). [4] Sur les groupes semi-simples déployés sur un anneau principal. *C. R. Acad. Sci. Paris* **262**, 1040–1042 (1966). [5] Sur les sous-groupes arithmétiques des groupes semi-simples déployés. *Ann. Sci. École Norm. Sup.* (4) **2**, 1–62 (1969). [6] Fonctions sphériques sur un groupe semi-simple p-adique. *C. R. Acad. Sci. Paris* **269**, 829–832 (1969).

Y. Matsushima, [1] Espaces homogènes de Stein des groupes de Lie complexes. *Nagoya Math. J.* **16**, 205–218 (1960).

F. Minbashian, [1] Pro-affine algebraic groups. *Amer. J. Math.* **95**, 174–192 (1973).

M. Miyanishi, [1] On the algebraic fundamental group of an algebraic group. *J. Math. Kyoto Univ.* **12**, 351–367 (1972).

C. Moore, [1] Group extensions of p-adic and adelic linear groups. *Inst. Hautes Études Sci. Publ. Math.* **35**, 5–70 (1968).

G. D. Mostow, [1] Self-adjoint groups. *Ann. of Math.* **62**, 44–55 (1955). [2] Fully reducible subgroups of algebraic groups. *Amer. J. Math.* **78**, 200–221 (1956).

D. Mumford, [1] *Geometric Invariant Theory.* Berlin–Heidelberg–New York: Springer 1965. [2] *Abelian Varieties.* London: Oxford Univ. Press 1970. [3] *Introduction to Algebraic Geometry*, preliminary version of Chapters I–III. Cambridge: Harvard Univ. Math. Dept. [4] *Introduction to Algebraic Geometry.* Berlin–Heidelberg–New York: Springer, in preparation.

M. Nagata, [1] Complete reducibility of rational representations of a matric group. *J. Math. Kyoto Univ.* **1**, 87–99 (1961). [2] Invariants of a group in an affine ring. *J. Math. Kyoto Univ.* **3**, 369–377 (1964). [3] *Lectures on the Fourteenth Problem of Hilbert.* Bombay: Tata Inst. of Fundamental Research 1965.

T. Nakamura, [1] A remark on unipotent groups of characteristic $p > 0$. *Kodai Math. Sem. Rep.* **23**, 127–130 (1971). [2] On commutative unipotent groups defined by Seligman. *J. Math. Soc. Japan* **23**, 377–383 (1972).

T. Ono [1] Arithmetic of orthogonal groups. *J. Math. Soc. Japan* **7**, 79–91 (1955). [2] Sur les groupes de Chevalley. *J. Math. Soc. Japan* **10**, 307–313 (1958). [3] Sur la reduction modulo p des groupes linéaires algébriques. *Osaka J. Math.* **10**, 57–73 (1958). [4] Arithmetic of algebraic tori. *Ann. of Math.* **74**, 101–139 (1961). [5] On the field of definition of Borel subgroups of semi-simple algebraic groups. *J. Math. Soc. Japan* **15**, 392–395 (1963). [6] On the Tamagawa number of algebraic tori. *Ann. of Math.* **78**, 47–73 (1963). [7] On the relative theory of Tamagawa numbers. *Ann. of Math.* **82**, 88–111 (1965). [8] On algebraic groups and discontinuous groups. *Nagoya Math. J.* **27**, 279–322 (1966). [9] A remark on Gaussian sums and algebraic groups *J. Math. Kyoto Univ.* **13**, 139–142 (1973).

F. Oort, [1] *Commutative Group Schemes. Lect. Notes in Math.* **15**. Berlin–Heidelberg–New York: Springer 1966. [2] Algebraic group schemes in characteristic zero are reduced. *Invent. Math.* **2**, 79–80 (1966).

B. Parshall, [1] Regular elements in algebraic groups of prime characteristic. *Proc. Amer. Math. Soc.* **39**, 57–62 (1973).

V. P. Platonov, [1] Automorphisms of algebraic groups. *Dokl. Akad. Nauk SSSR* **168**, 1257–1260 (1966) = *Soviet Math. Dokl.* **7**, 825–829 (1966). [2] Theory of algebraic linear groups and periodic groups. *Izv. Akad. Nauk SSSR Ser. Mat.* **30**, 573–620 (1966) = *Amer. Math. Soc. Transl. Ser.* **2**, Vol. **69**, 61–110 (1968). [3] Invariance theorems and algebraic groups with an almost-regular automorphism. *Izv. Akad. Nauk SSSR Ser. Mat.* **31**, 687–696 (1967) = *Math. USSR–Izv.* **1**, 667–676 (1967). [4] Proof of the finiteness hypothesis for the solvable subgroups of algebraic groups. *Sibirski Mat. Ž.* **10**, 1084–1090 (1969) = *Siberian Math. J.* **10**, 800–804 (1969). [5] The problem of strong approximation and the Kneser-Tits conjecture for algebraic groups. *Izv. Akad. Nauk SSSR Ser. Mat.* **33**, 1211–1219 (1969) = *Math. USSR–Izv.* **3**, 1139–1147 (1969). Addendum. *Izv. Akad. Nauk SSSR Ser. Mat.* **34**, 775–777 (1970) = *Math. USSR–Izv.* **4**, 784–786 (1970). [6] The congruence problem for solvable integral groups. *Dokl. Akad. Nauk BSSR* **15**, 869–872 (1971). [Russian] MR 44 #4009.

V. P. Platonov, V. I. Jančevskiĭ, [1] The structure of unitary groups and the commutator group of a simple algebra over global fields. *Dokl. Akad. Nauk SSSR* **208**, 541–544 (1973) = *Soviet Math. Dokl.* **14**, 132–136 (1973).

V. P. Platonov, M. V. Milovanov, [1] Determination of algebraic groups by arithmetic subgroups. *Dokl. Akad. Nauk SSSR* **209**, 43–46 (1973) = *Soviet Math. Dokl.* **14**, 331–335 (1973).

V. L. Popov, [1] Criteria for the stability of the action of semisimple group on a factorial variety. *Izv. Akad. Nauk SSSR Ser. Mat.* **34**, 523–531 (1970) = *Math. USSR–Izv.* **4**, 527–535 (1970). [2] On the stability of the action of an algebraic group on an algebraic variety. *Izv. Akad. Nauk SSSR Ser. Mat.* **36**, 371–385 (1972) = *Math. USSR–Izv.* **6**, 367–379 (1972).

G. Prasad, M. S. Raghunathan, [1] Cartan subgroups and lattices in semi-simple groups. *Ann. of Math.* **96**, 296–317 (1972).

M. S. Raghunathan, [1] Cohomology of arithmetic subgroups of algebraic groups. I. *Ann. of Math.* **86**, 409–424 (1967). II. ibid. **87**, 279–304 (1968). [2] A note on quotients of real algebraic groups by arithmetic subgroups. *Invent. Math.* **4**, 318–335 (1968). [3] *Discrete Subgroups of Lie Groups.* Berlin–Heidelberg–New York: Springer 1972. [4] On the congruence subgroup problem. To appear.

R. Ree, [1] On some simple groups defined by C. Chevalley. *Trans. Amer. Math. Soc.* **84**, 392–400 (1957). [2] Construction of certain semi-simple groups. *Canad. J. Math.* **16**, 490–508 (1964). [3] Commutators in semi-simple algebraic groups. *Proc. Amer. Math. Soc.* **15**, 457–460 (1964).

R. W. Richardson, Jr., [1] Conjugacy classes in Lie algebras and algebraic groups. *Ann. of Math.* **86**, 1–15 (1967). [2] Principal orbit types for algebraic transformation spaces in characteristic zero. *Invent. Math.* **16**, 6–14 (1972). [3] Conjugacy classes in parabolic subgroups of semi-simple algebraic groups. *Bull. London Math. Soc.* **6**, 21–24 (1974).

C. Riehm, [1] The congruence subgroup problem over local fields. *Amer. J. Math.* **92**, 771–778 (1970).

M. Rosenlicht, [1] Some basic theorems on algebraic groups. *Amer. J. Math.* **78**, 401–443 (1956). [2] Some rationality questions on algebraic groups. *Ann. Mat. Pura Appl.* **43**, 25–50 (1957).

[3] Toroidal algebraic groups. *Proc. Amer. Math. Soc.* **12**, 984–988 (1961). [4] On quotient varieties and the affine embedding of certain homogeneous spaces. *Trans. Amer. Math. Soc.* **101**, 211–223 (1961). [5] On a result of Baer. *Proc. Amer. Math. Soc.* **13**, 99–101 (1962). [6] Questions of rationality for solvable algebraic groups over nonperfect fields. *Annali di Mat.* (*IV*) **61**, 97–120 (1963).

I. Satake, [1] On the theory of reductive algebraic groups over a perfect field. *J. Math. Soc. Japan* **15**, 210–235 (1963). [2] Theory of spherical functions on reductive algebraic groups over p-adic fields. *Inst. Hautes Études Sci. Publ. Math.* **18**, 5–69 (1963). [3] Symplectic representations of algebraic groups satisfying a certain analyticity condition. *Acta Math.* **117**, 215–279 (1967). [4] On a certain invariant of the groups of type E_6 and E_7. *J. Math. Soc. Japan* **20**, 322–335 (1968). [5] *Classification Theory of Semi-Simple Algebraic Groups.* New York: Marcel Dekker 1971.

D. Schattschneider, [1] On restricted roots of semi-simple algebraic groups. *J. Math. Soc. Japan* **21**, 94–115 (1969).

C. Schoeller, [1] Groupes affines, commutatifs, unipotents sur un corps non parfait. *Bull. Soc. Math. France* **100**, 241–300 (1972).

G. B. Seligman, [1] *Algebraic Groups, mimeographed lecture notes.* New Haven: Yale Univ. Math. Dept. 1964. [2] *Modular Lie Algebras.* Berlin–Heidelberg–New York: Springer 1967. [3] Algebraic Lie algebras. *Bull. Amer. Math. Soc.* **74**, 1051–1065 (1968). [4] On some commutative unipotent groups. *Invent. Math.* **5**, 129–137 (1968). [5] On two-dimensional algebraic groups. *Scripta Math.* **29**, 453–465 (1973).

J. -P. Serre, [1] *Groupes algébriques et corps de classes.* Paris: Hermann 1959. [2] *Algèbres de Lie semi-simples complexes.* New York: W. A. Benjamin 1966. [3] *Groupes de congruence. Sém. Bourbaki*(*1966–67*),*Exp.330.* New York: W. A. Benjamin 1968. [4] Groupes de Grothendieck des schémas en groupes réductifs déployés. *Inst. Hautes Etudes Sci. Publ. Math.* **34**, 37–52 (1968). [5] Le problème des groupes de congruence pour SL_2. *Ann. of Math.* **92**, 489–527 (1970). [6] Cohomologie des groupes discrets. In: *Prospects in Mathematics, Ann. of Math. Studies* **70**, pp. 77–169. Princeton: Princeton Univ. Press 1971.

C. S. Seshadri, [1] Mumford's conjecture for GL(2) and applications. In: *Algebraic Geometry,* ed. S. Abhyankar, pp. 347–371. London: Oxford Univ. Press 1969. [2] Quotient spaces modulo reductive algebraic groups. *Ann. of Math.* **95**, 511–556 (1972). [3] Theory of moduli. Proc. A. M. S. Summer Inst. (Arcata), to appear.

SGAD = M. Demazure, A. Grothendieck [1]

I. R. Shafarevich, [1] Foundations of algebraic geometry. *Russian Math. Surveys* **24**, No. 6, 1–178 (1969). [2] *Basic Algebraic Geometry.* Berlin–Heidelberg–New York: Springer 1974.

A. -M. Simon, [1] Structure des groupes unipotents commutatifs. *C. R. Acad. Sci. Paris* **276**, 347–349 (1973).

T. A. Springer, [1] Some arithmetical results on semi-simple Lie algebras. *Inst. Hautes Études Sci. Publ. Math.* **30**, 115–141 (1966). [2] A note on centralizers in semi-simple groups. *Indag. Math.* **28**, 75–77 (1966). [3] Weyl's character formula for algebraic groups. *Invent. Math.* **5**, 85–105 (1968). [4] The unipotent variety of a semisimple group. In: *Algebraic Geometry,* ed. S. Abhyankar, pp. 373–391. London: Oxford Univ. Press 1969. [5] *Jordan Algebras and Algebraic Groups.* Berlin–Heidelberg–New York: Springer 1973.

T. A. Springer, R. Steinberg, [1] Conjugacy classes. In: *Seminar on Algebraic Groups and Related Finite Groups, Lect. Notes in Math.* **131**, pp. 167–266. Berlin–Heidelberg–New York: Springer 1970.

R. Steinberg, [1] Prime power representations of finite linear groups. I. *Canad. J. Math.* **8**, 580–591 (1956). II. ibid. **9**, 347–351 (1957). [2] Variations on a theme of Chevalley. *Pacific J. Math.* **9**, 875–891 (1959). [3] The simplicity of certain groups. *Pacific J. Math.* **10**, 1039–1041 (1960). [4] Automorphisms of finite linear groups. *Canad. J. Math.* **12**, 606–615 (1960). [5] Automorphisms of classical Lie algebras. *Pacific J. Math.* **11**, 1119–1129 (1961). [6] Générateurs, relations et revêtements de groupes algébriques. In: *Colloq. Théorie des Groupes Algébriques* (*Bruxelles, 1962*), pp. 113–127. Paris: Gauthier-Villars 1962. [7] Representations of algebraic groups. *Nagoya Math. J.* **22**, 33–56 (1963). [8] Regular elements of semisimple algebraic groups. *Inst. Hautes Études Sci. Publ. Math.* **25**, 49–80 (1965). [9] Classes of elements of semisimple algebraic groups. In: *Proc. Intl. Congress of Mathematicians* (*Moscow 1966*), pp. 277–284. Moscow: Mir 1968. [10] *Lectures on Chevalley Groups, mimeographed lecture notes.* New Haven: Yale Univ. Math. Dept. 1968. [11] Endomorphisms of linear algebraic groups. *Mem. Amer. Math. Soc.* **80** (1968). [12] Algebraic groups and finite groups. *Illinois J. Math.* **13**, 81–86 (1969). [13] *Conjugacy Classes in Algebraic Groups. Lect. Notes in Math.* **366.** Berlin–Heidelberg–New York: Springer 1974. [14] Abstract homomorphisms of simple algebraic groups. *Sém. Bourbaki* (1972–73), Exp. 435. *Lect. Notes in Math.* **383.** Berlin–Heidelberg–New York: Springer 1974.

U. Stuhler, [1] Unipotente und nilpotente Klassen in einfachen Gruppen und Liealgebren vom Typ G_2. *Indag. Math.* **33**, 365–378 (1971).

J. B. Sullivan, [1] Automorphisms of affine unipotent groups in positive characteristic. *J. Algebra* **26**, 140–151 (1973). [2] A decomposition theorem for pro-affine solvable algebraic groups over algebraically closed fields. *Amer. J. Math.* **94**, 221–228 (1973).

M. E. Sweedler, [1] Conjugacy of Borel subgroups, an easy proof. To appear.

M. Takeuchi, [1] A note on geometrically reductive groups. *J. Fac. Sci Univ. Tokyo* **20**, 387–396 (1973).

T. Tasaka, [1] On the quasi-split simple algebraic groups defined over an algebraic number field. *J. Fac. Sci. Univ. Tokyo* **15**, 147–168 (1968). [2] On the second cohomology groups of the fundamental groups of simple algebraic groups over perfect fields. *J. Math. Soc. Japan* **21**, 244–258 (1969).

J. Tits, [1] Sur la classification des groupes algébriques semisimples. *C. R. Acad. Sci. Paris* **249**, 1438–1440 (1959). [2] Sur la trialité et certains groupes qui s'en déduisent. *Inst. Hautes Études Sci. Publ. Math.* **2**, 13–60 (1959). [3] Théorème de Bruhat et sous-groupes paraboliques. *C. R. Acad. Sci. Paris* **254**, 2910–2912 (1962). [4] Groupes semi-simples isotropes. In: *Colloq. Théorie des Groupes Algébriques (Bruxelles, 1962)*, pp. 137–147. Paris: Gauthier-Villars 1962. [5] Groupes simples et géométries associées. In: *Proc. Intl. Congress of Mathematicians*, pp. 197–221. Stockholm 1962. [6] Algebraic and abstract simple groups. *Ann. of Math.* **80**, 313–329 (1964). [7] *Structures et groupes de Weyl. Sém. Bourbaki (1964–65), Exp. 288.* New York: W. A. Benjamin 1966. [8] Normalisateurs de tores. I. Groupes de Coxeter étendues. *J. Algebra* **4**, 96–116 (1966). [9] *Lectures on Algebraic Groups, mimeographed lecture notes.* New Haven: Yale Univ. Math. Dept. 1967. [10] Représentations linéaires irréductibles d'un groupe réductif sur un corps quelconque. *J. Reine Angew. Math.* **247**, 196–220 (1971). [11] Free subgroups in linear groups. *J. Algebra* **20**, 250–270 (1972). [12] *Buildings of Spherical Type and Finite BN-pairs. Lect. Notes in Math.* **386**. Berlin–Heidelberg–New York: Springer 1974.

B. Ju. Veĭsfeĭler, [1] A certain property of semisimple algebraic groups. *Funkcional. Anal. i Priložen.* **2**, no. 2, 84–85 (1968) = *Functional Anal. Appl.* **2**, 257–258 (1968). [2] Certain properties of singular semisimple algebraic groups over non-closed fields. *Trudy Moskov. Mat. Obšč.* **20**, 111–136 (1969) = *Trans. Moscow Math. Soc.* **20**, 109–134 (1971). [3] A remark on certain algebraic groups. *Funkcional. Anal. i Priložen.* **4**, no. 1, 91 (1970) = *Functional Anal. Appl.* **4**, 81–82 (1970). [4] Semisimple algebraic groups which are split over a quadratic extension. *Izv. Akad. Nauk SSSR Ser. Mat.* **35**, 56–71 (1971) = *Math. USSR–Izv.* **5**, 57–72 (1971). [5] Hasse principle for algebraic groups split over a quadratic extension. *Funkcional. Anal. i Priložen.* **6**, no. 2, 21–23 (1972) = *Functional Anal. Appl.* **6**, 102–104 (1972).

B. Ju Veĭsfeĭler, V. G. Kac, [1] Exponentials in Lie algebras of characteristic p. *Izv. Akad. Nauk SSSR Ser. Mat.* **35**, 762–788 (1971) = *Math. USSR–Izv.* **5**, 777–803 (1971).

F. D. Veldkamp, [1] Representations of algebraic groups of type F_4 in characteristic 2. *J. Algebra* **16**, 326–339 (1970). [2] The center of the universal enveloping algebra of a Lie algebra in characteristic p. *Ann. Sci. École Norm. Sup.* (4) **5**, 217–240 (1972). [3] Roots and maximal tori in finite forms of semisimple algebraic groups. *Math. Ann.* **207**, 301–314 (1974).

D. -N. Verma, [1] Role of affine Weyl groups in the representation theory of algebraic Chevalley groups and their Lie algebras. In: *Lie Groups and their Representations*, ed. I. M. Gel'fand. Budapest: Akad. Kiado 1974.

V. E. Voskresenskiĭ, [1] Two-dimensional algebraic tori. *Izv. Akad. Nauk SSSR Ser. Mat.* **29**, 239–244 (1965) = *Amer. Math. Soc. Transl. Ser.* 2, Vol. **73**, 190–195 (1968). [2] On two-dimensional algebraic tori. II. *Izv. Akad. Nauk SSSR Ser. Mat.* **31**, 711–716 (1967) = *Math. USSR–Izv.* **1**, 691–696 (1967). [3] Picard groups of linear algebraic groups. In: *Studies in Number Theory*, No. 3, pp. 7–16. Saratov: Izdat. Saratov. Univ. 1969 [Russian]. MR 42 #5999. [4] On the birational equivalence of linear algebraic groups. *Dokl. Akad. Nauk SSSR* **188**, 978–981 (1969) = *Soviet Math. Dokl.* **10**, 1212–1215 (1969). [5] Birational properties of linear algebraic groups. *Izv. Akad. Nauk SSSR Ser. Mat.* **34**, 3–19 (1970) = *Math. USSR–Izv.* **4**, 1–18 (1970). [6] Rationality of certain algebraic tori. *Izv. Akad. Nauk SSSR Ser. Mat.* **35**, 1037–1046 (1971) = *Math. USSR–Izv.* **5**, 1049–1056 (1971).

B. A. F. Wehrfritz, [1] *Infinite Linear Groups.* Berlin–Heidelberg–New York: Springer 1973.

A. Weil, [1] *Variétés abéliennes et courbes algébriques.* Paris: Hermann 1948. [2] On algebraic groups of transformations. *Amer. J. Math.* **77**, 355–391 (1955). [3] On algebraic groups and homogeneous spaces. *Amer. J. Math.* **77**, 493–512 (1955). [4] Algebras with involutions and the classical groups. *J. Indian Math. Soc.* **24**, 589–623 (1961). [5] *Adeles and algebraic groups.* Princeton: Inst. Advanced Study 1961.

H. Weyl, [1] *The Classical Groups.* Princeton: Princeton Univ. Press 1946.

D. J. Winter, [1] On automorphisms of algebraic groups. *Bull. Amer. Math. Soc.* **72**, 706–708 (1966). [2] Algebraic group automorphisms having finite fixed point sets. *Proc. Amer. Math. Soc.* **18**, 371–377 (1967). [3] Fixed points and stable subgroups of algebraic group automorphisms. *Proc. Amer. Math. Soc.* **18**, 1107–1113 (1967). [4] On groups of automorphisms of Lie algebras. *J. Algebra* **8**, 131–142 (1968).

Z. E. Zalesskiĭ, [1] A remark on the triangular linear group. *Vesci Akad. Navuk BSSR Ser. Fiz.—Mat. Navuk* 1968, no. 2, 129–131. [Russian]. MR 38 #5944.

Index of Terminology